RILEM 252-CMB Symposium

RILEM BOOKSERIES

Volume 20

RILEM, The International Union of Laboratories and Experts in Construction Materials, Systems and Structures, founded in 1947, is a non-governmental scientific association whose goal is to contribute to progress in the construction sciences, techniques and industries, essentially by means of the communication it fosters between research and practice. RILEM's focus is on construction materials and their use in building and civil engineering structures, covering all phases of the building process from manufacture to use and recycling of materials. More information on RILEM and its previous publications can be found on www.RILEM.net.

More information about this series at http://www.springer.com/series/8781

Lily D. Poulikakos · Augusto Cannone Falchetto
Michael P. Wistuba · Bernhard Hofko
Laurent Porot · Hervé Di Benedetto
Editors

RILEM 252-CMB Symposium

Chemo-Mechanical Characterization of Bituminous Materials

Springer

Editors
Lily D. Poulikakos
Empa
Dübendorf, Switzerland

Augusto Cannone Falchetto
Braunschweig Pavement Engineering Center
Technische Universität Braunschweig
Braunschweig, Niedersachsen, Germany

Michael P. Wistuba
Braunschweig Pavement Engineering Center
Technische Universität Braunschweig
Braunschweig, Niedersachsen, Germany

Bernhard Hofko
Institute of Transportation
Technische Universität Wien
Vienna, Wien, Austria

Laurent Porot
Kraton Chemical B.V.
Almere, The Netherlands

Hervé Di Benedetto
University of Lyon
Vaulx en Velin, France

ISSN 2211-0844 ISSN 2211-0852 (electronic)
RILEM Bookseries
ISBN 978-3-030-00475-0 ISBN 978-3-030-00476-7 (eBook)
https://doi.org/10.1007/978-3-030-00476-7

Library of Congress Control Number: 2018954493

This Springer imprint is published by the registered company Springer Nature Switzerland AG
The registered company address is: Gewerbestrasse 11, 6330 Cham, Switzerland

Preface

The RILEM Technical Committee 252 CMB on Chemo-Mechanical Characterization of Bituminous Materials was launched in Stockholm in June 2013 by Niki Kringos and Lily Poulikakos. The TC was active for 5 years (2013–2018) and intended to combine chemical and mechanical characterization of bituminous materials in order to gain a better understanding of the behavior of this complex material. Through the cooperative effort facilitated through Rilem, it was possible to bring more visibility to this field of research which in a traditionally construction field is not always easy to do. Applying chemo-mechanics allows us to gain a better understanding of the long-term behavior of traditional and new materials such as high RAP mixtures and reduced temperature mixtures. Having a fundamental understanding of the combined chemo-mechanical properties can greatly enhance the tools used to improve the material's sustainability and functionality.

The TC work was primarily focused on the characterization of aging and the relevance of the current short-term (RTFOT) and long-term aging (PAV) techniques for new types of mixtures. Two TGs were actively led by Bernhard Hofko and James Grenfell.

The editors would like to thank all active members of the TC for their continuous support through participation, supply of materials, and discussions during the TC meetings. The steering and organizing committees would like to gratefully acknowledge the contribution of Di Wang, Nina Eßmann, Anna-Lena Gertig, and Tess Sigwarth in helping to prepare the present collection of papers. The cooperation of the scientific committee in reviewing the contributions to the symposium is greatly appreciated.

Augusto Cannone Falchetto would like to personally acknowledge the German Research Foundation (Deutsche Forschungsgemeinschaft—DFG) for the grant he was awarded to financially support the RILEM 252-CMB symposium.

As is apparent by the wide range of topics that are presented in forty-eight contributions in the symposium, chemo-mechanical characterization is gaining more acceptance within our community. The inputs were divided among seven parts. Bitumen Aging Mechanisms and Characterization is the topic of Part I. Part II addresses Chemo-Mechanical Coupling. In Part III, Low, Intermediate and High Temperature Behavior is discussed. Part IV addresses Microstructure and Micro-Mechanics. Part V covers the topic of Recycling and Rejuvenation. In Part VI, Multiphase Analysis of Binders is discussed, and Part VII is devoted to other approaches in the field.

Looking into the future, as conventional materials become scarce, using alternative materials is inevitable and we hope that this type of characterization will continue to flourish in the asphalt field.

July 2018

Lily Poulikakos
Augusto Cannone Falchetto
Bernhard Hofko
Laurent Porot
Hervé Di Benedetto
Michael P. Wistuba

Organization

Steering Committee

Chairs

Lily D. Poulikakos	EMPA, Switzerland
Augusto Cannone Falchetto	TU Braunschweig, Germany

Co-chairs

Bernhard Hofko	TU Wien, Austria
Laurent Porot	Kraton Corporation, the Netherlands
Hervé Di Benedetto	Université de Lyon, France
Michael P. Wistuba	TU Braunschweig, Germany

International Scientific Committee

Gordon Airey	The University of Nottingham, UK
Hassan Baaj	University of Waterloo, Canada
Krishna Prapoorna Biligiri	Indian Institute of Technology Tirupati, India
Björn Birgisson	Texas A&M University, USA
Francesco Canestrari	Università Politecnica delle Marche, Italy
Augusto Cannone Falchetto	TU Braunschweig, Germany
Emmanuel Chailleux	IFSTTAR, France
Hervé Di Benedetto	University of Lyon, ENTPE, France
Elham Fini	North Carolina A&T State University, USA
Zhen Fu	Chang'an University, China
Meng Guo	Beijing University of Technology, China
Yuejie Han	Chang'an University, China
Bernhard Hofko	TU Wien, Austria

Chichun Hu	South China University of Technology, China
Luis Guillermo Loría Salazar	Universidad de Costa Rica, Costa Rica
Xiaohu Lu	Nynas, Sweden
Feng Ma	Chang'an University, China
Wangyu Ma	West Texas Paving Inc., USA
Mihai Marasteanu	University of Minnesota, USA
Ki Hoon Moon	Korea Expressway Corporation, South Korea
Jianzhong Pei	Chang'an University, China
Laurent Porot	Kraton Corporation, the Netherlands
Lily D. Poulikakos	EMPA, Switzerland
Yu Qin	CREEC, China
Cédric Sauzéat	University of Lyon, ENTPE, France
Hilde Soenen	Nynas Belgium, Belgium
Yichang Tsai	Georgia Institute of Technology, USA
Lucia Tsantilis	Politecnico di Torino, Italy
Di Wang	TU Braunschweig, Germany
Hainian Wang	Chang'an University, China
Hao Wang	Rutgers, The State University of New Jersey, USA
Weina Wang	Chongqing Jiaotong University, China
Michael P. Wistuba	TU Braunschweig, Germany
Adam Zofka	Road and Bridge Research Institute (IBDiM), Poland

Contents

RILEM Publications

The following list is presenting the global offer of RILEM Publications, sorted by series. Each publication is available in printed version and/or in online version.

RILEM Proceedings (PRO)

PRO 1: Durability of High Performance Concrete (ISBN: 2-912143-03-9; e-ISBN: 2-351580-12-5; e-ISBN: 2351580125); *Ed. H. Sommer*
PRO 2: Chloride Penetration into Concrete (ISBN: 2-912143-00-04; e-ISBN: 2912143454); *Eds. L.-O. Nilsson and J.-P. Ollivier*
PRO 3: Evaluation and Strengthening of Existing Masonry Structures (ISBN: 2-912143-02-0; e-ISBN: 2351580141); *Eds. L. Binda and C. Modena*
PRO 4: Concrete: From Material to Structure (ISBN: 2-912143-04-7; e-ISBN: 2351580206); *Eds. J.-P. Bournazel and Y. Malier*
PRO 5: The Role of Admixtures in High Performance Concrete (ISBN: 2-912143-05-5; e-ISBN: 2351580214); *Eds. J. G. Cabrera and R. Rivera-Villarreal*
PRO 6: High Performance Fiber Reinforced Cement Composites - HPFRCC 3 (ISBN: 2-912143-06-3; e-ISBN: 2351580222); *Eds. H. W. Reinhardt and A. E. Naaman*
PRO 7: 1st International RILEM Symposium on Self-Compacting Concrete (ISBN: 2-912143-09-8; e-ISBN: 2912143721); *Eds. Å. Skarendahl and Ö. Petersson*
PRO 8: International RILEM Symposium on Timber Engineering (ISBN: 2-912143-10-1; e-ISBN: 2351580230); *Ed. L. Boström*
PRO 9: 2nd International RILEM Symposium on Adhesion between Polymers and Concrete ISAP '99 (ISBN: 2-912143-11-X; e-ISBN: 2351580249); *Eds. Y. Ohama and M. Puterman*
PRO 10: 3rd International RILEM Symposium on Durability of Building and Construction Sealants (ISBN: 2-912143-13-6; e-ISBN: 2351580257); *Ed. A. T. Wolf*

PRO 11: 4th International RILEM Conference on Reflective Cracking in Pavements (ISBN: 2-912143-14-4; e-ISBN: 2351580265); *Eds. A. O. Abd El Halim, D. A. Taylor and El H. H. Mohamed*

PRO 12: International RILEM Workshop on Historic Mortars: Characteristics and Tests (ISBN: 2-912143-15-2; e-ISBN: 2351580273); *Eds. P. Bartos, C. Groot and J. J. Hughes*

PRO 13: 2nd International RILEM Symposium on Hydration and Setting (ISBN: 2-912143-16-0; e-ISBN: 2351580281); *Ed. A. Nonat*

PRO 14: Integrated Life-Cycle Design of Materials and Structures - ILCDES 2000 (ISBN: 951-758-408-3; e-ISBN: 235158029X); (ISSN: 0356-9403); *Ed. S. Sarja*

PRO 15: Fifth RILEM Symposium on Fibre-Reinforced Concretes (FRC) - BEFIB'2000 (ISBN: 2-912143-18-7; e-ISBN: 291214373X); *Eds. P. Rossi and G. Chanvillard*

PRO 16: Life Prediction and Management of Concrete Structures (ISBN: 2-912143-19-5; e-ISBN: 2351580303); *Ed. D. Naus*

PRO 17: Shrinkage of Concrete – Shrinkage 2000 (ISBN: 2-912143-20-9; e-ISBN: 2351580311); *Eds. V. Baroghel-Bouny and P.-C. Aïtcin*

PRO 18: Measurement and Interpretation of the On-Site Corrosion Rate (ISBN: 2-912143-21-7; e-ISBN: 235158032X); *Eds. C. Andrade, C. Alonso, J. Fullea, J. Polimon and J. Rodriguez*

PRO 19: Testing and Modelling the Chloride Ingress into Concrete (ISBN: 2-912143-22-5; e-ISBN: 2351580338); *Eds. C. Andrade and J. Kropp*

PRO 20: 1st International RILEM Workshop on Microbial Impacts on Building Materials (CD 02) (e-ISBN: 978-2-35158-013-4); *Ed. M. Ribas Silva*

PRO 21: International RILEM Symposium on Connections between Steel and Concrete (ISBN: 2-912143-25-X; e-ISBN: 2351580346); *Ed. R. Eligehausen*

PRO 22: International RILEM Symposium on Joints in Timber Structures (ISBN: 2-912143-28-4; e-ISBN: 2351580354); *Eds. S. Aicher and H.-W. Reinhardt*

PRO 23: International RILEM Conference on Early Age Cracking in Cementitious Systems (ISBN: 2-912143-29-2; e-ISBN: 2351580362); *Eds. K. Kovler and A. Bentur*

PRO 24: 2nd International RILEM Workshop on Frost Resistance of Concrete (ISBN: 2-912143-30-6; e-ISBN: 2351580370); *Eds. M. J. Setzer, R. Auberg and H.-J. Keck*

PRO 25: International RILEM Workshop on Frost Damage in Concrete (ISBN: 2-912143-31-4; e-ISBN: 2351580389); *Eds. D. J. Janssen, M. J. Setzer and M. B. Snyder*

PRO 26: International RILEM Workshop on On-Site Control and Evaluation of Masonry Structures (ISBN: 2-912143-34-9; e-ISBN: 2351580141); *Eds. L. Binda and R. C. de Vekey*

PRO 27: International RILEM Symposium on Building Joint Sealants (CD03; e-ISBN: 235158015X); *Ed. A. T. Wolf*

PRO 28: 6th International RILEM Symposium on Performance Testing and Evaluation of Bituminous Materials - PTEBM'03 (ISBN: 2-912143-35-7; e-ISBN: 978-2-912143-77-8); *Ed. M. N. Partl*

PRO 29: 2nd International RILEM Workshop on Life Prediction and Ageing Management of Concrete Structures (ISBN: 2-912143-36-5; e-ISBN: 2912143780); *Ed. D. J. Naus*

PRO 30: 4th International RILEM Workshop on High Performance Fiber Reinforced Cement Composites - HPFRCC 4 (ISBN: 2-912143-37-3; e-ISBN: 2912143799); *Eds. A. E. Naaman and H. W. Reinhardt*

PRO 31: International RILEM Workshop on Test and Design Methods for Steel Fibre Reinforced Concrete: Background and Experiences (ISBN: 2-912143-38-1; e-ISBN: 2351580168); *Eds. B. Schnütgen and L. Vandewalle*

PRO 32: International Conference on Advances in Concrete and Structures 2 vol. (ISBN (set): 2-912143-41-1; e-ISBN: 2351580176); *Eds. Ying-shu Yuan, Surendra P. Shah and Heng-lin Lü*

PRO 33: 3rd International Symposium on Self-Compacting Concrete (ISBN: 2-912143-42-X; e-ISBN: 2912143713); *Eds. Ó. Wallevik and I. Níelsson*

PRO 34: International RILEM Conference on Microbial Impact on Building Materials (ISBN: 2-912143-43-8; e-ISBN: 2351580184); *Ed. M. Ribas Silva*

PRO 35: International RILEM TC 186-ISA on Internal Sulfate Attack and Delayed Ettringite Formation (ISBN: 2-912143-44-6; e-ISBN: 2912143802); *Eds. K. Scrivener and J. Skalny*

PRO 36: International RILEM Symposium on Concrete Science and Engineering – A Tribute to Arnon Bentur (ISBN: 2-912143-46-2; e-ISBN: 2912143586); *Eds. K. Kovler, J. Marchand, S. Mindess and J. Weiss*

PRO 37: 5th International RILEM Conference on Cracking in Pavements – Mitigation, Risk Assessment and Prevention (ISBN: 2-912143-47-0; e-ISBN: 2912143764); *Eds. C. Petit, I. Al-Qadi and A. Millien*

PRO 38: 3rd International RILEM Workshop on Testing and Modelling the Chloride Ingress into Concrete (ISBN: 2-912143-48-9; e-ISBN: 2912143578); *Eds. C. Andrade and J. Kropp*

PRO 39: 6th International RILEM Symposium on Fibre-Reinforced Concretes - BEFIB 2004 (ISBN: 2-912143-51-9; e-ISBN: 2912143748); *Eds. M. Di Prisco, R. Felicetti and G. A. Plizzari*

PRO 40: International RILEM Conference on the Use of Recycled Materials in Buildings and Structures (ISBN: 2-912143-52-7; e-ISBN: 2912143756); *Eds. E. Vázquez, Ch. F. Hendriks and G. M. T. Janssen*

PRO 41: RILEM International Symposium on Environment-Conscious Materials and Systems for Sustainable Development (ISBN: 2-912143-55-1; e-ISBN: 2912143640); *Eds. N. Kashino and Y. Ohama*

PRO 42: SCC'2005 - China: 1st International Symposium on Design, Performance and Use of Self-Consolidating Concrete (ISBN: 2-912143-61-6; e-ISBN: 2912143624); *Eds. Zhiwu Yu, Caijun Shi, Kamal Henri Khayat and Youjun Xie*

PRO 43: International RILEM Workshop on Bonded Concrete Overlays (e-ISBN: 2-912143-83-7); *Eds. J. L. Granju and J. Silfwerbrand*

PRO 44: 2nd International RILEM Workshop on Microbial Impacts on Building Materials (CD11) (e-ISBN: 2-912143-84-5); *Ed. M. Ribas Silva*

PRO 45: 2nd International Symposium on Nanotechnology in Construction, Bilbao (ISBN: 2-912143-87-X; e-ISBN: 2912143888); *Eds. Peter J. M. Bartos, Yolanda de Miguel and Antonio Porro*

PRO 46: ConcreteLife'06 - International RILEM-JCI Seminar on Concrete Durability and Service Life Planning: Curing, Crack Control, Performance in Harsh Environments (ISBN: 2-912143-89-6; e-ISBN: 291214390X); *Ed. K. Kovler*

PRO 47: International RILEM Workshop on Performance Based Evaluation and Indicators for Concrete Durability (ISBN: 978-2-912143-95-2; e-ISBN: 9782912143969); *Eds. V. Baroghel-Bouny, C. Andrade, R. Torrent and K. Scrivener*

PRO 48: 1st International RILEM Symposium on Advances in Concrete through Science and Engineering (e-ISBN: 2-912143-92-6); *Eds. J. Weiss, K. Kovler, J. Marchand, and S. Mindess*

PRO 49: International RILEM Workshop on High Performance Fiber Reinforced Cementitious Composites in Structural Applications (ISBN: 2-912143-93-4; e-ISBN: 2912143942); *Eds. G. Fischer and V. C. Li*

PRO 50: 1st International RILEM Symposium on Textile Reinforced Concrete (ISBN: 2-912143-97-7; e-ISBN: 2351580087); *Eds. Josef Hegger, Wolfgang Brameshuber and Norbert Will*

PRO 51: 2nd International Symposium on Advances in Concrete through Science and Engineering (ISBN: 2-35158-003-6; e-ISBN: 2-35158-002-8); *Eds. J. Marchand, B. Bissonnette, R. Gagné, M. Jolin and F. Paradis*

PRO 52: Volume Changes of Hardening Concrete: Testing and Mitigation (ISBN: 2-35158-004-4; e-ISBN: 2-35158-005-2); *Eds. O. M. Jensen, P. Lura and K. Kovler*

PRO 53: High Performance Fiber Reinforced Cement Composites - HPFRCC5 (ISBN: 978-2-35158-046-2; e-ISBN: 978-2-35158-089-9); *Eds. H. W. Reinhardt and A. E. Naaman*

PRO 54: 5th International RILEM Symposium on Self-Compacting Concrete (ISBN: 978-2-35158-047-9; e-ISBN: 978-2-35158-088-2); *Eds. G. De Schutter and V. Boel*

PRO 55: International RILEM Symposium Photocatalysis, Environment and Construction Materials (ISBN: 978-2-35158-056-1; e-ISBN: 978-2-35158-057-8); *Eds. P. Baglioni and L. Cassar*

PRO 56: International RILEM Workshop on Integral Service Life Modelling of Concrete Structures (ISBN: 978-2-35158-058-5; e-ISBN: 978-2-35158-090-5); *Eds. R. M. Ferreira, J. Gulikers and C. Andrade*

PRO 57: RILEM Workshop on Performance of cement-based materials in aggressive aqueous environments (e-ISBN: 978-2-35158-059-2); *Ed. N. De Belie*

PRO 58: International RILEM Symposium on Concrete Modelling - CONMOD'08 (ISBN: 978-2-35158-060-8; e-ISBN: 978-2-35158-076-9); *Eds. E. Schlangen and G. De Schutter*

PRO 59: International RILEM Conference on On Site Assessment of Concrete, Masonry and Timber Structures - SACoMaTiS 2008 (ISBN set:

978-2-35158-061-5; e-ISBN: 978-2-35158-075-2); *Eds. L. Binda, M. di Prisco and R. Felicetti*

PRO 60: Seventh RILEM International Symposium on Fibre Reinforced Concrete: Design and Applications - BEFIB 2008 (ISBN: 978-2-35158-064-6; e-ISBN: 978-2-35158-086-8); *Ed. R. Gettu*

PRO 61: 1st International Conference on Microstructure Related Durability of Cementitious Composites 2 vol., (ISBN: 978-2-35158-065-3; e-ISBN: 978-2-35158-084-4); *Eds. W. Sun, K. van Breugel, C. Miao, G. Ye and H. Chen*

PRO 62: NSF/RILEM Workshop: In-situ Evaluation of Historic Wood and Masonry Structures (e-ISBN: 978-2-35158-068-4); *Eds. B. Kasal, R. Anthony and M. Drdácký*

PRO 63: Concrete in Aggressive Aqueous Environments: Performance, Testing and Modelling, 2 vol., (ISBN: 978-2-35158-071-4; e-ISBN: 978-2-35158-082-0); *Eds. M. G. Alexander and A. Bertron*

PRO 64: Long Term Performance of Cementitious Barriers and Reinforced Concrete in Nuclear Power Plants and Waste Management - NUCPERF 2009 (ISBN: 978-2-35158-072-1; e-ISBN: 978-2-35158-087-5); *Eds. V. L'Hostis, R. Gens, C. Gallé*

PRO 65: Design Performance and Use of Self-consolidating Concrete - SCC'2009 (ISBN: 978-2-35158-073-8; e-ISBN: 978-2-35158-093-6); *Eds. C. Shi, Z. Yu, K. H. Khayat and P. Yan*

PRO 66: 2nd International RILEM Workshop on Concrete Durability and Service Life Planning - ConcreteLife'09 (ISBN: 978-2-35158-074-5; ISBN: 978-2-35158-074-5); *Ed. K. Kovler*

PRO 67: Repairs Mortars for Historic Masonry (e-ISBN: 978-2-35158-083-7); *Ed. C. Groot*

PRO 68: Proceedings of the 3rd International RILEM Symposium on 'Rheology of Cement Suspensions such as Fresh Concrete (ISBN: 978-2-35158-091-2; e-ISBN: 978-2-35158-092-9); *Eds. O. H. Wallevik, S. Kubens and S. Oesterheld*

PRO 69: 3rd International PhD Student Workshop on 'Modelling the Durability of Reinforced Concrete (ISBN: 978-2-35158-095-0); *Eds. R. M. Ferreira, J. Gulikers and C. Andrade*

PRO 70: 2nd International Conference on 'Service Life Design for Infrastructure' (ISBN set: 978-2-35158-096-7, e-ISBN: 978-2-35158-097-4); *Eds. K. van Breugel, G. Ye and Y. Yuan*

PRO 71: Advances in Civil Engineering Materials - The 50-year Teaching Anniversary of Prof. Sun Wei' (ISBN: 978-2-35158-098-1; e-ISBN: 978-2-35158-099-8); *Eds. C. Miao, G. Ye, and H. Chen*

PRO 72: First International Conference on 'Advances in Chemically-Activated Materials – CAM'2010' (2010), 264 pp, ISBN: 978-2-35158-101-8; e-ISBN: 978-2-35158-115-5, *Eds. Caijun Shi and Xiaodong Shen*

PRO 73: 2nd International Conference on 'Waste Engineering and Management - ICWEM 2010' (2010), 894 pp, ISBN: 978-2-35158-102-5; e-ISBN: 978-2-35158-103-2, *Eds. J. Zh. Xiao, Y. Zhang, M. S. Cheung and R. Chu*

PRO 74: International RILEM Conference on 'Use of Superabsorsorbent Polymers and Other New Additives in Concrete' (2010) 374 pp., ISBN: 978-2-35158-104-9; e-ISBN: 978-2-35158-105-6; *Eds. O. M. Jensen, M. T. Hasholt, and S. Laustsen*
PRO 75: International Conference on 'Material Science - 2nd ICTRC - Textile Reinforced Concrete - Theme 1' (2010) 436 pp., ISBN: 978-2-35158-106-3; e-ISBN: 978-2-35158-107-0; *Ed. W. Brameshuber*
PRO 76: International Conference on 'Material Science - HetMat - Modelling of Heterogeneous Materials - Theme 2' (2010) 255 pp., ISBN: 978-2-35158-108-7; e-ISBN: 978-2-35158-109-4; *Ed. W. Brameshuber*
PRO 77: International Conference on 'Material Science - AdIPoC - Additions Improving Properties of Concrete - Theme 3' (2010) 459 pp., ISBN: 978-2-35158-110-0; e-ISBN: 978-2-35158-111-7; *Ed. W. Brameshuber*
PRO 78: 2nd Historic Mortars Conference and RILEM TC 203-RHM Final Workshop – HMC2010 (2010) 1416 pp., e-ISBN: 978-2-35158-112-4; *Eds. J. Válek, C. Groot, and J. J. Hughes*
PRO 79: International RILEM Conference on Advances in Construction Materials Through Science and Engineering (2011) 213 pp., ISBN: 978-2-35158-116-2, e-ISBN: 978-2-35158-117-9; *Eds. Christopher Leung and K. T. Wan*
PRO 80: 2nd International RILEM Conference on Concrete Spalling due to Fire Exposure (2011) 453 pp., ISBN: 978-2-35158-118-6, e-ISBN: 978-2-35158-119-3; *Eds. E. A. B. Koenders and F. Dehn*
PRO 81: 2nd International RILEM Conference on Strain Hardening Cementitious Composites (SHCC2-Rio) (2011) 451 pp., ISBN: 978-2-35158-120-9, e-ISBN: 978-2-35158-121-6; *Eds. R. D. Toledo Filho, F. A. Silva, E. A. B. Koenders and E. M. R. Fairbairn*
PRO 82: 2nd International RILEM Conference on Progress of Recycling in the Built Environment (2011) 507 pp., e-ISBN: 978-2-35158-122-3; *Eds. V. M. John, E. Vazquez, S. C. Angulo and C. Ulsen*
PRO 83: 2nd International Conference on Microstructural-related Durability of Cementitious Composites (2012) 250 pp., ISBN: 978-2-35158-129-2; e-ISBN: 978-2-35158-123-0; *Eds. G. Ye, K. van Breugel, W. Sun and C. Miao*
PRO 84: CONSEC13 - Seventh International Conference on Concrete under Severe Conditions – Environment and Loading (2013) 1930 pp., ISBN: 978-2-35158-124-7; e-ISBN: 978-2- 35158-134-6; *Eds. Z. J. Li, W. Sun, C. W. Miao, K. Sakai, O. E. Gjorv and N. Banthia*
PRO 85: RILEM-JCI International Workshop on Crack Control of Mass Concrete and Related issues concerning Early-Age of Concrete Structures – ConCrack 3 – Control of Cracking in Concrete Structures 3 (2012) 237 pp., ISBN: 978-2-35158-125-4; e-ISBN: 978-2-35158-126-1; *Eds. F. Toutlemonde and J.-M. Torrenti*
PRO 86: International Symposium on Life Cycle Assessment and Construction (2012) 414 pp., ISBN: 978-2-35158-127-8, e-ISBN: 978-2-35158-128-5; *Eds. A. Ventura and C. de la Roche*

PRO 87: UHPFRC 2013 – RILEM-fib-AFGC International Symposium on Ultra-High Performance Fibre-Reinforced Concrete (2013), ISBN: 978-2-35158-130-8, e-ISBN: 978-2-35158-131-5; *Ed. F. Toutlemonde*

PRO 88: 8th RILEM International Symposium on Fibre Reinforced Concrete (2012) 344 pp., ISBN: 978-2-35158-132-2, e-ISBN: 978-2-35158-133-9; *Ed. Joaquim A. O. Barros*

PRO 89: RILEM International workshop on performance-based specification and control of concrete durability (2014) 678 pp, ISBN: 978-2-35158-135-3, e-ISBN: 978-2-35158-136-0; *Eds. D. Bjegović, H. Beushausen and M. Serdar*

PRO 90: 7th RILEM International Conference on Self-Compacting Concrete and of the 1st RILEM International Conference on Rheology and Processing of Construction Materials (2013) 396 pp., ISBN: 978-2-35158-137-7, e-ISBN: 978-2-35158-138-4; *Eds. Nicolas Roussel and Hela Bessaies-Bey*

PRO 91: CONMOD 2014 - RILEM International Symposium on Concrete Modelling (2014), ISBN: 978-2-35158-139-1; e-ISBN: 978-2-35158-140-7; *Eds. Kefei Li, Peiyu Yan and Rongwei Yang*

PRO 92: CAM 2014 - 2nd International Conference on advances in chemically-activated materials (2014) 392 pp., ISBN: 978-2-35158-141-4; e-ISBN: 978-2-35158-142-1; *Eds. Caijun Shi and Xiadong Shen*

PRO 93: SCC 2014 - 3rd International Symposium on Design, Performance and Use of Self-Consolidating Concrete (2014) 438 pp., ISBN: 978-2-35158-143-8; e-ISBN: 978-2-35158-144-5; *Eds. Caijun Shi, Zhihua Ou and Kamal H. Khayat*

PRO 94 (online version): HPFRCC-7 - 7th RILEM conference on High performance fiber reinforced cement composites (2015), e-ISBN: 978-2-35158-146-9; *Eds. H. W. Reinhardt, G. J. Parra-Montesinos and H. Garrecht*

PRO 95: International RILEM Conference on Application of superabsorbent polymers and other new admixtures in concrete construction (2014), ISBN: 978-2-35158-147-6; e-ISBN: 978-2-35158-148-3; *Eds. Viktor Mechtcherine and Christof Schroefl*

PRO 96 (online version): XIII DBMC: XIII International Conference on Durability of Building Materials and Components(2015), e-ISBN: 978-2-35158-149-0; *Eds. M. Quattrone and V. M. John*

PRO 97: SHCC3 – 3rd International RILEM Conference on Strain Hardening Cementitious Composites (2014), ISBN: 978-2-35158-150-6; e-ISBN: 978-2-35158-151-3; *Eds. E. Schlangen, M. G. Sierra Beltran, M. Lukovic and G. Ye*

PRO 98: FERRO-11 – 11th International Symposium on Ferrocement and 3rd ICTRC - International Conference on Textile Reinforced Concrete (2015), ISBN: 978-2-35158-152-0; e-ISBN: 978-2-35158-153-7; *Ed. W. Brameshuber*

PRO 99 (online version): ICBBM 2015 - 1st International Conference on Bio-Based Building Materials (2015), e-ISBN: 978-2-35158-154-4; *Eds. S. Amziane and M. Sonebi*

PRO 100: SCC16 - RILEM Self-Consolidating Concrete Conference (2016), ISBN: 978-2-35158-156-8; e-ISBN: 978-2-35158-157-5; *Ed. Kamal H. Kayat*

PRO 101 (online version): III Progress of Recycling in the Built Environment (2015), e-ISBN: 978-2-35158-158-2; *Eds. I. Martins, C. Ulsen and S. C. Angulo*

PRO 102 (online version): RILEM Conference on Microorganisms-Cementitious Materials Interactions (2016), e-ISBN: 978-2-35158-160-5; *Eds. Alexandra Bertron, Henk Jonkers and Virginie Wiktor*

PRO 103 (online version): ACESC'16 - Advances in Civil Engineering and Sustainable Construction (2016), e-ISBN: 978-2-35158-161-2; *Eds. T. Ch. Madhavi, G. Prabhakar, Santhosh Ram and P. M. Rameshwaran*

PRO 104 (online version): SSCS'2015 - Numerical Modeling - Strategies for Sustainable Concrete Structures (2015), e-ISBN: 978-2-35158-162-9

PRO 105: 1st International Conference on UHPC Materials and Structures (2016), ISBN: 978-2-35158-164-3, e-ISBN: 978-2-35158-165-0

PRO 106: AFGC-ACI-fib-RILEM International Conference on Ultra-High-Performance Fibre-Reinforced Concrete – UHPFRC 2017 (2017), ISBN: 978-2-35158-166-7, e-ISBN: 978-2-35158-167-4; *Eds. François Toutlemonde and Jacques Resplendino*

PRO 107 (online version): XIV DBMC – 14th International Conference on Durability of Building Materials and Components (2017), e-ISBN: 978-2-35158-159-9; *Eds. Geert De Schutter, Nele De Belie, Arnold Janssens and Nathan Van Den Bossche*

PRO 108: MSSCE 2016 - Innovation of Teaching in Materials and Structures (2016), ISBN: 978-2-35158-178-0, e-ISBN: 978-2-35158-179-7; *Ed. Per Goltermann*

PRO 109 (2 volumes): MSSCE 2016 - Service Life of Cement-Based Materials and Structures (2016), ISBN Vol. 1: 978-2-35158-170-4, Vol. 2: 978-2-35158-171-4, Set Vol. 1&2: 978-2-35158-172-8, e-ISBN: 978-2-35158-173-5; *Eds. Miguel Azenha, Ivan Gabrijel, Dirk Schlicke, Terje Kanstad and Ole Mejlhede Jensen*

PRO 110: MSSCE 2016 - Historical Masonry (2016), ISBN: 978-2-35158-178-0, e-ISBN: 978-2-35158-179-7; *Eds. Inge Rörig-Dalgaard and Ioannis Ioannou*

PRO 111: MSSCE 2016 - Electrochemistry in Civil Engineering (2016), ISBN: 978-2-35158-176-6, e-ISBN: 978-2-35158-177-3; *Ed. Lisbeth M. Ottosen*

PRO 112: MSSCE 2016 - Moisture in Materials and Structures (2016), ISBN: 978-2-35158-178-0, e-ISBN: 978-2-35158-179-7; *Eds. Kurt Kielsgaard Hansen, Carsten Rode and Lars-Olof Nilsson*

PRO 113: MSSCE 2016 - Concrete with Supplementary Cementitious Materials (2016), ISBN: 978-2-35158-178-0, e-ISBN: 978-2-35158-179-7; *Eds. Ole Mejlhede Jensen, Konstantin Kovler and Nele De Belie*

PRO 114: MSSCE 2016 - Frost Action in Concrete (2016), ISBN: 978-2-35158-182-7, e-ISBN: 978-2-35158-183-4; *Eds. Marianne Tange Hasholt, Katja Fridh and R. Doug Hooton*

PRO 115: MSSCE 2016 - Fresh Concrete (2016), ISBN: 978-2-35158-184-1, e-ISBN: 978-2-35158-185-8; *Eds. Lars N. Thrane, Claus Pade, Oldrich Svec and Nicolas Roussel*

PRO 116: BEFIB 2016 – 9th RILEM International Symposium on Fiber Reinforced Concrete (2016), ISBN: 978-2-35158-187-2, e-ISBN: 978-2-35158-186-5; *Eds. N. Banthia, M. di Prisco and S. Soleimani-Dashtaki*

PRO 117: 3rd International RILEM Conference on Microstructure Related Durability of Cementitious Composites (2016), ISBN: 978-2-35158-188-9, e-ISBN: 978-2-35158-189-6; *Eds. Changwen Miao, Wei Sun, Jiaping Liu, Huisu Chen, Guang Ye and Klaas van Breugel*

PRO 118 (4 volumes): International Conference on Advances in Construction Materials and Systems (2017), ISBN Set: 978-2-35158-190-2, Vol. 1: 978-2-35158-193-3, Vol. 2: 978-2-35158-194-0, Vol. 3: ISBN: 978-2-35158-195-7, Vol. 4: ISBN: 978-2-35158-196-4, e-ISBN: 978-2-35158-191-9; *Ed. Manu Santhanam*

PRO 119 (online version): ICBBM 2017 - Second International RILEM Conference on Bio-based Building Materials, (2017), e-ISBN: 978-2-35158-192-6; *Ed. Sofiane Amziane*

PRO 120 (2 volumes): EAC-02 - 2nd International RILEM/COST Conference on Early Age Cracking and Serviceability in Cement-based Materials and Structures, (2017), Vol. 1: 978-2-35158-199-5, Vol. 2: 978-2-35158-200-8, Set: 978-2-35158-197-1, e-ISBN: 978-2-35158-198-8; *Eds. Stéphanie Staquet and Dimitrios Aggelis*

PRO 121 (2 volumes): SynerCrete18: Interdisciplinary Approaches for Cement-based Materials and Structural Concrete: Synergizing Expertise and Bridging Scales of Space and Time, (2018), Set: 978-2-35158-202-2, Vol. 1: 978-2-35158-211-4, Vol. 2: 978-2-35158-212-1, e-ISBN: 978-2-35158-203-9; *Eds. Miguel Azenha, Dirk Schlicke, Farid Benboudjema and Agnieszka Knoppik*

PRO 122: SCC'2018 China - Fourth International Symposium on Design, Performance and Use of Self-Consolidating Concrete, (2018), ISBN: 978-2-35158-204-6, e-ISBN: 978-2-35158-205-3; *Eds. C. Shi, Z. Zhang and K. H. Khayat*

PRO 123: Final Conference of RILEM TC 253-MCI: Microorganisms-Cementitious Materials Interactions (2018), Set: 978-2-35158-207-7, Vol. 1: 978-2-35158-209-1, Vol. 2: 978-2-35158-210-7, e-ISBN: 978-2-35158-206-0; *Ed. Alexandra Bertron*

PRO 124 (online version): Fourth International Conference Progress of Recycling in the Built Environment (2018), e-ISBN: 978-2-35158-208-4; *Eds. Isabel M. Martins, Carina Ulsen and Yury Villagran*

PRO 125 (online version): SLD4 - 4th International Conference on Service Life Design for Infrastructures (2018), e-ISBN: 978-2-35158-213-8; *Eds. Guang Ye, Yong Yuan, Claudia Romero Rodriguez, Hongzhi Zhang and Branko Savija*

PRO 126: Workshop on Concrete Modelling and Material Behaviour in honor of Professor Klaas van Breugel (2018), ISBN: 978-2-35158-214-5, e-ISBN: 978-2-35158-215-2; *Ed. Guang Ye*

PRO 127 (online version): CONMOD2018 - Symposium on Concrete Modelling (2018), e-ISBN: 978-2-35158-216-9; *Eds. Erik Schlangen, Geert de Schutter, Branko Savija, Hongzhi Zhang and Claudia Romero Rodriguez*

PRO 128: SMSS2019 - International Conference on Sustainable Materials, Systems and Structures (2019), ISBN: 978-2-35158-217-6, e-ISBN: 978-2-35158-218-3

RILEM Reports (REP)

Report 19: Considerations for Use in Managing the Aging of Nuclear Power Plant Concrete Structures (ISBN: 2-912143-07-1); *Ed. D. J. Naus*
Report 20: Engineering and Transport Properties of the Interfacial Transition Zone in Cementitious Composites (ISBN: 2-912143-08-X); *Eds. M. G. Alexander, G. Arliguie, G. Ballivy, A. Bentur and J. Marchand*
Report 21: Durability of Building Sealants (ISBN: 2-912143-12-8); *Ed. A. T. Wolf*
Report 22: Sustainable Raw Materials - Construction and Demolition Waste (ISBN: 2-912143-17-9); *Eds. C. F. Hendriks and H. S. Pietersen*
Report 23: Self-Compacting Concrete state-of-the-art report (ISBN: 2-912143-23-3); *Eds. Å. Skarendahl and Ö. Petersson*
Report 24: Workability and Rheology of Fresh Concrete: Compendium of Tests (ISBN: 2-912143-32-2); *Eds. P. J. M. Bartos, M. Sonebi and A. K. Tamimi*
Report 25: Early Age Cracking in Cementitious Systems (ISBN: 2-912143-33-0); *Ed. A. Bentur*
Report 26: Towards Sustainable Roofing (Joint Committee CIB/RILEM) (CD 07) (e-ISBN: 978-2-912143-65-5); *Eds. Thomas W. Hutchinson and Keith Roberts*
Report 27: Condition Assessment of Roofs (Joint Committee CIB/RILEM) (CD 08) (e-ISBN: 978-2-912143-66-2); *Ed. CIB W 83/RILEM TC166-RMS*
Report 28: Final report of RILEM TC 167-COM 'Characterisation of Old Mortars with Respect to Their Repair (ISBN: 978-2-912143-56-3); *Eds. C. Groot, G. Ashall and J. Hughes*
Report 29: Pavement Performance Prediction and Evaluation (PPPE): Interlaboratory Tests (e-ISBN: 2-912143-68-3); *Eds. M. Partl and H. Piber*
Report 30: Final Report of RILEM TC 198-URM 'Use of Recycled Materials' (ISBN: 2-912143-82-9; e-ISBN: 2-912143-69-1); *Eds. Ch. F. Hendriks, G. M. T. Janssen and E. Vázquez*
Report 31: Final Report of RILEM TC 185-ATC 'Advanced testing of cement-based materials during setting and hardening' (ISBN: 2-912143-81-0; e-ISBN: 2-912143-70-5); *Eds. H. W. Reinhardt and C. U. Grosse*
Report 32: Probabilistic Assessment of Existing Structures. A JCSS publication (ISBN: 2-912143-24-1); *Ed. D. Diamantidis*
Report 33: State-of-the-Art Report of RILEM Technical Committee TC 184-IFE 'Industrial Floors' (ISBN: 2-35158-006-0); *Ed. P. Seidler*
Report 34: Report of RILEM Technical Committee TC 147-FMB 'Fracture mechanics applications to anchorage and bond' Tension of Reinforced Concrete Prisms – Round Robin Analysis and Tests on Bond (e-ISBN: 2-912143-91-8); *Eds. L. Elfgren and K. Noghabai*

Report 35: Final Report of RILEM Technical Committee TC 188-CSC 'Casting of Self Compacting Concrete' (ISBN: 2-35158-001-X; e-ISBN: 2-912143-98-5); *Eds. Å. Skarendahl and P. Billberg*

Report 36: State-of-the-Art Report of RILEM Technical Committee TC 201-TRC 'Textile Reinforced Concrete' (ISBN: 2-912143-99-3); *Ed. W. Brameshuber*

Report 37: State-of-the-Art Report of RILEM Technical Committee TC 192-ECM 'Environment-conscious construction materials and systems' (ISBN: 978-2-35158-053-0); *Eds. N. Kashino, D. Van Gemert and K. Imamoto*

Report 38: State-of-the-Art Report of RILEM Technical Committee TC 205-DSC 'Durability of Self-Compacting Concrete' (ISBN: 978-2-35158-048-6); *Eds. G. De Schutter and K. Audenaert*

Report 39: Final Report of RILEM Technical Committee TC 187-SOC 'Experimental determination of the stress-crack opening curve for concrete in tension' (ISBN: 978-2-35158-049-3); *Ed. J. Planas*

Report 40: State-of-the-Art Report of RILEM Technical Committee TC 189-NEC 'Non-Destructive Evaluation of the Penetrability and Thickness of the Concrete Cover' (ISBN: 978-2-35158-054-7); *Eds. R. Torrent and L. Fernández Luco*

Report 41: State-of-the-Art Report of RILEM Technical Committee TC 196-ICC 'Internal Curing of Concrete' (ISBN: 978-2-35158-009-7); *Eds. K. Kovler and O. M. Jensen*

Report 42: 'Acoustic Emission and Related Non-destructive Evaluation Techniques for Crack Detection and Damage Evaluation in Concrete' - Final Report of RILEM Technical Committee 212-ACD (e-ISBN: 978-2-35158-100-1); *Ed. M. Ohtsu*

Report 45: Repair Mortars for Historic Masonry - State-of-the-Art Report of RILEM Technical Committee TC 203-RHM (e-ISBN: 978-2-35158-163-6); *Eds. Paul Maurenbrecher and Caspar Groot*

Report 46: *S*urface delamination of concrete industrial floors and other durability related aspects guide - Report of RILEM Technical Committee TC 268-SIF (e-ISBN: 978-2-35158-201-5); *Ed. Valerie Pollet*

Chemo-Mechanical Characterization of Bituminous Materials: Bitumen Aging Mechanisms and Characterization

A Mechanism Based Reaction-Diffusion Model for Spurt Oxidation of Bitumen

Uwe Mühlich(✉)

University of Antwerp, Antwerp, Belgium
uwe.muehlich@uantwerpen.be

Abstract. Focusing on the spurt reaction, related primarily to the formation of sulfoxides, we derive a reaction-diffusion model within the frame of thermodynamics of irreversible processes (TIP). It can account for bitumen composition and, it has been derived with the objective of minimizing the number of purely phenomenological parameters by considering underlying mechanisms at molecular level. Key aspects of the model are discussed by means of film ageing simulations.

Keywords: Bitumen · Oxidative ageing · Reaction-diffusion model TIP

1 Introduction

Information about oxidative aging is typically obtained from film aging in combination with infrared spectroscopy, see, e.g., Petersen (2009), and papers cited therein. These experiments identify sulfoxides and carbonyl groups as the main reaction products of the oxidation process. Furthermore, experimental results indicate the existence of at least two reactions of different reaction rate: a fast "spurt" reaction and a much slower "long term" reaction. Spectroscopy results suggest in addition, that the spurt reaction produces almost exclusively sulfoxides, whereas the fast reaction generates both, sulfoxides and carbonyl groups. Peterson and co-workers discuss in several papers possible chemical mechanisms for these reactions, see, e.g., Petersen (2009) and Petersen and Glaser (2011).

A purely phenomenological reaction-diffusion model for simulating oxidative aging which considers two reaction mechanisms is proposed by Herrington (2012). Here, we present a reaction-diffusion model able to account for bitumen composition considering underlying mechanisms at molecular level.

2 Continuum Thermodynamics Framework

We consider a mixture of N individual species occupying at time τ a region B in space with boundary ∂B. The local mass balance for species α, valid everywhere in B, is given by

$$\partial_\tau \rho_\alpha + \operatorname{div} (\rho_\alpha \, \mathrm{v} + J_\alpha) = r_\alpha \tag{1}$$

L. D. Poulikakos et al. (Eds.): RILEM 252-CMB 2018, RILEM Bookseries 20, pp. 3–8, 2019.
https://doi.org/10.1007/978-3-030-00476-7_1

with partial density ρ_α, reference velocity vector v and diffusion flux vector $J_\alpha = J_{\alpha 1} e_1 + J_{\alpha 2}\ e_2 + J_{\alpha 3}\ e_3$, where the base vectors of a cartesian coordinate system are denoted by e_k, k = 1,2,3. Production or consumption of species α, e.g., due to chemical reactions, is accounted for by r_α and ∂_τ indicates the partial derivative with respect to time τ. The components of the individual flux vectors are given by

$$J_{\alpha j} = -\frac{1}{T}\sum_{\beta=1}^{N-1} L_{\alpha\beta}\left[\mu_\beta - \mu_N\right]_{,j} \tag{2}$$

with absolute temperature T and purely phenomenological coefficients $L_{\alpha\beta}$. In addition, the short hand notation $f_{,j}$ indicates the partial derivative of some function f with respect to the coordinate x_j. Provided, that there is only one chemical reaction, the individual source/ sink terms r_α possess the following general form

$$r_\alpha = \Lambda s_\alpha M_\alpha \tag{3}$$

with reaction rate density Λ, stoichiometric coefficients s_α and the molar mass of the individual representatives of species α, M_α. Regarding the chemical potentials of the individual species, we use a standard approach

$$\mu_\alpha = \mu_\alpha^0 + \ln a_\alpha = \mu_\alpha^0 + \ln \chi_\alpha + \ln \gamma_\alpha \tag{4}$$

where μ_α^0 is the chemical potential of the pure species and the activities a_α account for the deviation from μ_α^0 in the case of a mixture. The activity of species α is supposed to be proportional to its molar fraction in the mixture, i.e., $a_\alpha = \gamma_\alpha \chi_\alpha$ with activity coefficient γ_α. The absence of terms related to heat conduction indicates that we are limiting our self to isothermal conditions.

3 A Reaction-Diffusion Model for Spurt Reaction

Classification of bitumen in terms of composition can be accomplished, for instance, by means of so-called SARA fractions, Read and Whiteoak (2003). In addition, we consider the spurt mechanism proposed by Petersen (2009). However, we assume a simplified reaction path. The corresponding reaction can be described as follows. A dioxygen molecule comes close to a saturated ring, situated between two aromatic rings. Two hydrogen atoms are taken off from the saturated ring, which converts the latter into an aromatic ring. This process is called "aromatizing" in the following. In addition, the O_2 and the two hydrogen atoms find somewhere else a sulfur atom. A sulfoxide is formed and the remaining oxygen forms together with the two hydrogen atoms a water molecule.

Considering, that saturates do not participate at all in the chemical reaction, we define the following five species: 1-dioxygen, 2-water, 3-aromatized compounds with one S = O (ARA*), 4-aromatizable compounds, which means aromatics, resins and asphaltenes with traces of sulfur (ARA) and 5-saturates, numbered in the following by α = 1,…,5.

We assume that every ARA member can be affected only once, which implies that the ARA species consists of average compounds. Furthermore, the simplified chemical reaction equation for short term aging reads

$$1O_2 + 1ARA \Leftrightarrow 1ARA^* + 1H_2O.$$

All molar fractions can be expressed by means of χ_1, χ_2 and the initial molar fraction of saturates χ_5^0, as follows

$$\chi_3 = \chi_2 \tag{5}$$

$$\chi_5 = \chi_5^0[1 - \chi_1 - \chi_2] \tag{6}$$

$$\chi_4 = \left[1 - \chi_5^0\right]\left[1 - \chi_1 - \chi_2\right] - \chi_2 \tag{7}$$

For the system considered here, dioxygen is the only species which diffuses. Therefore, v = *0* is a natural choice regarding Eq. 1. The essential mass balances in our model, rewritten in molar concentrations, read therefore

$$\partial_\tau c_1 + \operatorname{div} \overline{J}_1 = \overline{r}_1 = -\Lambda$$

$$\partial_\tau c_1 = \overline{r}_2 = +\Lambda$$

with $\overline{J} = J_1/M_1, \overline{r}_1 = r_1/M_1$ and $\overline{r}_2 = r_2/M_2$. For deriving explicitly diffusion flux and reaction rate density, we prefer, however, to work with molar fractions.

The activity model discussed in Vidal (2003) in the context of regular solution theory reads

$$\ln \gamma_\alpha = \frac{v_\alpha^*}{RT}\left[\delta_\alpha - \overline{\delta}\right] \text{ with } \overline{\delta} = \sum_{\alpha=1}^{N} \delta_\alpha \Phi_\alpha, \tag{8}$$

where v_α^* is the molar volume of the pure substance α in liquid state, δ_α is the solubility parameter and Φ_α is the corresponding volumetric fraction

$$\Phi_\alpha = \frac{v_\alpha^* \chi_\alpha}{\sum_{\beta=1}^{N} v_\beta^* \chi_\beta}. \tag{9}$$

Since dioxygen is the only species which diffuses, all gradients in chemical potential difference can be expressed by means of one particular gradient, e.g. $[\mu_1 - \mu_2]_{,j}$ and all phenomenological coefficients in Eq. 2 can be condensed in one single parameter L_{11}^*

$$J_{1j} = \frac{L_{11}^*}{T} [\mu_1 - \mu_2]_{,j}$$

and by using the definition of the chemical potentials, we obtain

$$J_{1j} = -L\left\{ \left[1 + \frac{\chi_1}{1 - \chi_1 - \chi_2} + \chi_1 \frac{\partial A_{15}}{\partial \chi_1} \right] \chi_{1,j} + \left[\frac{\chi_1}{1 - \chi_1 - \chi_2} + \chi_1 \frac{\partial A_{15}}{\partial \chi_2} \right] \chi_{2,j} \right\} \quad (10)$$

if, in addition, the concept of mobility is used. Furthermore, it follows from Eq. 8, that

$$A_{15} = \ln \frac{\gamma_1}{\gamma_5} = \frac{v_1^* \left[\delta_1 - \bar{\delta} \right]^2}{v_5^* \left[\delta_5 - \bar{\delta} \right]^2} \quad (11)$$

To relate the reaction rate density as well to information about composition, a reaction model following Pekar and Samohyl (2014) is used here. The result for the reaction rate reads

$$\Lambda = B_{14} \chi_1 \left[1 - \chi_5^0 \right] \left[1 - \chi_1 - \chi_2 \right] \exp\left(\frac{v_1^*}{RT} \left[\delta_1 - \bar{\delta} \right]^2 + \frac{v_4^*}{RT} \left[\delta_4 - \bar{\delta} \right]^2 \right). \quad (12)$$

4 Film Oxidation Simulations

The material parameters used in the following are given in Table 1. Data for SARA-fractions are taken from Powers (2014) and Akbarzadeh et al. (2005), respectively. Data for water and dioxygen originate from standard tables Three different values for χ_5^0 are considered: $\chi_5^0 = 0.15, 0.25, 0.35$ referred to as composition 1,2 and 3, respectively, in the following. Film aging simulations were performed using a film of thickness h = 1 mm with only one free surface exposed to air. It is assumed here, that a one-dimensional model is sufficient for reproducing the situation. Following Herrington (2012), a fixed dioxygen concentration, $c_1^s = 8.10 \times 10^{-7}$ mol/cm^3 is prescribed at the free surface. Furthermore, the fluxes of oxygen and water vanish at the bottom of the film. In addition, the flux term of water vanishes as well at the free surface. The finite element scheme for the one-dimensional case has been implemented in FEniCS, Alnaes et al. (2015). Oxygen uptake $C_{O_2}(\tau)$ as a function of time τ is computed during simulations. For composition 3 ($\chi_5^0 = 0.35$) and two combinations of L and B_{14}, the predictions obtained by the corresponding simulations are shown in Fig. 1.

5 Discussion

The model proposed here provides consistent results. Firstly, the reaction rate density Λ (Eq. 12) vanishes in the absence of reactants, dioxygen or ARA. Secondly, final oxygen uptake decreases with increasing amount of saturates,χ_5^0. Finally, the values

Table 1. Parameters used in simulations. The density for oxygen refers to liquid oxigen. The conversion factor for solubilities is $1\sqrt{MPa} = 3.162(g/cm/s^2)^{0.5}$

Species		Density g/cm^3	Molecular weight g/mol	Molar volume cm^3/mol	Solubility MPa$^{0.5}$	(g/cm/s^2)$^{0.5}$
dioxygen		1.114	32	28.0	14.0	44272.2
water		0.988	18	18.2	48.0	151790.4
ARA*				1709.0	20.0	63246.0
ARA	aromatics	1.006	440	437.0	21.0	
	resins	1.054	990	940.0	19.0	
	asphaltenes	1.200	4500	3750.0	20.0	
	average			1709.0	20.0	63246.0
saturates		0.887	370	417.0	16.5	52177.9

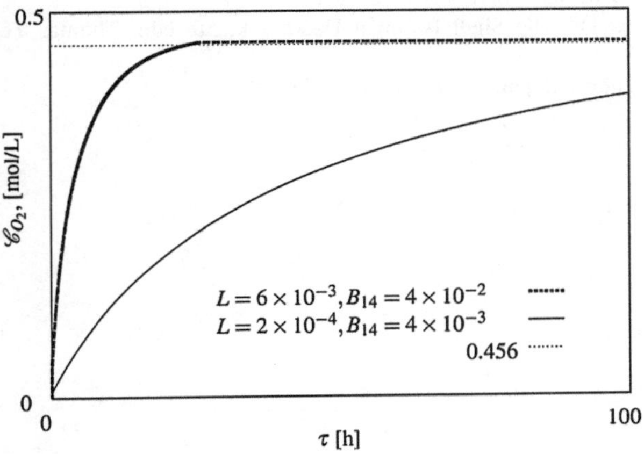

Fig. 1. Oxygen uptake in mol/L measured by integrating c_2 over film thickness as a function of time in hours for two combinations of diffusion and reaction parameters

obtained for final oxygen uptake are of same magnitude as the values reported in Petersen (2009).

On the other hand, diffusion is not significantly affected by the oxidation process in our model. In other words, our model cannot explain the change in diffusivity due to oxidation. The argument used, e.g., by Herrington (2012), that diffusivity depends on viscosity, which in turn depends on aging, implies, that a change in viscosity cannot be explained either just by accounting for the generation of sulfoxides. Our model therefore supports the hypothesis, that, simultaneously, there must be other processes at molecular level, causing eventually a stiffening of the material.

References

Akbarzadeh, K., Alboudwarej, H., Svrcek, W.Y., Yarranton, H.W.: A generalized regular solution model for asphaltene precipitation from n-alkane diluted heavy oils and bitumens. Fluid Phase Equilib. **232**(1), 159–170 (2005)

Alnaes, M., Blechta, J., Hake, J., Johansson, A., Kehlet, B., Logg, A., Richardson, C., Ring, J., Rognes, M., and Wells, G.: The fenics project version 1.5. Archive of Numerical Software, 3 (100) (2015)

Herrington, P.R.: Diffusion and reaction of oxygen in bitumen films. Fuel **94**, 86–92 (2012)

Pekar, M., Samohyl, I.: The Thermodynamics of Linear Fluids and Fluid Mixtures. Springer, Switzerland (2014)

Petersen, J.: A review of the fundamentals of asphalt oxidation. Technical report, Transportation Research Board (2009)

Petersen, J.C., Glaser, R.: Asphalt oxidation mechanisms and the role of oxidation products on age hardening revisited. Road Mater. Pavement Des. **12**(4), 795–819 (2011)

Powers, D.P.: Characterization and Asphaltene Precipitation Modeling of Native and Reacted Crude Oils. PhD thesis, University of Calgary (2014)

Read, J., Whiteoak, D.: The Shell Bitumen Handbook, 5th edn. Thomas Telford Publishing, London (2003)

Vidal, J. (2003). Thermodynamics. Editions OPHRYS

Aging Characterization of Biobinder Produced from Renewable Sources

Ingrid Gabrielle do Nascimento Camargo(✉),
Liedi Légi Bariani Bernucci, and Kamilla L. Vasconcelos

Department of Transportation Engineering, Polytechnic School,
University of São Paulo, São Paulo, Brazil
{ingridgnc,liedi,kamilla.vasconcelos}@usp.br

Abstract. This study aims to investigate the effect of the temperature in the aging of a biobinder obtained from vegetal source. In this sense, the Rolling Thin Film Oven Test test was performed according to the ASTM D2872 (2012), considering the standard test temperature (163 °C) and two others test temperatures (150 °C and 180 °C). The unaged and aged binders (Neat Asphalt Binder, AC 30/45, and the biobinder) had their rheological and chemical properties analyzed. Under the same aging conditions, it was found that the biobinder was more susceptible to aging than the neat asphalt binder (AC 30/45). Regarding the biobinder aged under different conditions, the incremental RTFOT temperature presented some beneficial aspects to high-temperature performance; however, it might compromise the performance at intermediate temperature which can be inferred from fatigue tests. Chemical results indicated that the carbonyl and sulphoxide indexes were not adequate for the analysis of the oxidation and aging of the biobinder.

Keywords: Biobinder · Aging · Rheology · Chemical analysis

1 Introduction

The demand for alternatives binders to asphalt binders is fostered by a number of political, social and economic issues, many of which are related to oil price instability, environmental policies to reduce emissions, and the search for reduction of dependence on oil and its derivatives. It is possible to obtain alternative binders, named generically of biobinders, from wide range of renewable sources and different production processes.

Manure from swine, wood residues, algae, vegetable and cooking oil residues are some of the possible renewable sources studied in previous researches (Fini et al. 2011; Yang et al. 2014; Dhasmana et al. 2015; Sun et al. 2017). According to Raouf and Williams (2010), the biobinder may be employed for different purposes: (i) as substitutes (100% replacement), (ii) as extenders (75% to 25% replacement), and (ii) as modifiers (less than 10%).

The feasibility analysis of employing biobinders as pavement construction material has been based on the current methodologies used to investigate the properties of asphalt binders and asphalt mixtures. However, it is still uncertain how effective are

L. D. Poulikakos et al. (Eds.): RILEM 252-CMB 2018, RILEM Bookseries 20, pp. 9–14, 2019.
https://doi.org/10.1007/978-3-030-00476-7_2

these methodologies to characterize important material properties and performance in the field such as: workability, water solubility, odor, adhesiveness, among others (Kluttz 2012).

In that sense, it is important to better understand the aging mechanism of the biobinders since it may interfere in the performance and durability of this material in the field. The aging of asphalt pavements is usually subdivided in short-term and long-term ageing related to, respectively: (i) The aging due to the high temperatures which the asphalt mixture is submitted during its manufacture, transport and compaction; (ii) Aging occurred during the useful life of the asphalt pavement, mainly related with the binder degradation given by its oxidation mechanism (Abbas et al. 2002; Read and Whiteoak 2003).

The short-term aging can be simulated in the laboratory by means of the Rolling Thin Film Oven Test (RTFOT). This procedure consists in submitting the binder to specific pressure and temperature conditions, in order to simulate the oxidation of the material during the production of the asphalt pavement related to the high temperatures that it is subjected. In case of conventional asphalt binders, the RTFOT is performed according to the protocols of the ASTM D 2872 (2012). However, when certain types of modified binders are considered, such testing should be adapted for an adequate time and/or test temperature (Airey 2003).

In this perspective, some studies considering the biobinders, as the developed by Williams et al. (2015), recommended to change the RTFOT test temperature to lower value (120 °C) than the recommended by the ASTM D 2872 (163 °C) standard. The temperature restrictions might be related the possible volatilization of some components and change of the chemical structure of some types of biobinder.

2 Materials and Methods

In this study 2 types of binders were analyzed: (i) one Neat asphalt binder, penetration grade AC 30/45; and (ii) one Biobinder produced from renewable plant raw materials. The physical properties of the AC 30/45 and the Biobinder are follow, respectively: softening point (°C): 55.0 and 56.5; penetration at 25 °C (0.1 mm): 25 and 29; viscosity at 135 °C (Pa.s): 0.45 and 0.33.

The standard method used to simulate the short-term aging of the binders was the RTFOT, performed according to the ASTM D 2872 (2012). Initially, the aging test was performed at the temperature of 163 °C, for both binders. Additionally, two other temperatures (150 °C and 180 °C) were considered in order to verify the effect of the temperature on the short-term aging of the biobinder. The effect of aging on binder chemistry and rheology was investigated using infrared spectroscopy and dynamic mechanical analysis, as described below.

The frequency sweep test was performed on the DHR-3 rheometer from TA Instruments. During the test execution it was adopted a deformation rate of 0.01% (to ensure that the material was within the linear viscoelastic zone) and the frequency was ranged between 1 rad/s and 100 rad/s. The test temperature was varied from 40 °C to 76 °C. Based on time-temperature superposition principle (TTSP), the master curve was constructed at the reference temperature of 40 °C.

The multiple stress creep and recovery (MSCR) test was performed according to ASTM D 7405 (2015), at the temperature of 64 °C and considering aged samples of binders. From the initial and final deformation values obtained during the cycles (load-recovering), for each of the applied stress levels, it was possible to calculate a mean value for the non-recoverable compliance (J_{nr}) and recovery (R).

The linear amplitude sweep test (LAS) was performed according the provisional standard AASTHO TP 101/2014. The tests were carried out in aged binders at 25 °C. The rupture criterion used in this study was the maximum value of the shear stress supported by the binder and the results analysis was done under the principles of the Viscoelastic Continuous Damage theory.

The Chemical analysis was performed by Fourier transform infrared (FTIR) spectroscopy. The spectra were recorded on a Perkin Elmer Frontier ATR-FTIR spectrometer. For each spectrum, twenty scans were considered, with a resolution of 4 cm^{-1} and it varied from 4000 cm^{-1} to 600 cm^{-1}. From the obtained spectrum for each binder it was possible to calculate the carbonyl and sulfoxide indexes, related to the oxidation of functional groups present in the binder (Marsac et al. 2014).

3 Results and Discussion

The master curves of the dynamic shear modulus (|G*|) of the binders were obtained according the TTSP. It was found that before the aging the biobinder exhibited lower |G*| values than the asphalt binder (AC 30/45) (Fig. 1a). On the other hand, considering the aged binders at the temperature (163 °C), it is possible to verify that the biobinder and AC 30/45 have shown similar values of |G*|, that it indicates that the biobinder is more susceptible to ageing than the AC 30/45. The biobinder (Yang et al. 2014) has also shown to be more susceptible to aging than conventional asphalt binders.

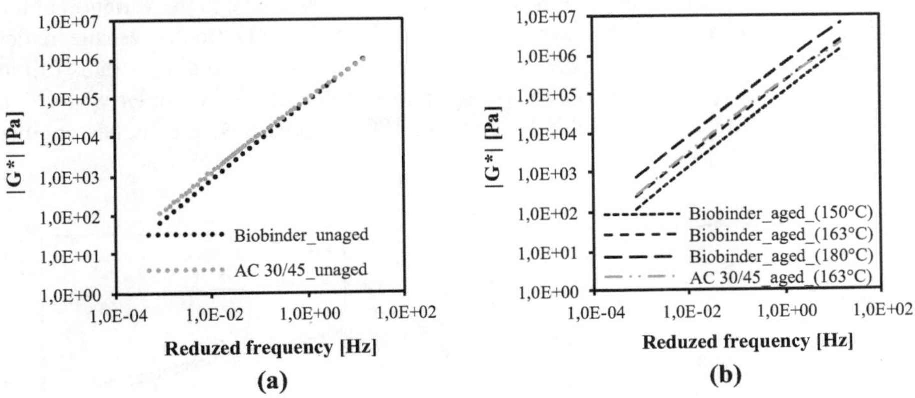

Fig. 1. Binders master curve at 40 °C: (a) Unaged condition, (b) RTFOT aged conditions

Regarding the effect of the test temperature on the biobinder stiffness, the binder aged at 150 °C and 180 °C, respectively, were shown to be less and more stiffer than the biobinder aged at 163 °C (Fig. 1b).

The MSCR test results indicate the similarity in the high temperature performance between the biobinder and AC 30/45, both aged at 163 °C. The increase in RTFOT temperature resulted in a biobinder less susceptible to permanent deformation, since the binder aged at 180 °C showed the highest recovery values (Fig. 2a) and the lowest value of non-recoverable compliance (J_{nr}) (Fig. 2b) (when compared to the biobinder aged in lower temperatures (150 °C and 163 °C).

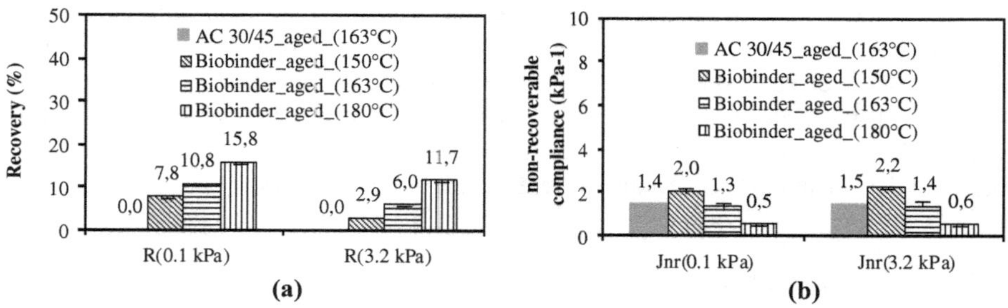

Fig. 2. MSCR test results: (a) Recovery values, (b) J_{nr} values

The LAS test results provide the curve that correlates the integrity and the damage suffered by the binder during the test. It can be inferred that the AC 30/45 binder presented higher resistance to damage than the biobinder, regardless of the degree of aging to which they were subjected (Fig. 3a). The fatigue curves of the binder (Fig. 3b) indicate that the AC 30/45 presented a longer fatigue life than the biobinder, for the deformation rates up to 10%. However, by the angular coefficient of each binder fatigue curve it is possible to infer that the asphalt binder is more sensitive to the variation of the applied strain rate than the biobinder. The more aged the biobinder (in the case the binder aged at 180 °C) the better the performance of the binder at low applied strains (up to YY%), however, aging tends to penalize the performance of the biobinder when high strains are considered. The biobinder aged at 180 °C presented the higher angular coefficient.

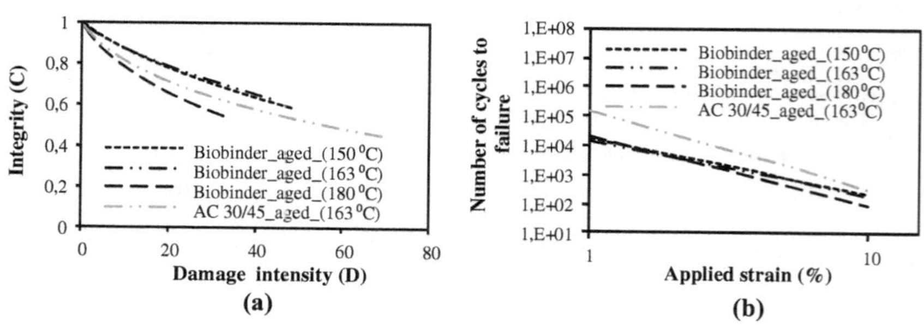

Fig. 3. Binders fatigue analysis: (a) Integrity curves (b) Fatigue life

The spectrum of the biobinder shows a high concentration of peaks between the 1800 and 1680 cm^{-1} bands, which indicates the high concentration of oxygen in its composition when compared to asphalt binder AC 30/45. This finding is in agreement with those obtained by Gong et al. (2017) and Zhang et al. (2017).

As expected, the aging of AC30/45 increased its carbonyl and sulfoxide indexes. However, the aged biobinders in general have exhibited lower carbonyl and indexes when compared to the biobinder in the unaged condition. The observations indicate that the carbonyl and sulfoxide indexes may not be adequate to evaluate the biobinder aging.

4 Conclusions

Rheological and chemical analysis of the effect of the short-term aging temperature showed that the fatigue and permanent deformation performance of the bi-obinders was significantly affected by test temperature at which the RTFOT was conducted. The sensitivity of the biobinder to temperature is probably related to its expressive effect on the oxidation and loss of volatiles.

The chemical results indicated that the carbonyl and sulphoxide indexes were not adequate for the analysis of the oxidation of the biobinder. In order to better understand the degradation mechanism of the biobinder it is necessary to development an adequate methodology for aging characterization.

References

Abbas, A., Choi, B.C., Masad, E., Papagiannakis, T.: The influence of laboratory aging method on the rheological properties of asphalt binders. J. Test. Eval. **30**, 171–176 (2002). https://doi.org/10.1520/JTE12304J

Airey, G.D.: State of the art report on ageing test methods for bituminous pavement materials. Int. J. Pavement Eng. **4**, 165–176 (2003). https://doi.org/10.1080/1029843042000198568

Dhasmana, H., Ozer, H., Al-qadi, I.L., Zhang, Y., Schideman, L., Sharma, B.K., Zhang, P.: Rheological and chemical characterization of biobinders from different biomass resources. Transp. Res. Rec. **2505**, 121–129 (2015). https://doi.org/10.3141/2505-16

Fini, E.H., Kalberer, E.W., Shahbazi, A., Basti, M., You, Z., Ozer, H., Aurangzeb, Q.: Chemical characterization of biobinder from swine manure: Sustainable modifier for asphalt binder. J. Mater. Civ. Eng. **23**, 1506–1513 (2011). https://doi.org/10.1061/(ASCE)MT.1943-5533.0000237

Gong, M., Zhu, H., Pauli, T., Yang, J., Wei, J., Yao, Z.: Evaluation of bio-binder modified asphalt's adhesion behavior using sessile drop device and atomic force microscopy. Constr. Build. Mater. **145**, 42–51 (2017). https://doi.org/10.1016/j.conbuildmat.2017.03.114

Kluttz, R.: Considerations for Use of Alternative Binders. In: Asphalt Pavements (2012). http://onlinepubs.trb.org/onlinepubs/circulars/ec165.pdf. Accessed 25 Apr 2018

Marsac, P., Piérard, N., Porot, L., Bergh, W.V.D., Grenfell, J., Mouillet, V., Pouget, S., Besamusca, J., Farcas, F., Gabet, T., Hugener, M.: Potential and limits of FTIR methods for reclaimed asphalt characterization. Mater. Struct. **47**, 1–14 (2014). https://doi.org/10.1617/s11527-014-0248-0

Raouf, M., Williams, R.: Temperature and shear susceptibility of a nonpetroleum binder as a pavement material. Transp. Res. Rec. **2180**, 9–18 (2010). https://doi.org/10.3141/2180-02

Read, J., Whiteoak, D.: The Shell Bitumen Handbook. Thomas Telford, London (2003)

Sun, D., Sun, G., Du, Y., Zhu, X., Lu, T., Pang, Q., Dai, Z.: Evaluation of optimized bio-asphalt containing high content waste cooking oil residues. Fuel **202**, 529–540 (2017). https://doi.org/10.1016/j.fuel.2017.04.069

Williams, R.C., Peralta, E.J.F., Puga, K.N.N.: Development of Non-Petroleum-Based Binders for Use in Flexible Pavements–Phase II (2015). Available via DIALOG. http://lib.dr.iastate.edu/cgi/viewcontent.cgi?article=1144&context=intrans_reports. Accessed 15 Apr 2018

Yang, X., You, Z., Dai, Q., Mills-beale, J.: Mechanical performance of asphalt mixtures modified by bio-oils derived from waste wood resources. Constr. Build. Mater. **51**, 424–431 (2014). https://doi.org/10.1016/j.conbuildmat.2013.11.017

Zhang, R., Wang, H., You, Z., Jiang, X., Yang, X.: Optimization of bio-asphalt using bio-oil and distilled water. J. Cleaner Prod. **165**, 281–289 (2017). https://doi.org/10.1016/j.jclepro.2017.07.154

Chemomap Imaging Microscopy Use to in Situ Assess Oxidative Ageing in Compacted Asphalt Mixtures

Sabine Vassaux[1(✉)], Vincent Gaudefroy[1], Laurence Boulangé[2], Audrey Pévère[3], and Virginie Mouillet[3]

[1] IFSTTAR MAST/MIT, Route de Bouaye, CS4, 44344 Bouguenais, France
sabinevassaux@yahoo.fr, vincent.gaudefroy@ifsttar.fr
[2] Eiffage Infrastructures, 3-7 Place de l'Europe, 78140 Vélizy-Villacoublay, France
laurence.boulange@eiffage.com
[3] Cerema Equipe-projet DIMA, 30 rue Albert Einstein, 13593 Aix-en-Provence, France
{audrey.pevere, virginie.mouillet}@cerema.fr

Abstract. Nowadays, the main challenge in the actual road infrastructure context is to guarantee an efficient, safe and durable road network for daily users. Without a regular maintenance, road pavement materials are ageing and lead to a local deterioration. To better predict the durability, it is also essential to assess road pavement materials properties after ageing. In literature, asphalt mixture ageing is often assessed by the measurement of consistency, oxidation and rheological properties on the binder, which is previously extracted with a chlorinated solvent. However, this method is solvent and time consuming; it also lies on the dissolution and possible modification of the bitumen structure. Therefore, the paper aims at developing an in-situ method to directly assess the ageing state in compacted road pavement asphalt mixtures. Based on a statistical chemomap Imaging ATR methodology, variations in the microscopic carbonyl absorbance are quantified in the analyzed bituminous mastic of recycled asphalt mixtures displaying different oxidation states. Results allow successfully validating the direct measurement of microscopic oxidation properties on asphalt mixture specimens. Through the calculation of a microscopic oxidation average, the degree of remobilization can be indirectly assessed in bituminous mixtures incorporating reclaimed asphalt. More largely, the statistical chemomap method opens new perspectives to link the measurement of local oxidative ageing properties with mechanical behavior of compacted asphalt mixtures specimens, for a better prediction of road materials performances and durability.

Keywords: Infrared imaging microscopy · No-destructive · Recycling Asphalt · Ageing · Chemical properties · Oxidation

L. D. Poulikakos et al. (Eds.): RILEM 252-CMB 2018, RILEM Bookseries 20, pp. 15–20, 2019.
https://doi.org/10.1007/978-3-030-00476-7_3

1 Motivation and Objective

The main challenge for the road construction companies is to produce durable and safe road pavement materials. However, during the long-term service life, road pavement is submitted to thermo-oxidative ageing which leads to a local surface layer deterioration including cracking, raveling and potholes. As ageing is one of the main factors affecting durability (Airey 2003; Lu and Isacsson 2002), it is necessary to gain knowledge on the evolution of materials properties after ageing to better predict durability and ageing resistance of road pavement materials for pavement design.

Bitumen ageing is quite complex and depends on several parameters like the void content (Al-Qadi et al. 2007), the pavement layer depth or the layer composition (Dony et al. 2016; Wang et al. 2014). On surface courses, bitumen is particularly exposed to climatic conditions (UV, temperature) and oxygen from air. This chemical ageing leads to the increase of carbonyl (C = O) and sulfoxide (S = O) functions in bitumen, to the increase of the asphaltenes content and of the complex modulus.

Therefore, to assess the bitumen properties after ageing, the common method consists in extracting the hydrocarbon binder from the asphalt mixture using a chlorinated solvent (Ziyani et al. 2017). Binder properties are then measured through several approaches including consistency approach, rheological approach as well as chemical approach (Hofko et al. 2017; Lopes et al. 2013; Lu and Isacsson 2002). Among these tests, the determination of carbonyl and sulfoxide indices is largely used (Mouillet et al. 2010) because carbonyl and sulfoxide functions are relevant infrared markers of the bitumen ageing using Fourier Transform Infra Red spectroscopy (Poulikakos et al. 2014). However, this method needs an extraction step that is solvent-consuming and may, by dissolution, modify binder internal structure (Stimilli et al. 2015) and induce structural modification in molecular weight (Themeli et al. 2017).

The objective of the paper is to develop a novel method to in situ assess the ageing state in road pavement asphalt mixtures. The selected approach consists in characterizing local chemical properties of compacted asphalt mixtures by measuring microscopic oxidation properties with infrared microscopy in imaging ATR (Attenuated Total Reflectance).

2 Materials and Method

2.1 Materials

To simulate different oxidation states, asphalt mixtures incorporating RA (Reclaimed Asphalt) were produced in laboratory using a BBmax MLPC® according to the EN 12697-35 standard (European Standard 2017) . A first asphalt mixture, H40, was manufactured according to a hot mix process and contains 40% RA. As a higher RA content leads to a higher oxidation, two other asphalt mixture formulations, H100 and W100, were then tested. 100% RA materials were mixed to a virgin binder amount according to hot (H) and warm (W) processes.

For all formulations, reclaimed asphalt were used in a 6/10 granular fraction and materials were mixed during 120 seconds to obtain a compacted rectangular slab sample with 180 × 500 × 50 mm^3 dimensions. Then, slab cuts were performed under a cold water, up to the core to obtain an asphalt mixture specimen displaying specific 20 × 5 × 20 mm^3 dimensions. More specially, refined cuts were realized on extremities of the slab height (50 mm^3) to obtain a sample thickness of 5 mm. The infrared analysis was performed on the asphalt mixture surface which wasn't directly exposed to air.

2.2 Infrared Microscopy Method

An infrared microscope (Spotlight 400 from Perkin Elmer) was used with an ATR (Attenuated Total Reflectance) accessory. A description of the apparatus is available in (Vassaux et al. 2018). The chemical characterization was performed on a refined zone (200 × 100 μm^2) of the intergranular bituminous mastic. For the analysis of the H40 sample, the bitumen spatial distribution (1A) and the aged binder spatial distribution (1B) were obtained by respectively integrating the ethylene function area (between 2995 and 2760 cm^{-1}) and the carbonyl function area (between 1720 and 1665 cm^{-1}). On Fig. 1A, spectra 2 and 3 confirm namely differences in bitumen concentration (Peak area = 0.042 > 0.003) whereas spectra 4 and 5 confirm the differences in the carbonyl function concentration (1B) (Peak area = 12.7 10^{-4} > 5.4 10^{-4}).

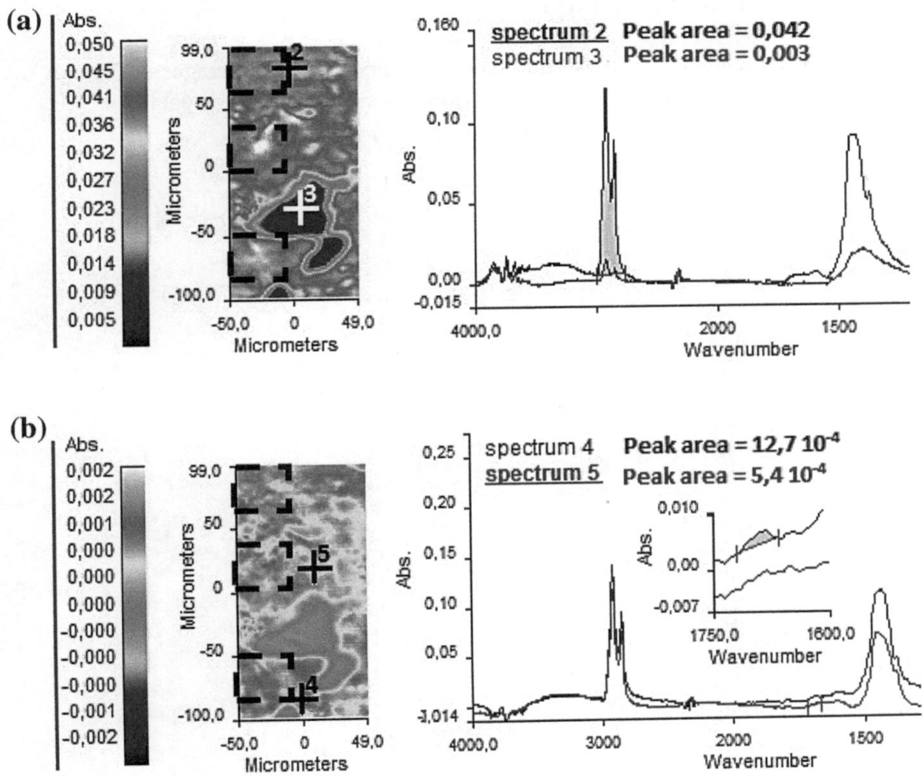

Fig. 1. Spatial distributions of the ethylene (A) and the carbonyl (B) functions for the infrared analysis of a H40 compacted asphalt mixture surface

To assess the bitumen oxidation on the 2D infrared map, a specific methodology has been developed to discuss oxidation variations for a same bitumen concentration: the bitumen map (Fig. 1A) was firstly divided into three zones of 20 × 20 μm^2 (black dotted lines) displaying a similar ethylene function absorbance (equals to 0.040 ± 0.002), these defined zones were similarly focused on the carbonyl spatial distribution (Fig. 1B).

Then, each reduced zone on Fig. 1B was converted into a micro carbonyl absorbance values matrix. The micro carbonyl absorbance value corresponds to the area absorbance of the carbonyl peak which was integrated on a spectrum defined on a surface of 1.56 × 1.56 μm^2 (pixel size) of the chemical mapping. As one reduced zone has arbitrary dimensions of 20 × 20 μm^2, each associated matrix has 196 numerical values defined on the reduced zone. From these data, a statistical treatment was performed to classify micro carbonyl absorbance values into defined intervals. Statistical curves were computed and compared. A CPA (for Curves Peaks Average) indicator was finally calculated from carbonyl absorbance maxima observed on the statistical curves. For the H40 analysis, CPA equals to (8.5 + 9.5 + 10.5) 10^{-4} divided by 3, that is to say $9.5.10^{-4}$. It allows characterizing the averaged oxidation state of analyzed zones.

3 Findings and Perspectives

Figure 2 shows statistical curves and respective curves peaks averages (CPA) for the analysis of H40 (2A), H100 (2B) and W100 (2C) asphalt mixture specimens.

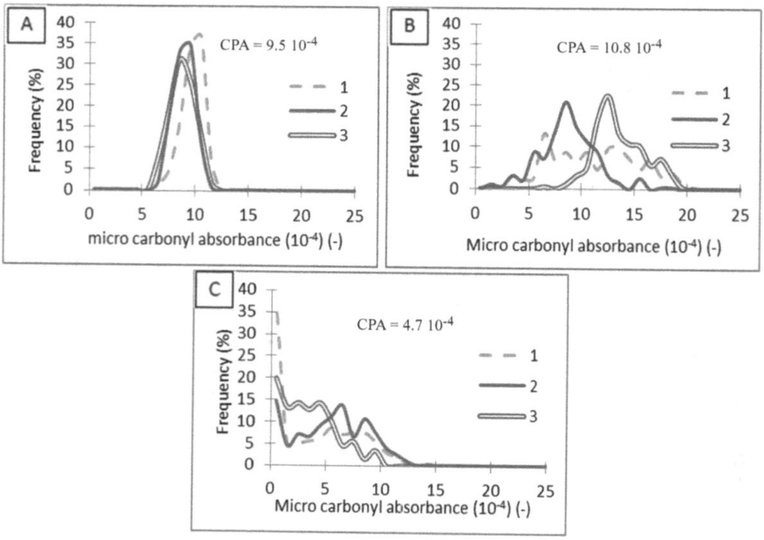

Fig. 2. Statistical curves comparison referring to the analysis of H40 (A), H100 (B) and W100 (C) asphalt mixture surfaces

On Fig. 2, some differences are observed in the quantification of microscopic carbonyl function absorbance in the mixes. For the H40 specimen (2A), all statistical curves are Gaussian and centered around a CPA value of 9.5 10^{-4}. Statistical curves are more enlarged for the H100 specimen (2B) which presents a similar average oxidation (CPA of 10.8 10^{-4}). For the 100% warm-mix asphalt mixture, W100, statistical curves are always enlarged and are more oriented to low micro carbonyl absorbance values than H100 ones, what is confirmed by the low CPA value of 4.7 10^{-4} (Fig. 2C). Based on curves shapes and CPA values, results show that three tested asphalt mixtures formulations aren't equivalent in term of oxidation properties.

For a similar incorporated high RA content (2B and 2C), the difference between respective CPA values may be explained either by a higher oxidation (due to high temperatures of the hot-mix process) or by a higher contribution of the RA aged binder in the intergranular mix due to a more efficient remobilization. Although identical incorporated RA content, a higher aged binder amount would be also available in the H100 (2B) mix because interactions between the aged and virgin binder are improved in hot conditions.

Then, the H40 asphalt mixture appears more oxidized than the W100 one. That may be explained too by the RA binder remobilization which was certainly total during the manufacture of the H40 hot-mix asphalt mixture containing the lowest incorporated RA content. Hence, according to the RA high aged binder amount to mix with the virgin binder and warm manufacturing temperatures, remobilization would be less efficient in the W100 asphalt mixture, what induces that the intergranular analyzed zone is mostly composed of the non-oxidized virgin binder.

This explanation is relevant with results (Fig. 2) and it is in total agreement with a previous study performed on model samples (Vassaux et al. 2018) which has highlighted that remobilization depends on the RA aged binder viscosity.

Thus, the CPA indicator also allows better understanding blending phenomena occurring between the virgin and aged binder. In recycled compacted asphalt mixtures, the oxidation parameter allows indirectly assessing the degree of remobilization of the aged RA binder which depends both on the RA content and on the manufacturing process.

Therefore, this statistical "chemomap" methodology leads to better assess the interaction degree between RA and virgin materials in recycled asphalt mixtures. As the method is solvent-free, the study opens also novel research perspectives to characterize the oxidative ageing state on fresh extracted compacted materials and to assess the influence of some additives (rejuvenators, bio-oils…) on the ageing resistance of asphalt mixtures. New research ways can be also developed concerning the coupling between microscopic oxidation properties (high oxidation gradient located in restricted area) and micro-cracks which may lead to a reduction of fatigue resistance and durability at long term.

References

Airey, G.D.: State of the art report on ageing test methods for bituminous pavement materials. Int. J. Pavement Eng. **4**(3), 165–176 (2003). https://doi.org/10.1080/1029843042000198568

Al-Qadi, I.L., Elseifi, M., Carpenter, S.H.: Reclaimed asphalt pavement - A literature review (2007). http://hdl.handle.net/2142/46007

Dony, A., Ziyani, L., Drouadaine, I., Pouget, S., Faucon Dumont, S., Simard, D., Mouillet, V., Poirier, J.E., Gabet, T., Boulangé, L., Nicolaï, A., Gueit, C.: MURE national project : FTIR spectroscopy study to assess ageing of asphalt mixtures. E&E Congr. (2016). https://doi.org/10.14311/ee.2016.154

European Standard: EN 12697-35 - Bituminous mixtures - Test methods for hot asphalt - Laboratory mixing (2017)

Hofko, B., Falchetto, A.C., Grenfell, J., Huber, L., Lu, X., Porot, L., Poulikakos, L.D., You, Z.: Effect of short-term ageing temperature on bitumen properties. Road Mater. Pavement Des. (2017). https://doi.org/10.1080/14680629.2017.1304268

Lopes, M.D.M., Zhao, D., Chailleux, E., Kane, M., Gabet, T., Petiteau, C.: Characterization of aging processes on the asphalt mixture surface. ISAP 2012 10 (2013)

Lu, X., Isacsson, U.: Effect of ageing on bitumen chemistry and rheology. Constr. Build. Mater. **16**, 15–22 (2002). https://doi.org/10.1016/S0950-0618(01)00033-2

Mouillet, V., Farcas, F., Valérie, B., Besson, S., Cédric, P., Le Cunff, F.: Identification et dosage des fonctions oxygénées présentes dans les liants bitumineux. Analyses par spectrométrie infrarouge à transformée de Fourier, Méthode d'essai du LCPC, no. 69 (2010)

Poulikakos, L.D., dos Santos, S., Bueno, M., Kuentzel, S., Hugener, M., Partl, M.N.: Influence of short and long term aging on chemical, microstructural and macro-mechanical properties of recycled asphalt mixtures. Constr. Build. Mater. **51**, 414–423 (2014). https://doi.org/10.1016/j.conbuildmat.2013.11.004

Stimilli, A., Virgili, A., Canestrari, F.: New method to estimate the "re-activated" binder amount in recycled hot-mix asphalt. Road Mater. Pavement Des. **16**, 442–459 (2015). https://doi.org/10.1080/14680629.2015.1029678

Themeli, A., Chailleux, E., Farcas, F., Chazallon, C., Migault, B., Buisson, N.: Molecular structure evolution of asphaltite-modified bitumens during ageing; comparisons with equivalent petroleum bitumens. Int. J. Pavement Res. Technol. **10**, 75–83 (2017). https://doi.org/10.1016/j.ijprt.2017.01.003

Vassaux, S., Gaudefroy, V., Boulangé, L., Soro, L.J., Pévère, A., Michelet, A., Barragan-montero, V., Mouillet, V.: Study of remobilization phenomena at reclaimed asphalt binder/virgin binder interphases for recycled asphalt mixtures using novel microscopic methodologies. Constr. Build. Mater. **165**, 846–858 (2018). https://doi.org/10.1016/j.conbuildmat.2018.01.055

Wang, Y.P., Wen, Y., Zhao, K., Chong, D., Wong, A.S.T.: Evolution and locational variation of asphalt binder aging in long-life hot-mix asphalt pavements. Constr. Build. Mater. **68**, 172–182 (2014). https://doi.org/10.1016/j.conbuildmat.2014.05.091

Ziyani, L., Boulangé, L., Nicolaï, A., Mouillet, V.: Bitumen extraction and recovery in road industry : a global methodology in solvent substitution from a comprehensive review. J. Clean. Prod. **161**, 53–68 (2017). https://doi.org/10.1016/j.jclepro.2017.05.022

Comparison of Short Term Laboratory Ageing on Virgin and Recovered Binder from HMA/WMA Mixtures

Gilda Ferrotti[1(✉)], Hassan Baaj[2], Jeroen Besamusca[3], Maurizio Bocci[1], Augusto Cannone Falchetto[4], James Grenfell[5], Bernhard Hofko[6], Laurent Porot[7], Lily D. Poulikakos[8], Zhanping You[9], and Zhanping You[9]

[1] Università Politecnica delle Marche, Ancona, Italy
{g.ferrotti,m.bocci}@univpm.it
[2] University of Waterloo, Waterloo, Canada
hassan.baaj@uwaterloo.ca
[3] Kuwait Petroleum Research and Technology, Rotterdam, The Netherlands
j.besamusca@Q8.com
[4] Technische Universität Braunschweig, Braunschweig, Germany
a.cannone-falchetto@tu-bs.de
[5] Australian Road Research Board, Port Melbourne, Australia
james.grenfell@arrb.com.au
[6] Technische Universität Wien, Vienna, Austria
bernhard.hofko@tuwien.ac.at
[7] Kraton Chemical B.V., Almere, The Netherlands
Laurent.Porot@Kraton.com
[8] Empa, Dübendorf, Switzerland
Lily.Poulikakos@empa.ch
[9] Michigan Technological University, Houghton, USA
Lily.Poulikakos@empa.ch, zyou@mtu.edu

Abstract. Oxidative ageing strongly affects asphalt mixture behavior. The Rolling Thin Film Oven Test (RTFOT) is currently used to simulate within a laboratory environment the binder short-term ageing, assuming that the mixture is produced at conventional Hot Mix Asphalt (HMA) temperatures (ca 160 °C). However, the introduction of Warm Mix Asphalts (WMAs), which are produced at lower temperatures than HMAs, could require adjustments in the short-term ageing simulation procedure as ageing is strongly influenced by the mixing and compaction temperatures. In this study, the physical properties of a straight-run bitumen, before ageing and after RTFOT ageing at two temperatures (123 °C and 163 °C), are compared to those of the same bitumen recovered from a HMA and a foamed WMA, both produced in laboratory. This comparison aims at determining the best RTFOT temperature for short-term ageing simulation for WMAs. To this end, all the binders were investigated through conventional (penetration value at 25 °C and softening point temperature) and rheological (frequency sweeps with dynamic shear rheometer) tests. Both conventional and rheological tests indicate that the WMA recovered binder is less aged than the binder aged at the standard conditioning temperature of 163 °C, whereas the HMA recovered binder is more aged than the artificially aged binder in the

L. D. Poulikakos et al. (Eds.): RILEM 252-CMB 2018, RILEM Bookseries 20, pp. 21–26, 2019.
https://doi.org/10.1007/978-3-030-00476-7_4

laboratory with RTFOT at 163 °C. These initial results support the idea that an appropriate ageing temperature for RTFOT short-term ageing simulation of WMA needs to be identified.

Keywords: WMA · RTFOT · Ageing · Foamed asphalt

1 Introduction

Environmental impact of asphalt pavement construction can be reduced by using Warm Mix Asphalt (WMA) technologies, that allow asphalt mixtures to be produced and placed at lower temperatures than traditional Hot Mix Asphalt (HMA) (D'Angelo et al. 2008). This results in several environmental and technical benefits such as reduction of energy consumption and CO_2 emission (Rubio et al. 2012, Hofko et al. 2017b), extended paving conditions (night-time and winter applications), extended workability time and reduction of binder aging (Bonaquist 2011, Prowell 2007).

Since bitumen ageing strongly influences asphalt mixture behaviour (Hofko et al. 2017a, Hofko and Hospodka 2016), it should be properly considered during ageing simulation in laboratory. The Rolling Thin Film Oven Test (RTFOT) (EN 12607-1, 2014) is commonly used for simulating the ageing occurring during manufacturing, transportation and placing of asphalt mixture (short-term ageing) and is conventionally performed at 163 °C. The use of WMAs technologies may require an adjustment in the RTFOT temperature as ageing is strongly influenced by temperature.

This work is part of a wider RILEM TC CMB 252 inter-laboratory study and aims to investigate the short-term ageing simulation of WMA mixtures in laboratory by performing RTFOT at two temperatures (123 and 163 °C). Conventional (Penetration value at 25 °C and Softening Point temperature) and rheological (through Dynamic Shear Rheometer, DSR) tests were performed on the neat bitumen (70/100 pen grade), on the RTFOT aged bitumens and on the same bitumen recovered from a HMA and from a foamed WMA, both prepared in laboratory.

2 Materials and Experimental Program

As above-mentioned, the results reported in this paper cover part of a much larger study, which looked at the ageing of four 70/100 pen bitumens. In this study, only the results of a 70/100 pen bitumen, labelled as B503, were considered.

The neat B503 bitumen was short term aged through the RTFOT at both 123 °C and 163 °C and the aged binders were labelled as B503B123 and B503B163, respectively. Moreover, the same bitumen was used to produce a traditional HMA mixture and a WMA mixture with foamed bitumen in the laboratory.

Asphalt mixtures were made up according to the AC 11 design as described in EN 13108 part 1. The coarse aggregates were porphyrite and the filler was limestone. The gradation of the asphalt mixtures used in this study is shown in Fig. 1.

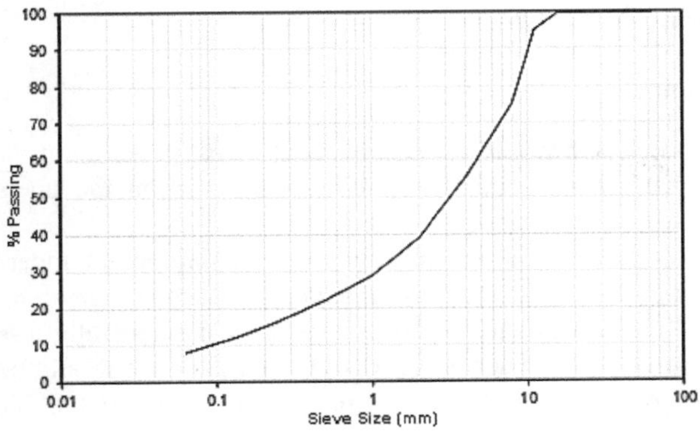

Fig. 1. Aggregate gradation according to AC11 mixture design

The HMA was prepared at lab #1. The aggregates and the filler were heated to 170 °C for 7 h while the bitumen was heated to the same temperature for 3 h. The mixing was performed at that temperature for 5 min to obtain a homogeneous asphalt mixture. The WMA mixture was produced in lab #2, by using foamed bitumen prepared in laboratory with a water content of 2.45% by weight of bitumen (Godenzoni et al. 2016) at a temperature of 150 °C. The filler and the aggregates were oven-dried at 110 °C for 5 h, whereas the bitumen was pre-heated at 150 °C for 2 h. The mixing was at 110 °C, immediately after the production of the foamed bitumen, for about 5 min.

After the production, both HMA and WMA were short-term laboratory aged and they were spread on a pan, achieving a homogeneous layer thickness of about 25 mm. They were then placed in a ventilated oven at 160 °C and 110 °C, respectively, and stored for 4 h. These temperatures reflect the mixing temperatures for these mixtures. Every 60 min, the mixtures were stirred to achieve homogeneous ageing. After the mixtures had cooled down to room temperature, the binders were recovered, according to EN 12697-3 (2013) using tetrachloroethylene as a solvent. The binders recovered from these mixtures after extraction were labelled as HMA_rec and WMA_rec, respectively.

The five binders were compared by performing conventional and rheological tests. Conventional tests consisted in the determination of the penetration value and the softening point temperature, according to EN 1426 (2015) and EN 1427 (2015), respectively. The rheological measurements were performed with the dynamic shear rheometer (DSR), by carrying out frequency sweep tests in the parallel plate configuration. For each binder, a frequency range between 0.1 and 10 Hz and a different number of temperatures, ranging from −10 to 80 °C, were selected. All the binders were tested in control strain conditions within the linear viscoelastic range of the materials. Two replicates were considered for each binder and each test.

3 Results and Analysis

3.1 Conventional Test Results

It is well known that age conditioning results in lower penetration value and higher softening point temperature and these, in turn, depend on the ageing temperature (Verhasselt 2000).

Figure 2 shows the penetration value and the softening point temperature results for all the tested binders. The values of HMA_rec indicate higher ageing than the standard conditioning of 163 °C, whereas WMA_rec seems to be generally less aged than the laboratory artificially-aged binder using RTFOT at 163 °C. It should be noted that the recovered binder behaviour could be partially influenced by the binder extraction procedure.

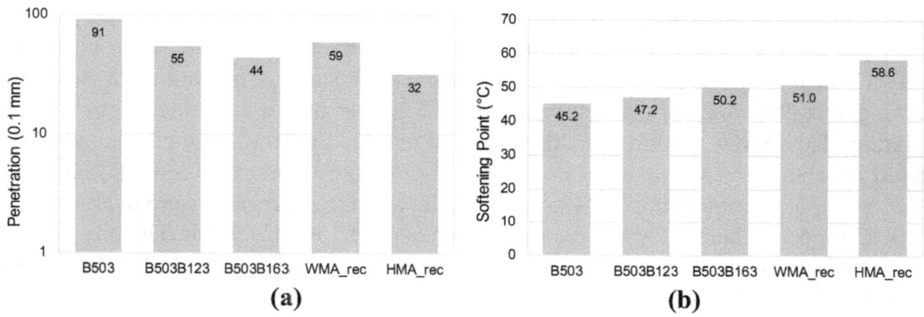

Fig. 2. Changes of B503 due to ageing conditioning: (a) penetration; (b) softening point

3.2 Rheological Test Results

The entire set of DSR data is shown in the Black diagram (Fig. 3a), where the norm of the phase angle δ is represented as a function of the complex modulus $|G^*|$. Data provide smooth curves for each binder, indicating their thermo-rheologically simple behavior.

As a first analysis, on the Black space, the curves for HMA_rec are flatter ("less concave") than the others. This is typically an indication of lower temperature susceptibility which is commonly observed for aged binders. Surprisingly, for WMA_rec and RTFOT-aged binders the curve shapes do not show significant difference with the virgin binder, assuming less impact on the visco-elasticity balance of the binder. Moreover, the WMA_rec binder shows a similar behavior to both B503B123 and B503B163, which are between the virgin binder B503 and the HMA_rec.

The raw data can also be used to determine the crossover temperature, defined as the temperature when the loss modulus (G") is equal to the storage modulus (G'), corresponding to a phase angle of 45° at a given frequency. In Fig. 3b, the norm of the complex modulus has been represented as a function of the crossover temperature, for the selected frequency of 1.59 Hz (10 rad/s). A linear correlation between $|G^*|$ and

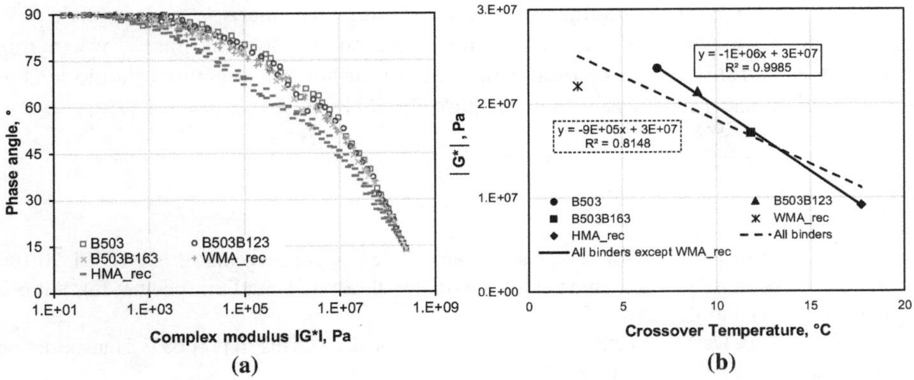

Fig. 3. DSR results: (a) Black diagram for all the binders tested, (b) Norm of the complex modulus vs. crossover temperature at the frequency of 1.59 Hz

the crossover temperature was observed ($R^2 = 0.8148$). If the WMA_rec result can be considered as an anomalous result and can be neglected, the R^2 value was improved to 0.9985. It has been shown that, due to ageing, the crossover temperature increased because of the hardening effect (Porot 2016). The higher the RTFOT temperature, the higher the change in properties. Moreover, in the case of laboratory produced HMA, the change is even larger than age conditioning on the binder in laboratory. This finding might indicate that the filler and the aggregates influence the system (Besamusca et al. 2012, Guo et al. 2017).

4 Conclusions

This work is part of an inter-laboratory study for the RILEM TC CMB. The objective of this paper is to evaluate the adequacy of the RTFOT to simulate short-term aging of Warm Mix Asphalt. To achieve this objective, a straight-run bitumen 70/100 was aged through RTFOT at two different temperatures (123 °C and 163 °C) and compared with a bitumen recovered from two laboratory produced asphalt mixtures: a traditional HMA and a WMA produced by using a foaming technique.

The analysis of the physical properties (Penetration value and Softening Point temperature) indicated that the lower the RTFOT temperature, the lower the change in the properties. At the same time, the binder recovered from the HMA has undergone more severe changes than that recovered from the WMA.

The analysis of the findings of the rheological tests by using Dynamic Shear Rheometer helped to better qualify the effect of short-term aging on the properties of the different tested binders and confirmed the previous conclusions. The change in rheological properties was higher with higher RTFOT temperature. Compared to lab-produced asphalt mixtures, the hot mix production generated more changes than the foamed WMA and also compared to the RTFOT binder conditioning at both temperatures.

As further steps in validating these initial findings, comparison with asphalt mixtures produced in the plant as well as investigation of other bitumens, warm mix technologies and mixtures is recommended. These further investigations should lead to an appropriate short-term ageing temperature for RTFOT simulation of WMA.

References

Besamusca, J., Volkers, A., Water, J., Gaarkeuken, B.: Simulating ageing of EN 12591 70/100 bitumen at laboratory conditioning compared to porous asphalt, 5th Eurasphalt & Eurobitume Congress, Istanbul (2012)

Bonaquist, R.: Mix Design Practices for Warm Mix Asphalt. NCHRP report 691. Transportation Research Board, Washington DC (2011)

D'Angelo, J., Harm, E., Bartoszek, J., Baumgardner, G.L., Corrigan, M., Cowsert, J., Harman, T., Jamshidi, M., Jones, W., Newcomb, D., Prowell, B., Sines, R., Yeaton, B.: Warm-Mix Asphalt: European Practice. FHWA-PL-08-007. US Federal Highway Administration, Alexandria (VA) (2008)

Godenzoni, C., Graziani, A., Perraton, D.: Complex modulus characterisation of cold recycled mixtures with foamed bitumen and different contents of reclaimed asphalt, Road Mater. Pavement Des. (2016). http://dx.doi.org/10.1080/14680629.2016.1142467

Guo, M., Bhasin, A., Tan, Y.: Effect of mineral fillers adsorption on rheological and chemical properties of asphalt binder. Constr. Build. Mater. **141**, 152–159 (2017)

Hofko, B., Hospodka, M.: Rolling thin film oven test and pressure ageing vessel conditioning parameters effect on viscoelastic behavior and binder performance grade. Transp. Res. Rec. **2574**, 111–116 (2016). https://doi.org/10.3141/2574-12

Hofko, B., Alavi, M.Z., Grothe, H., Jones, D., Harvey, J.: Repeatability and sensitivity of FTIR ATR spectral analysis methods for bituminous binders. Mater. Struct. **50**, 187 (2017a). https://doi.org/10.1617/s11527-017-1059-x

Hofko, B., Dimitrov, M., Schwab, O., Weiss, F., Rechberger, H., Grothe, H.: Technological and environmental performance of temperature-reduced mastic asphalt mixtures. Road Mater. Pavement **18**(1), 22–37 (2017b). https://doi.org/10.1080/14680629.2016.1141703

EN 12607-1: Bitumen and bituminous binders. Determination of the resistance to hardening under influence of heat and air. RTFOT method (2014)

EN12697-3: Bituminous mixtures. Test methods for hot mix asphalt. Part 3: Bitumen recovery: Rotary evaporator (2013)

EN 1426: Bitumen and bituminous binders. Determination of needle penetration (2015)

EN 1427: Bitumen and bituminous binders. Determination of the softening point. Ring and Ball method (2015)

Porot, L., Eduard, P.: Addressing asphalt binder aging through the viscous to elastic transition. In: ISAP Symposium, Jackson Hole, Wyoming (USA) (2016)

Prowell, B.: Warm mix asphalt. The international technology scanning program summary report Federal Highway Administration (2007)

Rubio, M.C., Martinez, G., Baena, L., Moreno, F.: Wam mix asphalt: an overview. J. Clean. Prod. **24**, 76–84 (2012). https://doi.org/10.1016/j.jclepro.2011.11.053

Verhasselt, F.A.: Kinetic approach to the ageing of bitumens. In: Yen and Chilingarian (ed.) Asphaltenes and Asphalts, 1st edn., vol. 2, pp. 475–497. Elsevier (2000). Chap.17

Effect of Artificial Ageing on Two Different Bitumen of Different Origin but Same Performance Grade

Alexandre Rogeaux, Alan Carter(✉), Daniel Perraton, and Abdeldjalil Daoudi

École de Technologie Supérieure (ÉTS), Montreal, Canada
alan.carter@etsmtl.ca

Abstract. Performance grade system for bitumen characterization is used to ensure that the selected bitumen has the required properties for the environmental conditions found in-situ. However, it has been shown that within a given PG grade, the properties can differ thru time. In this study the rheological characteristics evaluated with a DSR and a BBR are compared with the chemical composition of two different bitumen produced from different crude oil but with the same PG grade. The results have shown that even if both bitumens performs similarly with the usual ageing, RTFO and PAV, their characteristics get more different with intensive artificial laboratory ageing. The chemical characteristics of the bitumens were measured with infrared spectroscopy. It was shown that the change of their chemical composition with ageing is different. The links between their rheological properties and their chemical characteristics measured with FTIR that have been found is different for both bitumens.

Keywords: Bitumen · Artificial ageing · Performance grade · Rheology FTIR

1 Introduction

In Canada, bitumens are characterized mainly by the performance grade, PG, system. This system gives properties of the bitumen that is linked with hot mix asphalt properties. The high temperature in the PG H-L, represent the maximum temperature at which the material is cohesive enough to resist rutting. The L is the low temperature at which the material still resist thermal cracking. Bitumen produced to meet a specific PG grade do often have different chemical characteristics, and Speight (2006) have shown that the ageing sensitivity of bitumen is highly dependent of their chemical composition.

Based on laboratory tests done by the authors, it has been shown that two identical hot mix asphalt (HMA) made with different bitumen that has the same PG grade sometimes have different complex modulus, and fatigue resistance. This study was developed to better understand this problematic.

L. D. Poulikakos et al. (Eds.): RILEM 252-CMB 2018, RILEM Bookseries 20, pp. 27–32, 2019.
https://doi.org/10.1007/978-3-030-00476-7_5

2 Objective, Material and Methodology

The main objective of this research project is to evaluate the impact of ageing on the rheological and chemical characteristics of two bitumen of the same performance grade (PG), but prepared from different crude oil. More specifically, this project aims to verify if infrared spectroscopy and dynamic shear modulus of different bitumens from the same supplier that have the same performance grade evolve differently over time.

The two bitumens tested in this research program are non-modified PG58-28 which is the most common bitumen used in Quebec for base course. Both bitumen were produced by the same company in Canada and are sold as equivalent product. In this paper, the bitumens are labelled bitumen A and B.

The research program is separated into two main parts: (1) artificial ageing of the materials, and (2) rheological and chemical characterization of the bitumen. The laboratory artificial ageing of the bitumens were done with a rolling thin film oven (RTFO – ASTM D2872) and with a pressure ageing vessel (PAV – ASTM D6521). However, on top of the standardized RTFO and PAV combination used to define the bitumen performance grade, longer ageing period were used. The ageing protocol used is shown in Fig. 1.

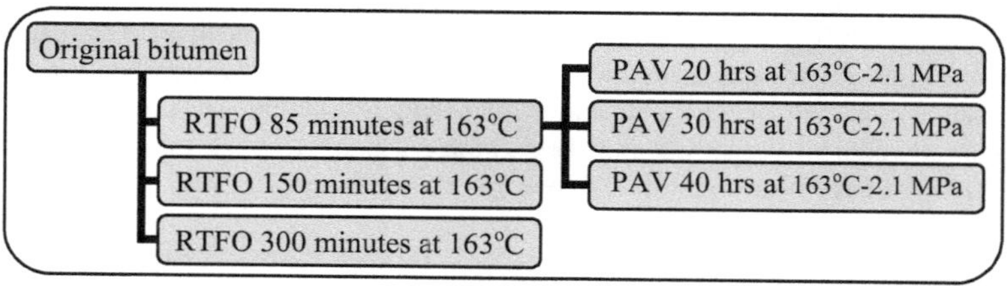

Fig. 1. Ageing protocol used on the bitumens

For the rheological characterization, a dynamic shear rheometer (DSR) and a bending beam rheometer (BBR) were used. For the DSR tests, a frequency sweep between 0,01 Hz and 15 Hz was done at 25, 35, 45 and 60 °C to obtain the complex shear modulus master curve, and the phase angle. Before doing those tests, the linear viscoelastic region limit was established at those temperatures and frequencies for all ageing steps. Three replicates were tested for every ageing steps for both bitumens.

The BBR tests were performed at −18 °C and −24 °C, and the creep stiffness, S(60), and the creep slope, m(60), were measured at both temperatures. Three replicates were tested for every temperature and every ageing steps for both bitumens.

Different chemical tests were performed, such as differential scanning calorimetry, but only the results from the infrared spectroscopy are presented here. The Fourier-transform infrared spectroscopy (FTIR) is often used to measure the change in the chemical composition of bitumens (Nivitha et al. 2016; Feng et al. 2015; Karlsson and Isacsson 2003; Edwards et al. 2005). In this project, the infrared spectrum was studied on wavelength between 450 and 4000 cm^{-1}, with a resolution of 0,5 cm^{-1}. Four replicates of each ageing steps for each bitumen were tested.

3 Results

Before showing the link between the rheological behaviour and the chemical characterization, it is important to look at the variation of the rheology with ageing.

3.1 Rheological and FTIR Results

As mentioned before, BBR tests were done at −18 °C and −24 °C for all ageing level. The results show that even if the creep stiffness, s(60), was very similar after the usual RTFO 85 min + PAV 20 h, the difference increases with ageing. For example, after a 40 h PAV ageing done after a RTFO, the creep stiffness of bitumen A (137 MPa) is about half of the one measured for bitumen B (244 MPa) at −18 °C. The same trend is observed at −24 °C. The difference in m-value between both bitumens is not as important, going from 0,238 for bitumen A after 40 h of PAV vs 0,274 for bitumen B.

It is more interesting to look at the results obtained with the DSR. In Fig. 2, the G* master curves at 25 °C are shown for bitumen A and bitumen B for two different ageing steps.

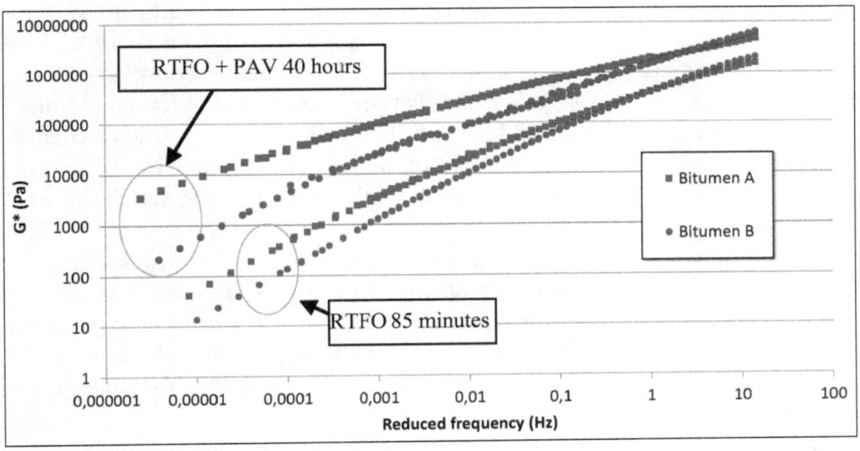

Fig. 2. Average complex shear modulus results are 25 °C for two different ageing levels for both bitumen tested

As it can be noted from Fig. 2, there is a bigger difference in the rheological behaviour of bitumen A and B after 40 h of PAV than after the initial usual ageing process. Even if it's complicated to state the age this artificial ageing process represents in the field, it is clear that pavement build with either of those two bitumens will behave differently. At very low frequency, or high temperature according the time temperature superposition principle, the modulus of bitumen A is more than 10 times more rigid than the bitumen B with RTFO + PAV 40 h. It is also important to note that the increase in rigidity between the two ageing steps is much more pronounced for bitumen A than for bitumen B. In order to completely evaluate the impact of this difference, it is required to also look at the phase angle. At 25 °C and the lowest reduced frequency, the phase angle is 89° for both bitumens after the usual ageing, and they reduce to 67° and 69° for bitumen A and bitumen B respectively after RTFO + PAV 40 h. A decrease in phase angle means a decrease in the viscous behaviour, which in turns means a decrease in the capacity to relax tension. So a higher modulus coupled with a lower phase angle means that both materials becomes more brittle with ageing (as expected), but the change is more significant for bitumen A than for bitumen B.

The same trend can be seen for all other ageing levels between those two extremes. Those results are not shown because of space limitation.

As mentioned before, when bitumen ages, its chemical composition changes. FTIR is a good tool to evaluate the change in chemical composition. As mentioned before, for FTIR tests, four replicates for every ageing levels for both bitumens were tested. The results shown here are the averages of those four results.

As stated previously, it has been shown that ageing mostly affects the carbonyl (around 1700 cm^{-1}) band and the sulfoxyde (around 1032 cm^{-1}) band. Because of this, it was decided to concentrate the study on the change of those two bands, even if a total of ten different bands were actually calculated and analyzed. The carbonyl and sulfoxyde index are calculated from the area of the peaks.

However, even if the peak itself can be easily identified, it is difficult to properly discriminate the total area of those peaks since many different methods exist to select their limits. In this study, fixed limits in term of wavelength were selected. For the carbonyl peak, the area was calculated between 1660 and 1760 cm^{-1}, and for sulphoxide, the area was calculated between 1010 and 1039 cm^{-1}. The calculation of the indexes was done by dividing the area of one of the two peaks by the sum of the area of the three aliphatic peaks. It's with this calculation that the variability was the lowest between replicates for both bitumens.

It is interesting to note that both bitumens have a different chemical signature; they have different indexes even before any ageing steps. This shows that it is possible to have two bitumens with similar rheological properties as measured in the performance grade system and be chemically different. This result can mean that those bitumen will age differently, or that the PG system is not characterizing the performance of the bitumen properly.

As expected, the results have shown an increase of the carbonyl and sulphoxide index with ageing. However, the analysis of the peaks by themselves is not very useful; it needs to be linked with other properties, like the rheological properties.

3.2 G* and FTIR Results Comparison

Many different correlations can be drawn between the rheology and the chemical characteristics measured with FTIR. In this study, the carbonyl index and the sulphoxide index were compared with the low temperature behaviour measured with BBR and the higher temperature behaviour measured with the DSR.

A linear relationship was found between the creep stiffness at −18 °C or −24 °C and the carbonyl or sulphoxide index. However, the relationship for both bitumens is quite different with bitumen B having a steeping slope meaning a greater change in creep stiffness according to either index. This higher rate change could mean that bitumen B ages faster than bitumen A, and, in the long run, would results in a somewhat fragile bituminous material.

Relationship between carbonyl and sulphoxide indexes and DSR results were also found. An example is shown in Fig. 3. In this case, it can be seen that there is a similar trend for both bitumens, but the trend is much clearer for Bitumen A. The relationship is just fair for bitumen B. This shows again a difference in the behaviour of both bitumens. When aging, the phase angle also changes. The results obtained (not shown here) between FTIR results and phase angle measured with the DSR show that bitumen A has a lower phase angle (more elastic) than bitumen B (more viscous) for every ageing steps, and the phase angle for bitumen B is less sensitive to aging than bitumen A, which is in contradiction with the rigidity. More tests are needed to better explain this situation.

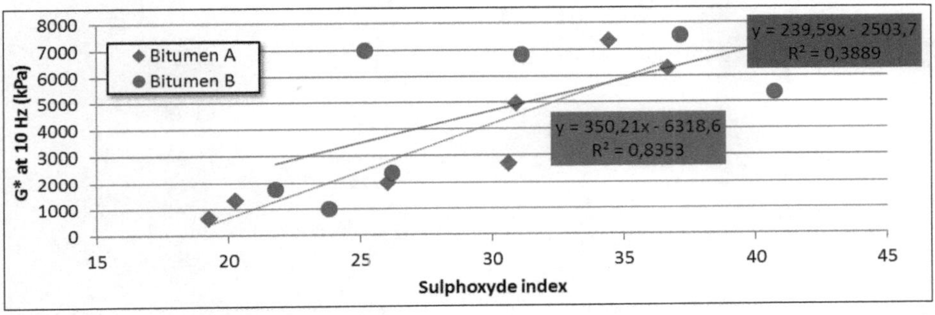

Fig. 3. G* at 10 Hz and 25 °C vs sulphoxide index

4 Conclusion

In this study, two bitumens were characterized with DSR and BBR tests before being analyzed with infrared spectroscopy. It was first observed that even if those two bitumens are considered equivalent in term of PG grade, they have different behaviour when they are aged for a longer period. Even if those ageing process are not standardized and it's complicated to link those artificial ageing with field conditions, we believe that it shows that it could be beneficial to age the bitumen more to better characterize them, since longer aging shows bigger differences.

Relationships were found between FTIR results and rheological properties, but they are different for both bitumens. This shows that the different observed rheological behaviour of the bitumen can be explained by the different chemical evolution with time. Even if they did have similar chemical composition before ageing according to FTIR results, their difference becomes more and more evident with time. More work is needed to better understand this aspect.

References

Edwards, Y., Tasdemir, Y., Isacsson, U.: Influence of commercial waxes on bitumen aging properties. Energy Fuels **19**(6), 2519–2525 (2005). https://doi.org/10.1021/ef050166r

Feng, Z.-G., Bian, H.-J., Li, X.-J., Yu, J.-Y.: FTIR analysis of UV aging on bitumen and its fractions. Mater. Struct. **49**, 1381–1389 (2015). https://doi.org/10.1617/s11527-015-0583-9

Karlsson, R., Isacsson, U.: Application of FTIR-ATR to characterization of bitumen rejuvenator diffusion. J. Mater. Civ. Eng. **15**(2), 157–165 (2003)

Nivitha, M.R., Prasad, E., Krishnan, J.M.: Ageing in modified bitumen using FTIR spectroscopy. Int. J. Pavement Eng. **17**(7), 565–577 (2016). https://doi.org/10.1080/10298436.2015.1007230

Speight, J.G.: The Chemistry and Technology of Petroleum, 4th edn. CRC Press, Boca Raton (2006)

Evaluation of Viscoelastic Properties and Cracking Behaviour of Asphalt Mixtures with Laboratory Aging

Runhua Zhang(✉), Jo Sias Daniel, and Eshan V. Dave

University of New Hampshire, Durham, NH, USA
rzl015@wildcats.unh.edu,
{jo.daniel, eshan.dave}@unh.edu

Abstract. Aging can significantly affect the performance of asphalt mixtures causing increase in stiffness, reduction in relaxation capability and increase in brittleness. These changes can be quantified through viscoelastic properties and cracking behaviour of asphalt mixtures. In this study, six mixtures are evaluated using different laboratory conditioning protocols (loose mixture conditioned at various temperatures and durations to represent different aging levels) to evaluate how the viscoelastic, fatigue, and fracture properties of the mixtures will change over time. Comparison of viscoelastic characterization between different aging levels and mixtures is conducted by using complex modulus testing. Semi Circular Bending (SCB) and Disk Shaped Compact Tension (DCT) tests are used to evaluate the fracture behaviour of the mixtures. The Simplified Viscoelastic Continuum Damage (S-VECD) testing is used to estimate the ability of the mixtures to resist fatigue cracking. The results indicate that the viscoelastic, fatigue and fracture properties of the mixtures change significantly with aging. The two long term conditioning protocols induce statistically similar changes in linear viscoelastic properties and fatigue properties but produce differences in fracture indices. In this study, two virgin mixtures generally have better fracture and fatigue performance than the four mixtures with RAP. Two mixtures which have the largest difference between the high and low performance grade temperatures show the largest change in fracture and fatigue properties with aging.

Keywords: Aging · Viscoelastic properties · Fatigue cracking
Thermal cracking

1 Introduction

Aging can significantly affect the performance of asphalt mixtures through an increased cracking susceptibility leading to potentially shorter pavement service life and lower serviceability. Several methods for laboratory conditioning of asphalt mixtures to simulate field aging are documented in the literature. The main objective of this study is to evaluate how the viscoelastic, fatigue, and fracture properties of asphalt mixtures evolve with different aging protocols and identify how mixture properties may influence the magnitude in change of properties with aging.

L. D. Poulikakos et al. (Eds.): RILEM 252-CMB 2018, RILEM Bookseries 20, pp. 33–38, 2019.
https://doi.org/10.1007/978-3-030-00476-7_6

In this study, several laboratory conditioning protocols are used. Protocols that use loose mix in the conditioning process were chosen to avoid differential aging gradients on compacted specimens of different geometry. The Asphalt Institute recommends conditioning of loose mixture for 24 h at 135 °C to simulate 7 to 10 years of aging in the field (Blankenship et al. 2010). The recent findings of the National Cooperative Highway Research Program (NCHRP) 09-54 project on long term aging of asphalt mixtures suggests 95 °C as an optimal temperature for aging loose mix (Kim et al. (2013) & Elwardany et al. 2017). In this study, periods of 5 and 12 days were chosen to represent the New Hampshire climate based on preliminary results from the NCHRP project (Rahbar-Rastegar et al. 2018).

2 Materials and Testing

This study includes laboratory testing on six typical New Hampshire plant mixed, lab compacted mixtures. Table 1 below shows the mixture information in this study. Recycled binder content is the percentage of the weight of recycled binder to the total binder weight.

Table 1.

Mixture ID	Virgin Binder Grade	NMAS (mm)	Total Binder Content (%)	Recycled Binder Content (%)
5828 M	PG 58-28	9.5	5.9	16.9
6428 M	PG 64-28	9.5	6.3	18.9
7628 M	PG 76-28	9.5	6.1	14.8
7034LV	PG 70-34	12.5	5.8	0
6428MV	PG 64-28	9.5	6.4	0
6428L	PG 64-28	12.5	5.8	18.5

Complex modulus testing was conducted following AASHTO T 342 to compare the linear viscoelastic properties of asphalt mixtures at different aging levels. Uniaxial direct tension cyclic fatigue testing was conducted following AASHTO TP 107 to evaluate the fatigue behavior of the asphalt mixtures.

To evaluate the fracture characteristics of asphalt mixtures, Semi Circular Bending (SCB) testing (AASHTO TP 124) was performed at 25 °C to evaluate intermediate temperature behavior using flexibility index (FI) (Ozer et al. 2016). The Disk Shaped Compact Tension (DCT) testing (ASTM D 7313) was conducted to evaluate the thermal cracking propensity using fracture strain tolerance (FST) (Zhu et al. 2017). The DCT test temperature for each mixture is based on the winter-time pavement in-service temperature for the location where mix is being used; it is calculated as 10 °C warmer than 98% reliability low pavement temperature from the LTPPBind database for the nearest weather station to the actual project site.

3 Results and Discussion

Dynamic modulus and phase angle master curves constructed from complex modulus testing results for different aging levels are presented for one of the study mixture in Fig. 1; each series represents the average of three replicates. The dynamic modulus increases as the aging level increases. The peak phase angle decreases and occurs at a lower frequency as materials age. The two higher levels of aging (24 h. at 135 °C and 12 days at 95 °C) show statistically similar dynamic modulus and phase angle values. These trends are similar for all mixtures evaluated in this study. As expected, the complex modulus testing results for the aged materials show higher stiffness (dynamic modulus) and lower relaxation capability (phase angle), which, in combination, can result in higher cracking susceptibility.

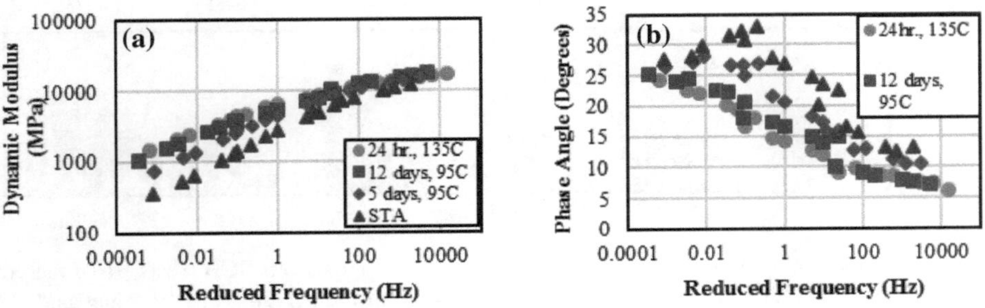

Fig. 1. Example (6428 M, PG 64-28, 9.5 mm, 18.9% RAP) (a) Dynamic Modulus, (b) Phase Angle Master Curves at 21.1 °C (STA: Short term aging)

Figure 2a shows the flexibility index (FI) parameter from SCB tests with the error bars representing one standard deviation; the FI aging ratios (LTOA divided by STA) are shown above each bar. Generally, FI and FI aging ratio decrease with increase of aging levels. There is a statistically significant difference in FI between the STA and the three long term aging levels. Intermediate aging causes the FI to drop to 20–40% of the STA condition while long term aging drops the FI to 5–30% of the STA condition. This indicates that the mixtures lose cracking resistance very quickly with aging. The 24 h. at 135 °C condition causes a larger drop of FI aging ratio than the 12 days at 95 °C.

The two mixtures without RAP (6428 MV and 7034 LV) generally have higher FI values at each aging level than other mixtures with RAP (7628 M also has good performance at STA condition). Another interesting observation is that even though the 7628 M and 7034 LV have good performance in terms of FI at the STA condition, the FI aging ratio drops more than other mixtures. Except for these two mixtures, the FI aging ratio for the other four mixtures is relatively consistent at each aging level (30–40% after 5 days at 95 °C; 20–30% after 12 days at 95 °C; 10–15% after 24 h. at 135 °C). This means that the 7628 M and 7034 LV mixtures, which have the largest difference between high and low performance grade temperatures, show the most impact from aging on FI values. However, the FI value for 7034 LV after aging is still higher than other mixtures.

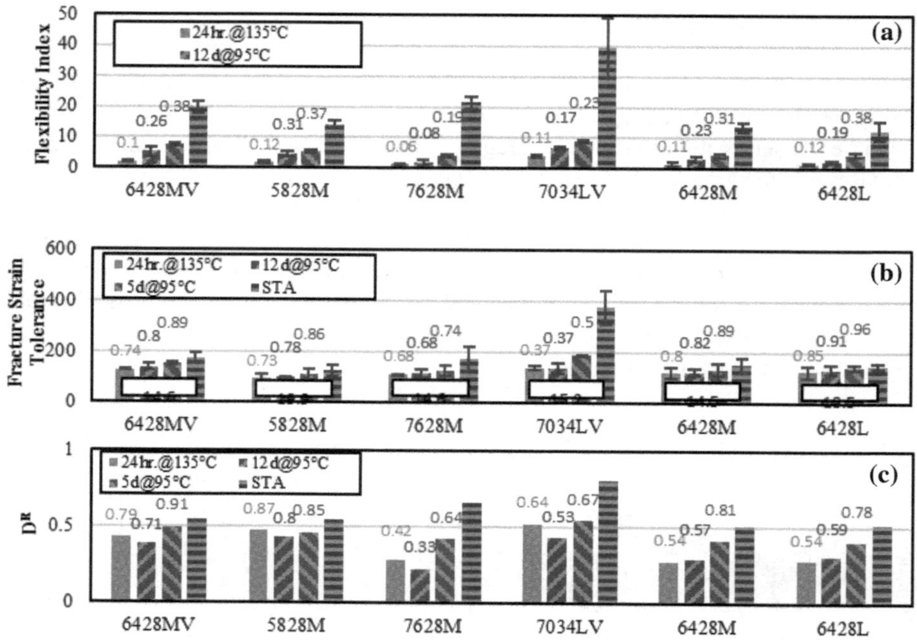

Fig. 2. (a) Flexibility Index (FI) Values and FI Aging Ratios from SCB Tests; (b) Fracture Strain Tolerance (FST) Values and FST Aging Ratios from DCT Tests, (c) DR Values and DR Aging Ratios from S-VECD Fatigue Tests

Figure 2b shows the FST values and aging ratios from the DCT testing. The DCT testing temperature is based on the in-service location, as shown in Fig. 2b. Generally, FST and FST aging ratio decrease when aging level increases. There is a statistically significant difference in FST between the STA and the two long term aging levels. Intermediate aging causes the FST to drop to 50–95% of the STA condition while long term aging drops the FST to 37–90% of the STA condition. The 24 h. at 135 °C condition causes a larger drop of FST aging ratio than the 12 days at 95 °C.

Like SCB results, the virgin mixtures (6428 MV and 7034 LV) usually have higher FST values at each aging level than the mixtures with RAP. The 7628 M and 7034 LV mixtures have the largest decrease in FST with aging. Except for these two mixtures, the FST aging ratio for other mixtures is relatively consistent at each aging level (85–95% after 5days at 95 °C; 75–90% after 12days at 95 °C; 70–85% after 24 h. at 135 °C). However, the FST value for 7034LV after aging is still higher than other mixtures.

Figure 2c shows the D^R values and D^R aging ratio from the S-VECD fatigue testing. Generally, a higher D^R value indicates better fatigue behavior. The D^R values and D^R aging ratio show a consistent trend with decreasing value with aging level, which is similar to the SCB and DCT results. However, the two long term aging levels are different, with the trend changing for each mixture. Statistical analysis shows that there is no significant difference in D^R values between these two long term aging levels.

One possible explanation is that D^R value is related to the stiffness of mix. Also, intermediate aging causes the D^R to drop to 65–95% of the STA condition while long term aging drops the D^R to 35–85% of the STA condition.

Similar to the SCB and DCT results, the 6428 MV and 7034 LV mixtures generally show the best fatigue performance as measured by D^R value. The D^R aging ratio for the 7628 M and 7034 LV mixtures shows the largest drop with aging as compared to the other mixtures. However, the D^R value for 7034 LV after aging is still higher than other mixtures.

4 Summary and Conclusion

The major objective of this study was to investigate the effects of aging on the viscoelastic, fatigue, and fracture properties of asphalt mixtures. The following conclusions can be drawn from the results of the testing and analysis:

1. Linear viscoelastic properties and fatigue behavior of mixtures with 24 h. at 135 °C and 12 days at 95 °C aging are statistically similar.
2. Both flexibility index (FI) and fracture strain tolerance (FST) after 24 h. at 135 °C condition typically show the most severe decline.
3. In this study, the two virgin mixtures generally have better fracture and fatigue performance than the mixtures with RAP.
4. The two mixtures with the largest difference between the high and low performance grade temperatures show the most impact from aging based on the fracture and fatigue results. However, the high PG spread virgin mixture still has better fracture and fatigue performance after aging than the other mixtures.
5. Except for the 7628 M and 7034 LV mixtures, the aging ratio of the fracture index (FI and FST) for other mixtures is relatively consistent.
6. As aging level increases from STA to 5 days at 95 °C, and 12 days at 95 °C, the general trend of the results from SCB, DCT and S-VECD fatigue testing, which are used to evaluate the craking behavior of the mix, is very similiar.

Acknowledgement. The authors would like to acknowledge New Hampshire Department of Transportation (NHDOT) for sponsoring this study and the University of New Hampshire Center for Infrastructure Resilience to Climate (UCIRC).

References

Blankenship, P.B., et al.: Laboratory and Field Investigation to Develop Test Procedures for Predicting Non-Load Associated Cracking of Airfield HMA Pavements, Airfield Asphalt pavement technology Program (2010)

Elwardany, M.D., et al.: Evaluation of Asphalt Mixture Laboratory Long-Term Aging Methods for Performance Testing and Prediction, Road Materials and Pavement Design, vol. 18 (2017)

Kim, R.Y., et al.: Long-term Aging of Asphalt Mixtures for Performance Testing and Prediction, Interim Report, NCHRP 09-54 (2013)

Ozer, H., et al.: Development of the fracture-based flexibility index for asphalt concrete cracking potential using modified semi-circle bending test parameters. Constr. Build. Mater. **115**, 390–401 (2016)

Rahbar-Rastegar, R., et al.: Evaluation of Viscoelastic and Fracture Properties of Asphalt Mixtures with Long-Term Laboratory Conditioning, Transportation Research Record: Journal of Transportation Research Board (2018)

Zhu, Y., et al.: Comprehensive Evaluation of Low Temperature Cracking Fracture Indices for Asphalt Mixtures, Road Materials and Pavement Design (2017)

Microstructural Investigation of Reclaimed Asphalt Binder with Bio-Based Rejuvenators

Maria Chiara Cavalli[1,2(✉)], Martins Zaumanis[1,2], and Lily D. Poulikakos[1,2]

[1] EMPA, Swiss Federal Laboratories for Materials Science and Technology, Überland Str. 129, 8600 Dübendorf, Switzerland
MariaChiara.Cavalli@empa.ch

[2] ETH Zurich, Institute for Mechanical Systems, 8092 Zurich, Switzerland

Abstract. Reclaimed asphalt pavement (RAP) has become an important source for road materials for energy consumption in asphalt pavement production. In this work, three different bio based rejuvenators were added to the RAP binder: a vegetable oil, a cashew nut shell oil and a tall oil. To better understand the microstructural morphology, atomic force microscopy (AFM) was used with quantitative nano-mechanical mapping (QNM) showing how the addition of rejuvenators decreased the elastic moduli of the RAP binder before and after aging procedure. Aim of this work is to enlighten the effect of rejuvenators at the morphology of RAP's binder.

Keywords: Reclaimed asphalt binder · Rejuvenators
Atomic force microscopy · Elastic moduli

1 Introduction

Recycling of asphalt pavements is an important step for sustainable roads construction and energy savings. Reclaimed asphalt pavement (RAP) is known for being harder and more brittle than standard asphalt concrete (Cavalli et al. 2017) thus, it is common to add chemical products called "rejuvenators" to the RAP mixtures in order to restore the properties of the aged binder. To do so, it is important to understand the bitumen's microstructure and how it is affected by rejuvenators (Das et al. 2016). Atomic force microscopy (AFM) is capable of giving information such information (dos Santos et al. 2015). The results of this work can be important for furthering understanding of the aging process in bitumen and the relation between composition and mechanical properties.

L. D. Poulikakos et al. (Eds.): RILEM 252-CMB 2018, RILEM Bookseries 20, pp. 39–43, 2019.
https://doi.org/10.1007/978-3-030-00476-7_7

2 Materials and Methods

The bituminous binder from RAP from a Swiss supplier was extracted using toluene (EN 12697-1). The binder content was 4.60% by weight of the mixture (EN 12697-1), having a penetration of 22×10^{-1} mm at 25 °C (EN 1426) and a softening point equal to 65.7 °C (EN 1427). 50/70 virgin binder used in this study showed a penetration of 62×10^{-1} mm at 25 °C and a softening point temperature of 48.75 °C. Three commercial bio-based rejuvenators were used: rejuvenator defined as "A" from natural seed oil, rejuvenator designated as "B" from cashew nut shell oil and rejuvenator "C" from tall based oil. The rejuvenator dosage was set at 5% by mass of RAP binder in order to get a penetration close to that of the virgin binder. Subsequently 100 grams of RAP binder were placed in an oven at a temperature equal to the softening point plus 80° C for 20 min (EN 12697-1). The amount of time was found to be the minimum required for the RAP binder to become liquid. Rejuvenators were added to the hot binder and the mixture was placed in the Speed Mixer™ for one minute at 3500 rpm. One minute was visually found suitable for homogenization. Samples were prepared by using the heat casting method following (dos Santos et al. 2015). As all specimens were prepared by using the same procedure, it can be assumed that oxidation during sample preparation can considered constant amongst all specimens (Das et al. 2013). The AFM force mapping was performed on a Bruker Icon3 AFM in PeakForce Quantitative Nanomechanical property Mapping (QNM) mode with a Nanoscope V controller and software Nanoscope 8.15. The sample's preparation followed what was described in (Soenen et al. 2014). The cantilever was Bruker RTESPA-150 silicon probe with a resonant frequency of 150 kHz and a spring constant of 6 N/m, a cantilever length equal to 125 μm and width of 35 μm.

3 Results and Discussion

Quantitative nano-mechanical (QNM) mapping acquire information on nanomechanical properties such as elastic modulus at sample's surface while simultaneously imaging sample topography at micron scale resolution. It is non-destructive to both tip and sample since it directly controls the peak normal force and minimizes the lateral force on the probe. The main disadvantage of AFM is that it needs a perfectly smooth surface and it can only detect surface features. Therefore, it is fundamental be aware that the sample preparation could influence what has been measured at the surface (Das et al. 2016). As can be seen in Fig. 1, the virgin binder images present a softer paraphase surrounding the "bees" structures. On the contrary, RAP binder images appear generally darker with few darker spots. As overall, RAP binder complex moduli at the surface are higher than the virgin binder both at unaged and aged state. As in Fig. 1, after aging the complex moduli for both RAP binder and virgin binder increase. In particular, it's possible to observe how after aging, "bees" appear smaller and with higher elastic moduli. In addition, it's shown how the morphology of RAP binder doesn't change significantly after aging however, despite the fact that the RAP binder is already aged, its elastic modulus increase of 30%.

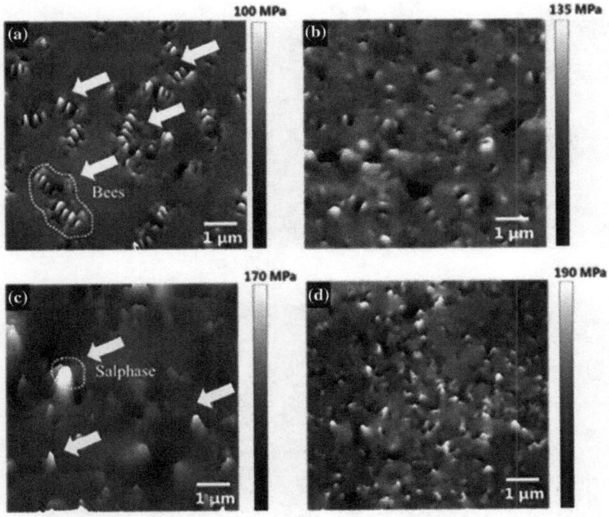

Fig. 1. AFM QNM images (10 × 10 μm). (a) Virgin binder 50/70 unaged; (b) RAP binder unaged; (c) Virgin binder 50/70 aged; (d) RAP binder aged

As in Fig. 2, RAP plus 5% A and RAP plus 5% C show the same branch formations. In addition, QNM mode demonstrates how branch formations are softer than the matrix in terms of complex moduli. In general, all the rejuvenated binders at unaged state, show lower complex moduli than the RAP binder. This can be an indication on the softening potential of rejuvenators in reducing the surface's moduli of the RAP binder. As shown in Fig. 2, RAP plus 5% B presents topography similar to the RAP binder with higher complex moduli than the other two modified binders. All binders increase their complex moduli at the surface after laboratory aging. In particular, RAP plus 5% B was more affected by aging than the other two binders. This corroborated the bulk moduli reported in (Cavalli et al. 2018).

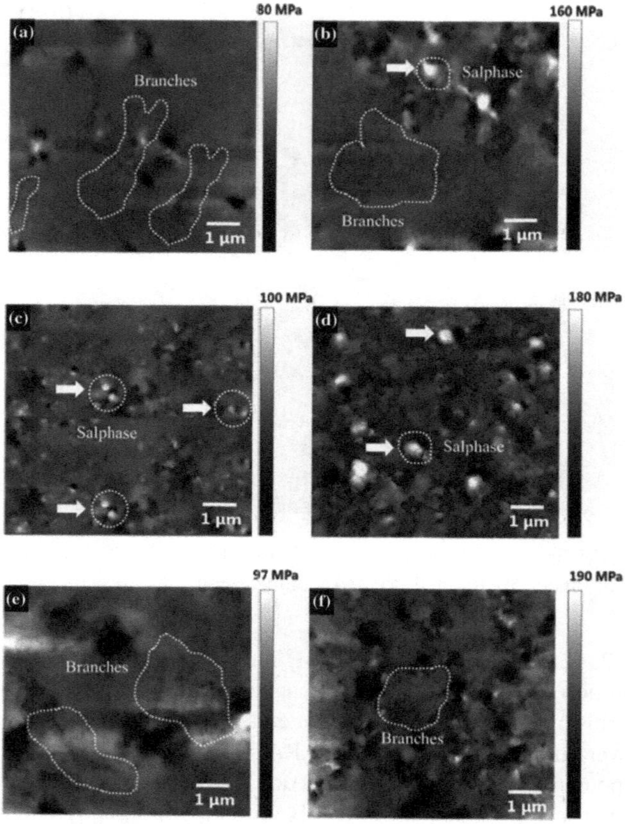

Fig. 2. AFM QNM images (10 × 10 µm). From top left clockwise: (a) RAP + 5% A (b) RAP + 5% A aged (c) RAP + 5% B (d) RAP + 5% B aged (e) RAP + 5% C (f) RAP + 5% C aged

4 Conclusions

In this study, a virgin binder, a RAP binder and modified RAP binder with three bio based rejuvenators (a seed oil, a cashew nut shell oil and a tall oil) were studied. Atomic force microscopy (AFM) has been utilized to characterize the elastic moduli. The addition of two types of rejuvenators could create new branches at the surface of the RAP binder however, the morphology of the RAP binder with rejuvenators could not replicate the one of the virgin binder. It has been shown how RAP binder was stiffer than the other binders and the addition of rejuvenators decreased the elastic moduli of the RAP binder. As a result of aging, the elastic moduli of all binder increased. In conclusion, atomic force microscopy has been shown as successful tool when studying elastic moduli of binders.

Acknowledgments. Authors would like to thank the Swiss federal office for the environment grant number UTF 489.19.14/IDM 2006.2423.487 for the financial support.

References

Cavalli, M.C., Partl, M.N., Poulikakos, L.D.: Measuring the binder film residues on black rock in mixtures with high amounts of reclaimed asphalt. J. Clean. Prod. **149**, 665–672 (2017). https://doi.org/10.1016/j.jclepro.2017.02.055

Cavalli, M.C., Zaumanis, M., Mazza, E., Partl, M.N., Poulikakos, L.D.: Effect of ageing on the mechanical and chemical properties of binder from RAP treated with bio-based rejuvenators. Compos. B Eng. **141**, 174–181 (2018). https://doi.org/10.1016/j.compositesb.2017.12.060

Das, P.K., Baaj, H., Tighe, S., Kringos, N.: Atomic force microscopy to investigate asphalt binders: a state-of-the-art review. Road Mater. Pavement Des. **17**(3), 693–718 (2016). https://doi.org/10.1080/14680629.2015.1114012

Das, P.K., Kringos, N., Wallqvist, V., Birgisson, B.: Micromechanical investigation of phase separation in bitumen by combining atomic force microscopy with differential scanning calorimetry results. Road Mater. Pavement Des. **14**, 25–37 (2013). https://doi.org/10.1080/14680629.2013.774744

dos Santos, S., Partl, M.N., Poulikakos, L.D.: From virgin to recycled bitumen: a microstructural view. Compos. B Eng. **80**, 177–185 (2015). https://doi.org/10.1016/j.compositesb.2015.05.042

EUROPEAN STANDARD 12697-1 Bituminous mixtures - Test methods for hot mix asphalt - Part 1: Soluble binder content. (2012)

EUROPEAN STANDARD EN 1426: Determination of the needle penetration (2012)

EUROPEAN STANDARD EN 1427: Determination of the softening point-Ring and Ball method (2012)

Soenen, H., Besamusca, J., Fischer, H.R., Poulikakos, L.D., Planche, J.-P., Das, P.K., Chailleux, E.: Laboratory investigation of bitumen based on round robin DSC and AFM tests. Mater. Struct. **47**(7), 1205–1220 (2014). https://doi.org/10.1617/s11527-013-0123-4

Recommendations of RILEM TC 252-CMB on the Effect of Short Term Aging Temperature on Long Term Properties of Asphalt Binder

Lily D. Poulikakos[1(✉)], Bernhard Hofko[2], Augusto Cannone Falchetto[3], Laurent Porot[4], Gilda Ferrotti[5], and Peter Mikhailenko[6]

[1] EMPA, Swiss Federal Laboratories for Materials Science and Technology, Dübendorf, Switzerland
lily.poulikakos@empa.ch
[2] Technische Universität Wien, Vienna, Austria
bernhard.hofko@tuwien.ac.at
[3] Technische Universität Braunschweig, Braunschweig, Germany
a.cannone-falchetto@tu-bs.de
[4] Kraton Chemical B.V., Paris, France
Laurent.Porot@Kraton.com
[5] Università Politecnica delle Marche, Ancona, Italy
g.ferrotti@univpm.it
[6] University of Waterloo, Waterloo, Canada
p2mikhai@uwaterloo.ca

Abstract. The Rilem Technical Committee on Chemo Mechanical Characterization of Bituminous Materials has investigated the effect of short term aging temperature on long term properties of asphalt binder, chemically, physically and microstructurally. The increased use of warm mix asphalt (WMA) technologies warrants such investigations in order to validate laboratory aging procedures. To this end, penetration, softening point, Fourier Transform Infrared Spectroscopy, dynamic shear rheology (DSR) and electron microscopy (ESEM) were used. The experimental results on binders and warm (WMA) and hot (HMA) mixtures from nine participating laboratories indicate that the binder source, as well as method of evaluation, result in different rankings and behaviors among the four binders used. The TC recommends the development of appropriate RTFOT aging temperatures for the simulation of binder aging in WMA.

Keywords: Aging · WMA · RTFOT · PAV · DSR · FTIR · ESEM
Asphalt · Bitumen

L. D. Poulikakos et al. (Eds.): RILEM 252-CMB 2018, RILEM Bookseries 20, pp. 44–49, 2019.
https://doi.org/10.1007/978-3-030-00476-7_8

1 Introduction

Warm Mix Asphalt (WMA) technologies allowing asphalt mixtures to be produced and placed at lower temperatures than traditional Hot Mix Asphalt (HMA) are increasingly popular as they reduce the environmental impacts of asphalt pavements (D'Angelo et al. 2008). In addition, as binder aging is strongly temperature related (Lu and Isacsson 2002), the binders of WMA are theoretically less aged. The Rolling Thin Film Oven Test (RTFOT) (EN 12607-1), conventionally performed at 163 °C, is used for laboratory simulation of aging during manufacturing, transportation, and placing of hot asphalt mixture (short-term aging). The present work summarizes the findings of the RILEM TC 252-CMB inter-laboratory study which aims at investigating the short-term aging simulation of WMA mixtures in laboratory.

2 Experimental Program

2.1 Materials

A detailed work program was developed (Hofko et al. 2017) to investigate both binder and mixtures materials. Four plain 70/100 penetration graded binders (EN 12591) from different sources were used as the core materials in this study. One of these binders was used to produce both HMA and WMA.

HMA was prepared in one of the participating laboratories by pre-heating the aggregates and the filler at 170 °C for 7 h and the binder at the same temperature for 3 h. Mixing was performed at the same temperature. The WMA mixture was produced in a different participating laboratory, using a foaming technique with water content of 2.45% by weight of binder (Godenzoni et al. 2016) at a temperature of 150 °C. Filler and aggregates were oven-dried at 110 °C for 5 h, whereas the binder was pre-heated at 150 °C for 2 h. Mixing was conducted at 110 °C, immediately after the production of the foamed binder.

After the production, both HMA and WMA were short term aged in laboratory as follows: they were spread on a pan to obtain a homogeneous layer thickness of about 25 mm. Thereafter, they were placed in a ventilated oven at 160 °C and 110 °C, respectively, and stored for 4 h. These temperatures reflect the mixing temperatures for these mixtures. Every 60 min, the mixtures were stirred for achieving homogeneous aging. After the mixtures had cooled down to room temperature, the binders were extracted and recovered, according to EN 12697-3.

2.2 Testing Program and Methods

The experimental plan for the four plain 70/100 penetration graded binders consisted in a combination of several laboratory tests. RTFOT aging was performed at three different temperatures (123C, 143C and 163 °C) and followed by conventional long-term aging with the Pressurized Aging Vessel PAV device (EN 14769). The aging on the set of binders used for the ESEM study was carried out at 123 °C and 163 °C at one of the participating laboratories. Due to the fact that ESEM is a very time consuming characterization method and as the significant difference was expected to be between 123C and 163C only those two temperatures were investigated with ESEM.

Conventional (Penetration value at 25 °C and Softening Point temperature) and rheological (Dynamic Shear Rheometer, DSR) tests, chemical characterizations (Fourier Transform Infrared Spectroscopy, FTIR) and microstructural characterizations (Environmental Scanning Electron Microscopy, ESEM) were performed on the original set of four binders (70/100 pen grade), on the RTFOT and PAV aged binders and, with the exception of ESEM, on the same binder recovered from short-term aged HMA and foamed WMA.

3 Results and Analysis

3.1 Conventional Test Results

The results of conventional tests showed that penetration value (EN 1427) decreased due to aging and softening point temperature (EN 1427) increased for all the tested binders. For example for B503 the penetration of virgin, RTFOT at 123 and RTFOT at 163 were 91, 55, 44 dmm respectively. The change in penetration and softening point is almost linear with temperature increase in short term conditioning (RTFOT) while the influence of long term conditioning (RTFOT 123 °C/PAV) shows a sharper variation. A combination of RTFOT at 143 °C and PAV was almost equivalent to the combined use of RTFOT at 163 °C and PAV. This might indicate a higher degree of chemical reaction, possibly with oxygen, that occurs from 143 °C. The values for the binder recovered from HMA indicate higher aging than the standard conditioning of 163 °C (Pen 32 vs 44 dmm for B503), whereas the binder extracted from WMA seemed to generally be less aged than the binder artificially aged in the laboratory with RTFOT at 163 °C (Pen 59 vs 44 dmm for B503).

3.2 Rheological Test Results

Rheological data using master curves show a global stiffening of the material that occurs across aging conditions, from the unaged to the apparently more severe combination of RTFOT aging at 163 °C and PAV. This effect is more significant at higher testing temperatures. Within the different RTFOT temperatures, it appears that when using 163 °C, a more severe increase in complex modulus occurs compared to the other two temperatures: 143 °C and 123 °C. A similar trend can also be observed for the three corresponding aging conditions when PAV is performed. Comparing the binder results to mixtures using binder B503 only, the DSR data shown in a Black diagram where the norm of the phase angle δ is represented as a function of the complex modulus $|G^*|$, form smooth curves for each binder, indicating their thermo-rheologically simple behavior (Fig. 1). The curves for HMA present a milder temperature susceptibility as commonly observed for aged binders. However, for WMA and RTFOT aged binders, the curve shapes do not show significant differences between the virgin binder, and thereby less impact on the visco-elasticity balance of the binder. Moreover, the G* values obtained from the binder extracted from the WMA mixture showed a similar behavior to the binder aged at both 123 and at 163 °C, which are between the virgin binder and the HMA (Fig. 1).

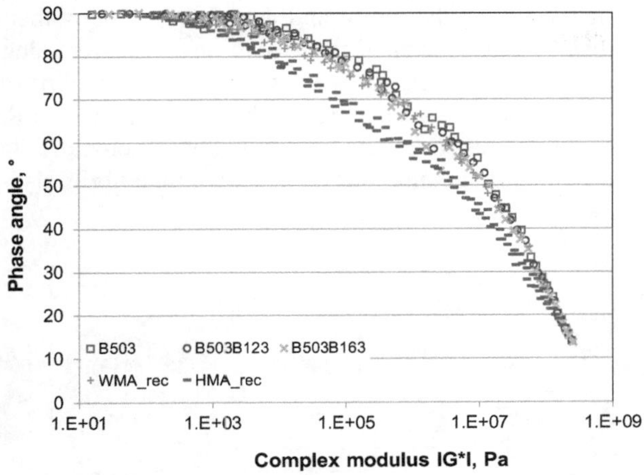

Fig. 1. Black Diagram showing a comparison between binder laboratory aging and binder extracted from aged WMA and HMA mixture (B503 = virgin; B503B123 = RTFOT aging at 123; B503B163 = RTFOT aging at 163C)

The crossover temperature, defined as the temperature when the loss modulus (G") is equal to the storage modulus (G') (phase angle is equal to 45° at a given frequency) increases for more severe aging as a result of hardening effect (Porot 2016). The higher the RTFOT temperature, the larger the change in properties. For the G* values obtained from the binder extracted from laboratory produced HMA, this change is even more dramatic than with the binder aging (Fig. 1).

3.3 FTIR Test Results

FTIR has been shown to be a promising tool for the characterization of aging of bitumen (Lamontagne et al. 2001). In these set of experiments, RTFOT ageing at 163 °C did not lead to a visible increase in the carbonyl area. Only after PAV ageing, a distinct band for carbonyl structures could be detected (3% for B504). As for the sulfoxide area, all binder samples show absorbance in this band already in the unaged state. In all cases, an RTFOT at 123 °C limited change of $I_{S=O}$ was seen, whereas $I_{S=O}$ stayed mostly constant at an RTFOT at 143 °C and increases for RTFOT at 163 °C (4% for B504).. As for PAV after the RTFOT, regardless of the RTFOT temperature, there was an increase in the indices, especially with higher RTFOT temperatures. Thus, the experimental results show that RTFOT temperature and therefore mix production temperature, has a stronger impact on the formation of sulfoxide structures than for carbonyl structures (7% vs 3% for B504). Also, the long-term aged state after PAV is affected by the short-term ageing temperature (Ferrotti et al. 2018).

3.4 Microstructural Test Results

The irradiation of asphalt binder samples by the ESEM electron beam over a period of time tend to produce fibril structures (Mikhailenko et al. 2017). The images of the resulting fibril structure for all the virgin and aged binders show that all the unaged binders have a fibril structure, which evolves through different degrees of aging. The RTFOT aging at 123 °C does not seem to induce much fibril evolution to the virgin binder, while the structure does appear to become denser with RTFOT aging at 163 °C (Fig. 2). The structures for all the binders evolve much more significantly with PAV aging, with the structure getting much denser, with the fibrils also getting smaller in diameter for most.

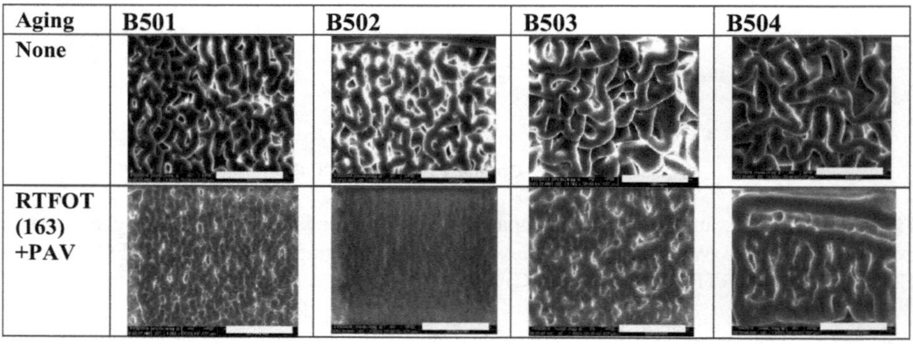

Fig. 2. Examples of ESEM images of virgin bitumen and aged bitumen (scale bar = 50 μm)

4 Conclusions and Recommendations

The analysis of conventional testing with penetration values and softening point temperatures showed that the lower the RTFOT temperature, the smaller the change in these properties. At the same time, the foamed WMA exhibited less changes compared to the HMA.

Rheological analysis using DSR confirmed the findings from the conventional evaluations. The change in rheological properties was higher with higher RTFOT temperatures. The hot mix production generated more changes than the foamed WMA and also compared to the RTFOT binder conditioning.

Due to aging, the cross over temperatures increased as a result of hardening effects. The higher the RTFOT temperature, the higher the change in properties. In the case of lab-produced HMA, the change is even higher than for laboratory binder aging.

The microstructures observed with ESEM for all the binders, evolve much more significantly with PAV aging, with the structure getting much denser, and the fibrils also getting smaller in diameter for most.

As a result of the chemo mechanical and microstructural investigations, the TC 252- CMB recommends the development of appropriate reduced short term aging temperature for RTFOT simulation of WMA. To this end, a wider sample of binders and mixtures need to be investigated and validated with in plant production or other warm mix technologies.

Acknowledgements. The authors would like to acknowledge the contributions of the active members of the Rilem 252-CMB technical committee in terms of experimental results and supplying bitumen samples.

References

D'Angelo, J., Harm, E., Bartoszek, J., Baumgardner, G.L., Corrigan, M., Cowsert, J., Harman, T., Jamshidi, M., Jones, W., Newcomb, D., Prowell, B., Sines, R., Yeaton, B.: Warm-Mix Asphalt: European Practice. FHWA-PL-08-007. US Federal Highway Administration, Alexandria (VA) (2008)

EN 12591: Bitumen and bituminous binders. Specifications for paving grade bitumen (2009)

EN 12607-1: Bitumen and bituminous binders. Determination of the resistance to hardening under influence of heat and air. RTFOT method (2014)

EN 1426: Bitumen and bituminous binders—determination of needle penetration. European Committee for Standardization (2007)

EN 1427 (2015) Bitumen and bituminous binders. Determination of the softening point. Ring and Ball method

EN 14769: Bitumen and bituminous binders - Accelerated long-term ageing conditioning by a Pressure Ageing Vessel (PAV). Brussels 2012

EN12697-3: Bituminous mixtures. Test methods for hot mix asphalt. Part 3: Bitumen recovery: Rotary evaporator (2013)

Ferrotti, G., Baaj, H., Besamusca, J., Bocci, M., Cannone-Falchetto, A., Grenfell, J., Hofko, B., Porot, L., Poulikakos, L., You, Z.: Comparison of short term laboratory ageing of bitumen and HMA and WMA bitumen aging. Submitted to the CMB Symposium (2018)

Godenzoni, C., Graziani, A., Perraton, D.: Complex modulus characterisation of cold recycled mixtures with foamed bitumen and different contents of reclaimed asphalt, Road Materials and Pavement Design (2016)

Hofko, B., Cannone Falchetto, A., Grenfell, J., Huber, L., Lu, X., Porot, L., Poulikakos, L.D., You, Z.: Effect of short-term ageing temperature on bitumen properties. RMPD, **18**(sup 2), 108-117 (2017)

Lamontagne, J., Dumas, P., Mouillet, V., Kister, J.: Comparison by Fourier transform infrared (FTIR) spectroscopy of different ageing techniques: application to road bitumens. Fuel **80**(4), 483–488 (2001)

Lu, X., Isacsson, U.: Effect of ageing on bitumen chemistry and rheology. Constr. Build. Mater. **16**, 15–22 (2002)

Mikhailenko, P., Kadhim, H., Baaj, H., Tighe, S.: Observation of asphalt binder microstructure with ESEM. J. Microsc. **267**, 347–355 (2017)

Porot, L., Eduard, P.: Addressing asphalt binder aging through the Viscous to Elastic Transition ISAP symposium 2016 Jacksonhole

Rheology and Bituminous Binder, A Review of Different Analyses

Laurent Porot(✉)

Kraton Chemical B.V., Almere, The Netherlands
Laurent.porot@kraton.com

Abstract. The use of Dynamic Shear Rheometer, DSR, has become a standard testing method to address the rheology of bituminous binder in various conditions. This test can generate a comprehensive set of data. Its interpretation can be done in different ways, depending on what will be the key features to address. This paper provides an overview and put in perspective the different parameters and approaches when analysing DSR data. It is based on a selection of different paving grade binders from soft to hard, one Polymer modified Binder and different aging conditionings of a standard binder. The analysis includes master curve of complex shear modulus, Black Space, Cole Cole diagram, as well single associated parameters such as Performance Grade criteria, cross-over parameters, Glover-Rowe parameter, amongst over. They all provide relevant and complementary information. While considered alone, they are not sufficient to address the whole range of behaviour. The combination of them can give a broader view of binder behaviour.

Keywords: Bitumen · Asphalt · Rheology · DSR · Master curve Black space

1 Introduction

Bituminous binder is a visco elastic material which is complex to describe and to characterise. Advanced characterisation, based on Dynamic Mechanical Analysis, DMA, helps having a broad overview of the behaviour of the materials through its rheology (Christensen 2003). For bituminous binder, Dynamic Shear Rheometer, DSR, is used now extensively either for purchase specifications, as in the US with the Performance Grade (PG) system (AASHTO T315) or for researches.

There are different ways to run DSR, and multiple ways to analyse and interpret the results, depending on the behaviour or conditions considered. Over the last two decades, multiple approaches have been introduced with new parameters studied, generating different trends.

This paper is reviewing some different approaches and parameters from analysing DSR measurements. It does not focus on modelling the DSR data to predict master curves.

L. D. Poulikakos et al. (Eds.): RILEM 252-CMB 2018, RILEM Bookseries 20, pp. 50–55, 2019.
https://doi.org/10.1007/978-3-030-00476-7_9

2 Experiment

In this study, DSR was run at a fixed frequency of 10 rad/s, in temperature sweep of 6 °C per minute between −30 °C and +90 °C when applicable, and with a single geometry of 10 mm plate and 2.5 mm gap. The lower temperature experiment was run in stress control, and the higher temperature was in strain control. While it may differ from other standardized teste methods, this experiment allows to run the measurement in one single test and generate continuous data set, avoiding the use of shifting function when building master curves.

The data came from a broad data base of various studies and binders. In order to highlight the main differences in behaviour a selection of 6 binders was made as described in Table 1.

Table 1. List of binders with main properties

Label	Binder type	Penetration value	Softening point temperature	Color code
Bit1 10/20	10/20 hard grade	14 × 0.1 mm	63.0 °C	Black
Bit2 50/70	50/70 medium grade	49 × 0.1 mm	50.4 °C	Orange
Bit3 160/220	160/220 soft grade	162 × 0.1 mm	39.8 °C	Yellow
PmB	PmB 25-55 65	30 × 0.1 mm	65.4 °C	Purple
AB cycle I	50/70 after 1 PAV*	16 × 0.1 mm	66.4 °C	Light blue
AB cycle III	5/70 after 3 PAV*	11 × 0.1 mm	79.9 °C	Medium blue

* Pressure Aging Vessel as run at 100 °C for 20 h

There are different approaches used to analyse the DSR data set depending on which temperature domain or main parameters are considered. Rheology enables to characterise the visco-elastic behaviour of material. The shear modulus |G ∗| is the direct response of the material to a solicitation, the strain deformation/displacement to an applied strain/force. It is composed of a storage modulus, G′, as a direct answer to the solicitation, an elastic response, and a loss modulus, G″, as a viscous response. The delta in response is characterised by the phase angle δ.

$$|G *| = \left(G'^2 + G''^2\right)^{1/2} \tan\delta = G''/G''$$

For asphaltic materials, stiffness and elasticity are required for the high temperature domain to limit permanent deformation. While at intermediate temperature or low temperature lower modulus and more viscous component will relax more stress and then limit cracking susceptibility. The balance between stiffness, as addressed by the absolute value of the shear modulus |G ∗|, and the visco elasticity, as addressed by the phase angle δ, is always foreseen as a dilemma to solve.

3 Complex Shear Modulus

Master curves, with $|G*|$ or phase angle/tanδ can be displayed vs. frequency or temperature; both are giving mirror figures. As measurements were made by varying temperature at a fixed frequency, there was no need for shifting temperature or frequency. Additionally the cross-over parameters were included. It relates to a phase angle equal to 45 ° (or tanδ equal to 1). It corresponds to the transition between a more predominantly elastic behaviour and a more predominantly viscous behaviour (Porot 2016). Figure 1 shows the data for the different binders. Some binders did not reach the 90 °C as they were too soft.

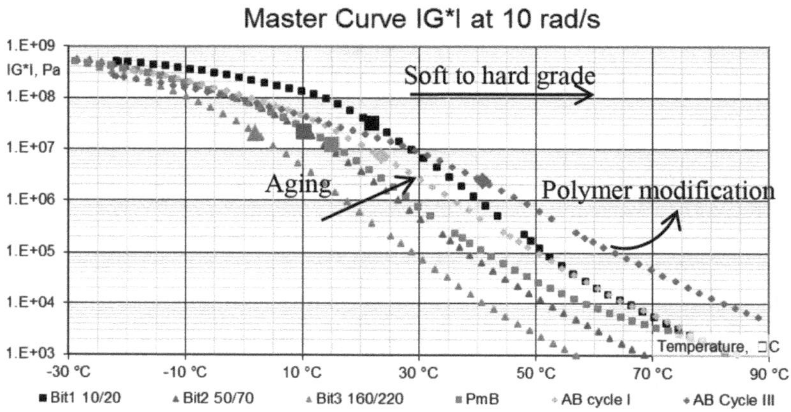

Fig. 1. Complex shear modulus for the different binders including cross over values

For the paving binder grades, soft to hard binder, the curves are shifted towards the left. For a same temperature, the shear modulus is higher while the binder is harder. The difference for the polymer modified binder is mostly seen in the high temperature domain with higher values than the medium paving grade. Over aging through one and three PAV cycles, the curves are turning right and up compared to the original binder. It means that the shear modulus increased but as well the overall slop of the curve became flatter, less temperature susceptible.

An analysis of the cross-over parameters confirms the trend. For the different pen grade binders, only the cross-over temperature is increasing with limited impact on the shear modulus. With harder grade, the binder remains elastic at higher temperature. The PmB shows both increase of cross-over temperature and decrease of cross-over shear modulus. To higher amplitude, with aging, the change is similar. The cross-over temperature is more related to stiffness, while the cross-over shear modulus to the temperature susceptibility.

4 Black Space

Black space plots the phase angle vs. complex shear modulus $|G*|$. It has no indication of temperature or frequency. However, the high temperatures are on the left and low temperatures on the right. The curvature of the curve can be characterised by the R-value, as the delta between the shear modulus and the glassy modulus (Christensen and Anderson 1992). The Glover-Rowe parameter (Rowe 2014) is included in the analysis. It is defined by G'/tanδ and was found to correlate with the ductility. The value is determined at 15 °C and 0.005 rads^{-1}, or, with time temperature superposition principle, it can be determined at 10 rads^{-1} for a shifted temperature of 44.7 °C. Figure 2 shows the data for the different binders, the dotted line are for the R-value and point-dotted lines the threshold values for the Glover-Rowe parameter. It is worth noticing that the points are more discriminant with higher phase angle, the viscous behaviour, corresponding to high temperature domain.

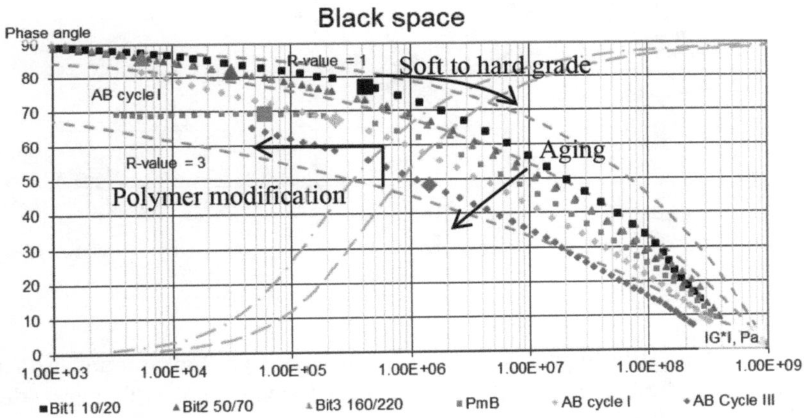

Fig. 2. Black Space for the different binders including Glover-Rowe parameters

For the different paving binder grades, soft to hard binder, the curves are almost overlapping, only a stretch can be observed while the binder is softer. With aging, the curves are shrinked, as a result of hardening, and becoming flatter, as a result of lower temperature susceptibility with higher R-value. Finally the PmB curve clearly displayed a different shape with more elasticity/lower phase angle and the clear presence of a rubber plateau with shear modulus below 10^5 Pa and phase angle of 70°. This highlights the benefits of PmB keeping elasticity at high temperature, while maintaining viscosity at intermediate and low temperatures.

The analysis of the Glover-Rowe parameter can correlate well with the stiffness of the binders. For the different paving grades, the harder the grade, the closer to the threshold it is, and even further for aged binder. It is worth noticing that the values correspond to high phase angle, more in the viscous behaviour, high temperature domain.

5 Other Analysis

Additional analyses of DSR data are also used. Amongst other are as follow.

Cole-Cole diagram plots G″ vs. G′ in linear scales as displayed in Fig. 3. There is no indication of temperature or frequency. This graph provides information on the low temperature behaviour where points are more discriminant. The shape is a bell curve with the indication of the peak as the maximum of the loss modulus, G″. The softer the higher the loss modulus, providing more relaxation while with aging the G″ is reduced consequently.

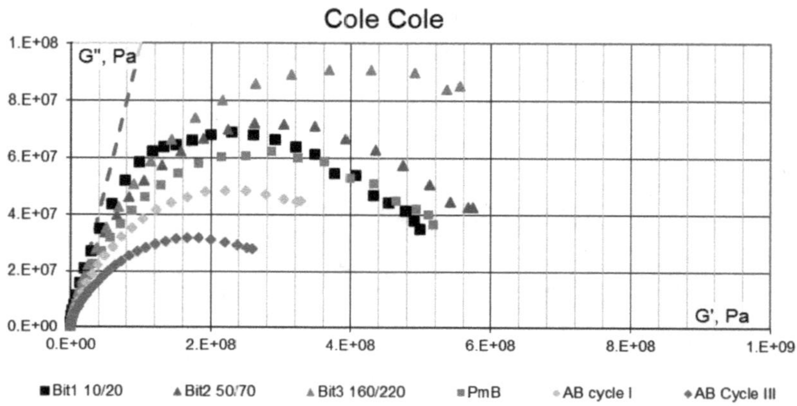

Fig. 3. Cole-Cole plot for the different binders

In the US, the SHRP Superpave approach is using the DSR measurements to assess the properties of the binder at high and intermediate temperatures, thus providing an indication of asphalt mix performance against rutting resistance, stiffness and cracking susceptibility. The G*sinδ = G″ and G*/sinδ, under different aging conditioning, are the criteria used in the PG grading purchase specifications as indicators of fatigue performance and rutting resistance.

And in Europe, a correlation is foreseen between the softening point temperature and the temperature at which the |G ∗| is equal to 15 kPa (Radenberg et al. 2016). As example in Germany, it is now part of bituminous binder specifications known as BTSV (Bitumen-Typisierungs-Schnell-Verfahren) temperature.

At a glance, Fig. 4 displays in a |G ∗| master curve, for one binder, the different parameters considered by researchers at this point of time. All of them are based on the same data source but with a different ways to analyse them. Clearly there are two set of points

- for the high temperature domain with the Glover-Rowe parameter, the high PG temperature or Temperature at |G ∗| equal to 15 kPa
- for the intermediate temperature domain with the PG intermediate temperature criteria and the cross-over temperature

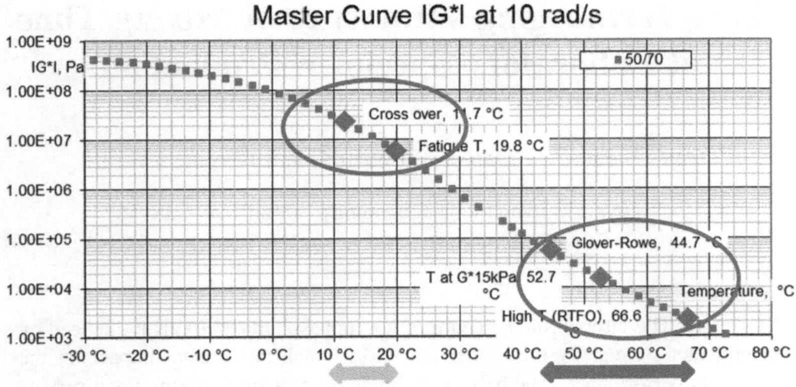

Fig. 4. Various DSR parameters in $|G*|$ vs temperature plot

6 Conclusion

There is a consensus today considering DSR as a useful tool to address the bituminous material behaviour in a wide range of temperature/frequency conditions. It leads to various ways interpreting the data. The shear modulus master curve or the Black space are ones of them. Single parameters, such as cross-over or Glover Rowe parameters are also considered. Based on different bituminous binder, this study has shown the limit and the complementarity of the different approaches.

The Black space, the Glover-Rowe, BTSV and IG*I/sinδ parameters address the high temperature domain. The master curve and the cross-over parameters, PG fatigue criteria address the intermediate temperature domain. The low temperature domain can be further addressed through the Cole Cole diagram.

References

AASHTO T315. Determining the Rheological Properties of Asphalt Binder Using a Dynamic Shear Rheometer (DSR). Washington (D.C.)

Christensen, D., Anderson, D.: Interpretation of dynamic mechanical test data for paving grade asphalt cements. Assoc. Asphalt Paving Technol. **61**, 67–116 (1992)

Christensen, R.: Theory of Viscoelasticity, 2nd edn. Dover, Mineola (N.Y.) (2003)

Radenberg, M., Flottmann, N., Koening, M., Holfert, T.: Extended bitumen testing in Germany, 6th Eurobitume & Eurasphalt congress. Prague dx (2016). https://doi.org/10.14311/EE.2016.403

Rowe, G.: Interelationships in Rheology for asphalt binder specifications, Canadian Technical Asphalt Association, 59th Annual Conference, Winnipeg Canada (2014)

Porot, L., Pieter, E.: Addressing asphalt binder aging through the Viscous to Elastic Transition. ISAP Symposium, Jacksonhole, US (2016)

Short Term Aging - Influence of Mixing Time at Laboratory Specimen Production

Daniel Steiner(✉), Daniel Maschauer, and Bernhard Hofko

Vienna University of Technology, Vienna, Austria
daniel.steiner@tuwien.ac.at

Abstract. The viscoelastic behavior of hot mix asphalt (HMA) is influenced by several factors, e.g. by the binder behavior, which has crucial impact on the pavement performance. The behavior changes due to aging during its in-service life within years, as well as during HMA production and compaction processes within a few hours. Therefore, aging processes are classified into long-term aging (LTA) and short-term aging (STA). This paper presents a study, analyzing STA processes at laboratory productions of HMA slabs. Six production cycles, with up to three slabs each were analyzed. The mix of binder and aggregates were prepared at once. This requires storing the hot and prepared mix inside the laboratory mixer while the first and the second slab are compacted, respectively. Bitumen was recovered for each produced slab. |G*| stiffness tests (DSR) are carried out afterwards at 52–82°C and 1.592 Hz. The results show that mixing times lead to a stiffness increase of 1.5 compared to the virgin bitumen, what is slightly below RTFO aging level. Recovered binder from Slab 2 and Slab 3 show increases of |G*| of 1.9 and 2.5. Therefore, for the used binder, longer mixing times have a significant impact on STA. The results can be used to back-calculate STA levels for specimens that are used for LTA aging studies. Furthermore, both binder and additional HMA stiffness results can serve as basis for multi-scale stiffness modelling by coupling the results.

Keywords: Hot mix asphalt · Reverse-Rotation compulsory mixer
Short term aging · DSR · Bitumen

1 Motivation and Objectives

Due to its organic origin, bitumen is undergoing constant change of its mechanical and chemical behavior. Talking about aging of bitumen and bituminous pavements a distinction is made between short-term aging (STA) and long-term aging (LTA). Reactions within a few hours during HMA production and compaction are defined as STA, while changes during its in-service life within years are defined as LTA. STA is characterized by fast oxidation by reason of elevated temperatures and a large specific surface, as well as the evaporation of remaining volatile components from the bitumen (Petersen 1994a, 1994b, Baek et al. 2012). LTA can be described as a slow oxidation starting from top of a pavement structure, since it is more exposed to oxidant gases available in the atmosphere (e.g. ozone, nitric oxides) (Morian et al. 2011). In the course of the years, the behavior of bitumen is transforming to stiffer and more brittle

L. D. Poulikakos et al. (Eds.): RILEM 252-CMB 2018, RILEM Bookseries 20, pp. 56–61, 2019.
https://doi.org/10.1007/978-3-030-00476-7_10

areas. As a result, pavements become more vulnerable to cracking by low-temperature and fatigue cracking (Teshale et al. 2011; Steiner et al. 2016a).

To assess bitumen aging in the laboratory the rolling thin film oven test (RTFOT) (CEN 2007) and the pressure aging vessel (PAV) (CEN 2012, Airey 2003) are standardized and widely accepted methods. To assess HMA aging in the laboratory, HMA slab production and oven conditioning of the mix are widely used to achieve STA. For simulating LTA of loose or compacted asphalt mix, far more procedures have been developed and published in the last decades. One of the latest developed methods is the Viennese Aging Procedure (VAPro) (Steiner et al. 2016b, Bell et al. 1994, Frigio et al. 2016).

In the laboratory, several factors (e.g. amount of HMA, temperature, mixing time) influence the materials properties at HMA slabs/specimens production for HMA LTA studies. Therefore, from a practical point of view, it is important to pay attention to some specific issues while carrying out laboratory HMA production to minimize STA differences between all produced samples. In many cases, slim slabs have to be produced in the laboratory and therefore only a small amount of material is required. Hence, a risk of segregation or stick of bitumen on the walls and stirrer of the mixer exists. Due to that, the mix design and the receipt can be changed, respectively. Furthermore, small amounts of materials lead to greater specific surfaces in large mixers, which can causes stronger STA. For these reasons, preparing and mixing of higher amounts of material for more than one slab at a time is preferred. Although the STA can be reduced with using greater amounts of materials, influences of additional storage times in the mixer due to the slab compacting has to be analyzed. Therefore, a small study was carried out within this paper. The results of the study can be used to estimate the sensitivity of the mixing time for slabs that are produces in series. More specifically, the binder stiffness results can serve as basis for multi-scale stiffness modelling with coupling them to HMA stiffness results, that where collated from specimen cored out of slab of this study.

2 Materials Test Methods and Experimental Program

The presented study looked into the impact of mixing time of HMA at slab production in the laboratory. Material from six different production days were analyzed using the same bitumen, HMA mix design and mixing temperatures. The amount of slabs varied for each production day between 1–3 slabs. Since normally only one roller compactor is available in common research and commercial laboratories, the mix has to kept within the mixer, while the first and the second slab are compacted, respectively. Changes in the mechanical behavior of the bitumen due to this additional storage time can occur because of STA.

To conduct the study, asphalt concrete with a maximum nominal aggregate size of 11 mm (AC 11) was employed. As a binder an unmodified 70/100 pen (PG 64-22/79 pen) was used. The binder content was set to 5.2% by mass with a target void content of 8.0% by volume. The mix was prepared in a laboratory reverse-rotation compulsory mixer, according to EN 12697-35, with a mixing temperature of +170 °C. HMA slabs (50 × 26 × 4 cm) were compacted in a roller compactor according to EN 12697-33.

At each slab production day, the mixture was prepared at once for all slabs. Therefore, the mixing time varied for each slab. For slab 1, the HMA was mixed and stored within the laboratory mixer for 5 min, HMA of slab 2 for 13 min and HMA of slab 3 for 21 min.

Eight specimens are cored from each slab with a diameter of 100 mm. Bitumen from slab remains was extracted according to EN 12697-3 with tetrachloroethylene (C2Cl4) as a solvent. The solvent-bitumen solution was distilled according to EN 12697-3 to recover the binder samples.

For analyzing the binder behavior, Dynamic Shear Rheometer (DSR) tests were carried out on bitumen samples recovered from all laboratory-STA HMA materials. The test conditions were chosen according to the SHRP procedure (Petersen et al. 1994a, 1994b) and EN 14770 with a temperature sweep from +52 °C to +82 °C using the large plate (diameter: 25 mm) and a 1 mm gap. A frequency of 1.592 Hz is employed. From test data the dynamic shear modulus |G*| and the phase angle φ vs. frequency are determined. Furthermore, the dynamic shear modulus |G*| of the same virgin and RTFOT -aged binder were determined as benchmarks.

3 Findings and Outlook

To investigate STA effects due to mixing time, bitumen from HMA productions were extracted and recovered from the HMA. Analysis of changes of the viscoelastic behavior were carried out. In addition, STA bitumen by RTFOT and LTA bitumen by RTFOT +PAV were tested to compare standardized bitumen aging procedures to the results.

The results of the DSR |G*| stiffness testing are shown in Fig. 1. The diagram shows the relative change in dynamic modulus |G*| of the binders in STA condition vs. unaged condition over the DSR testing temperature range. The data presented in the diagram are mean values (MV). Furthermore, the 95% confidence intervals for 4–6 single tests are stated. The solid lines represent data from the RTFOT aged bitumen. The results indicate significant changes in the mechanical behavior for slabs with longer mixing times. For all first slabs of a production cycle of three slabs the binder samples are 1.4–1.6 times stiffer than the virgin binder is. It can be seen, that these slabs are close to a RTFOT aging level, but they are slightly too soft to reach this level. The results of the second and third slabs are by far stiffer than RTFOT. The binder of the second slabs are 1.7–2.0, the third slabs are 2.1–2.5 times stiffer than the virgin binder is.

The results of the DSR phase angle φ stiffness tests are shown in Fig. 2. Changes due to the aging are shown by subtracting the phase angle in aged condition from phase angle in virgin condition. Therefore, this leads to the decrease of the phase angle due to STA. A similar picture as for the |G*| results can be seen. Here, the results of the first slabs of a productions cycle of three slabs perfectly match the RTFOT aging level. The phase angle was reduced between −4.7° at low testing temperatures and −1.5° at high testing temperatures. The MV of all second slabs reduces by the value of −1.9° to −6.1°. The phase angle of the third slabs is reduced between −3.0° and −8.5°.

In Fig. 3, a summary of the results is presented using DSR results at +64 °C. Here, the results are shown vs. the mixing times. To get a better overview, where the results are located between a standard RTFOT aging level and a standard RFTOT+PAV aging level, the results of a RTFOT and a RTFOT+PAV aged binder are illustrated. It can be

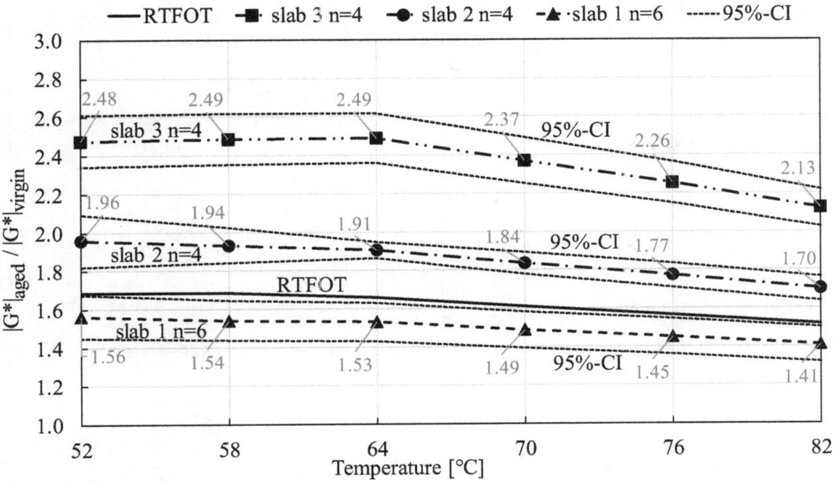

Fig. 1. DSR |G*| @ 52-82 °C, 1.592 Hz. Recovered bitumen from STA HMA slabs (n = 4–6) with different mixing times (5 min, 13 min, 21 min)

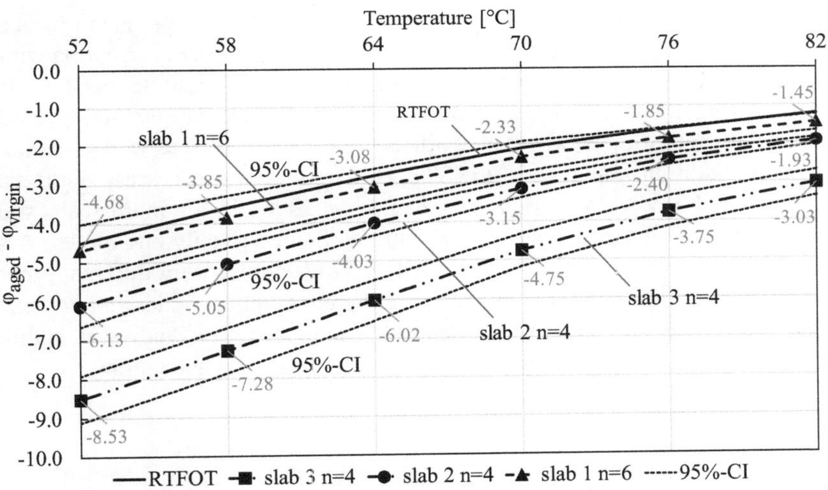

Fig. 2. DSR φ @ 52-82 °C, 1.592 Hz. Recovered bitumen from STA HMA slabs (n = 4-6) with different mixing times (5 min, 13 min, 21 min)

seen, that mixing time has significant influence on the mechanical behavior. Nevertheless, there is still room to reach a LTA PAV level. To get an idea about the scale of the influence, linear regressions are stated within the figures. Looking at the coefficient of determination, R^2 are equal or above 0.95, which indicates an excellent linear correlation of phase angel/dynamic modulus vs. mixing time.

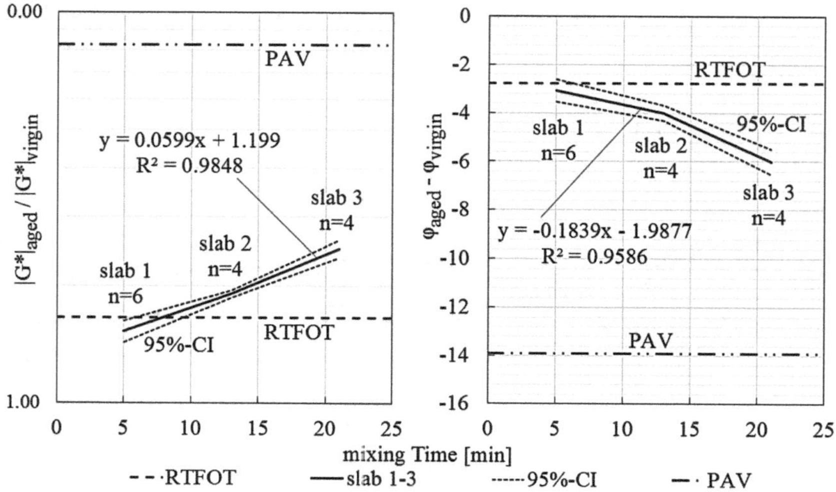

Fig. 3. DSR |G*| and φ @ 64 °C, 1.592 Hz vs mixing time. Recovered bitumen from slabs (n = 4–6)

The main drive for the research study presented within this paper is to give a better understanding of mixing time for STA of HMA in the laboratory. Summarizing the results, it can be stated, that longer mixing/storage times lead to changes in the mechanical behavior of the recovered binder. Since aging can be seen more clearly on binder level, the question is, whether significant difference can be as well seen on HMA level or not. Additional analysis of the STA HMA stiffness is necessary. Furthermore, both binder and HMA stiffness results can serve as basis for multi-scale stiffness modelling with coupling the results. This paper is limited to only one binder. A study of the aging behavior on HMA and binder level with binder from different provinces should give a better idea of aging in general. Different bitumen grades and provenances can lead to completely different behaviors. As the authors have already seen in different other unpublished work, the bitumen used in this paper is more prone to STA and LTA HMA aging than other bitumen.

References

Airey, G.D.: STAR on Ageing Test Meth. for Bit. Pavement Materials. I.J. of Pavem. Eng **4**, 165–176 (2003)

Baek, C., Underwood, B., Kim, Y.: Effects of oxidative aging on asphalt mixture properties. Transp. Res. Rec. J. Transp. Res. Board **2296**, 77–85 (2012)

Bell, C.A., Abwahab, Y., Cristi, M.E.: Selection of Laboratory Ageing Procedures for Asphalt Aggregate Mixtures (SHRP-A-383). SHRP, Washington, DC, National Research Council (1994)

CEN: EN 12607-1: Bitumen and bituminous binders – RTFOT. Brussels (2007)

CEN: EN 14769: Bitumen and bituminous binders - PAV. Brussels (2012)

Frigio, F., Raschia, S., Steiner, D., Hofko, B., Canestrari, F.: Aging effects on recycled WMA porous asphalt mixtures. Constr. Build. Mater. **123**, 712–718 (2016)

Morian, N., Hajj, E.Y., Glover, C.J., Sebaaly, P.E.: Oxidative aging of asphalt binders in hot-mix asphalt mixtures. Transp. Res. Rec. **2207**, 107–116 (2011)

Petersen, J.: Binder Characterization and Evaluation Volume 4: Test Methods SHRP-A-370. Washington, DC: Transportation Research Board. (Strategic Highway Research Program) (1994a)

Petersen, J.C., Robertson, T., Branthaver, J.F., Harnsberger, P.M., Duvall, J.J., Kim, S.S., Anderson, D.A., Christiansen, D.W., Bahia, H.U., Dongre, R., C.E., A., Sharma, M. G., Button, J. W., Glover, C.J.: Binder Characterization and Evaluation Volume 4: Test Methods (SHRP-A-370). SHRP. Washington, DC, National Reserach Counil (1994b)

Steiner, D., Hofko, B., Dimitrov, M., Blab, R.: Impact of Loading Rate and Temperature on Tensile Strength of Asphalt Mixtures at Low Temperatures. In: Chabot, A., Buttlar, G.W., Dave, V.E., Petit, C., Tebaldi, G. (eds.) 8th RILEM International Conference on Mechanisms of Cracking and Debonding in Pavements. Dordrecht. Springer, Netherlands (2016a)

Steiner, D., Hofko, B., Hospodka, M., Handle, F., Grothe, H., Fussl, J., Eberhardsteiner, L., Blab, R.: Towards an optimised lab procedure for long-term oxidative ageing of asphalt mix specimen. Int. J. Pavement Eng. **17**, 471–477 (2016b)

Teshale, E.Z., Moon, K.H., Turos, M., Marasteanu, M.: Pressure aging vessel and low-temperature properties of asphalt binders. Transp. Res. Rec. 117–124 (2011)

Viennese Aging Procedure – Behavior of Various Bitumen Provenances

Daniel Maschauer[1(✉)], Daniel Steiner[1], Johannes Mirwald[2], Bernhard Hofko[1], and Hinrich Grothe[2]

[1] Institute of Transportation, Research Center of Road Engineering, Vienna University of Technology, Gusshausstraße 28/230-3, 1040 Vienna, Austria
daniel.maschauer@tuwien.ac.at
[2] Institute of Materials Chemistry, Vienna University of Technology, Getreidemarkt 9/BC/01, 1060 Vienna, Austria

Abstract. Bitumen changes its properties in the course of time under natural and anthropogenic influences due to its organic origin. These processes are commonly called "aging". The material becomes stiffer and more brittle, resulting in less favorable low temperature and fatigue behavior. For this reason, it is important to simulate the aging of the material in the laboratory in an accelerated way to study the change in material behavior and minimize damage on the road. On the bitumen level, the standardized methods RTFOT (Rolling Thin Film Oven Test) and PAV (Pressure Aging Vessel) are used. Various methods have been developed in the past to simulate aging of asphalt mixes or compacted specimens. The study presented in this paper evaluated the aging method "Viennese Aging Procedure" (VAPro) for applicability with the aid of a parameter study with bitumen of different origin. VAPro uses realistic boundary conditions (temperature: +60 °C, pressure: ~0.3 bar) and increases the rate of aging of the compacted asphalt mix specimen by perfusing (1.0 l/min) with ozone and nitric oxides enriched air for three days. The state of aging of the extracted bitumen is assessed using the Dynamic Shear Rheometer (DSR). Significant differences between the employed bitumen are determined, which is probably caused by their different initial stiffness and their origin. The stiffness after VAPro of the extracted bitumen is between 1.2 and 2.6 times the RTFOT +PAV-aged state.

Keywords: VAPro · Ozone · NO_x · Long term aging · Hot mix asphalt Stiffness · DSR

1 Introduction

Organic materials, such as bituminous binders, change their properties over time due to natural and anthropogenic influences. Aging of bitumen and asphalt mixtures is divided in short-term aging (STA) during production and paving and long-term aging (LTA) during its service life on the road. STA is caused by fast chemical oxidation due to high temperatures and a high specific surface contacting with oxidant agents at mix

L. D. Poulikakos et al. (Eds.): RILEM 252-CMB 2018, RILEM Bookseries 20, pp. 62–67, 2019.
https://doi.org/10.1007/978-3-030-00476-7_11

production, as well as the vaporization of remaining volatile components (Baek et al. 2012). LTA is characterized by slow oxidation primarily in the upper layers of the asphalt by oxidant gases available in the atmosphere like ozone or nitric oxides (Morian et al. 2011). Due to aging, the bitumen becomes brittle and hard, which leads to a deterioration in the performance properties of asphalt pavements, especially, resulting in more likely low-temperature cracks and decreasing fatigue resistance (Teshale et al. 2011); (Hofko et al. 2014).

For this reason, it is important to accelerate the aging of the material in the laboratory in order to investigate the change in material behavior, as an input in mix design optimization, to minimize damage on the road by optimized pavements that are more durable. On bitumen level, the standardized methods RTFOT (Rolling Thin Film Oven Test) for STA and PAV (Pressure Aging Vessel) for LTA are commonly used (da Costa et al. 2010). For aging, loose or compacted asphalt mixtures in laboratory more than 30 procedures have been developed over the last decades. Many of these procedures use high temperature (above +100 °C) and/or high pressure that do not occur in the field (Bell et al. 1994). As a consequence, additional effects could be induced by these methods, such as evaporation of further volatile binder components that are not triggered in the field.

Therefore, a new procedure called 'Viennese Aging Procedure' (VAPro) was developed at Vienna University of Technology for aging compacted asphalt specimens under realistic boundary conditions concerning temperature and pressure (VAPro) (Steiner et al. 2016). The main objective of this paper is to test the applicability of VAPro. For this purpose, asphalt mixes with the same aggregate composition but with bitumen of different origins were examined and evaluated using the Dynamic Shear Rheometer on extracted binders after VAPro aging. The results show that the VAPro is easily applied and the aging level of the extracted bitumen is between 1.2 and 2.6 times the RTFOT+PAV-aged state. Moreover, it is evident that the origin of the bitumen has large impact on the achieved level of aging.

2 Materials and Methods

The study uses an asphalt concrete with a maximum grain size of 11 mm (AC 11). The coarse aggregates used for the mix is a porphyrite and the filler is powdered limestone. Six mixes were produced, four with paving grade bitumen (70/100) of different origins and two with polymer-modified bitumen (PmB 45/80-65) from the same source but different production years. Table 1 gives a list of the used bitumen. The bitumen content is set to 5.2% by mass for each mix with a target void content of 8% by volume.

For preparation of the mixes, a laboratory mixer, according to EN 12697-35 is used. The mixing temperatures are set at +170 °C for the paving grade bitumen and +185 °C for the polymer-modified bitumen, respectively. Subsequently, asphalt slabs (50 × 26 × 4 cm) are compacted in a roller compactor according to EN 12697-33. From these slabs, the specimens (eight per slab) with a diameter of 100 mm are cored. The air void content of the produced slabs ranges from 4.8 to 9.2% by volume. For each mix, three specimens with an air void content of about 7% by volume were selected for aging with VAPro.

Table 1. Bitumen types used

Bitumen type	Origin	Penetration in 1/10 mm	Dynamic shear modulus @ 1.59 Hz, 64 °C in MPa
70/100	A	88	1.34
70/100	B	79	1.09
70/100	C	64	1.77
70/100	D	55	1.57
PmB 45/80–65	A (2012)	67	5.58
PmB 45/80–65	A (2017)	65	4.97

Figure 1 shows the setup that is used for VAPro. Compressed air from the laboratory system is directed through the ozone generator. The generator enriches the compressed air with ozone and nitric oxides using a dielectric barrier discharge tube. A flow regulator ensures a constant flowrate of 0.9–1.1 l/min. The gas mixture is then heated to +70 °C (T_{liq}) via a heating coil, which is placed in a beaker glass filled with vegetable oil that is placed on a heatable magnetic stirrer. The specimen is mounted within a triaxial cell between two filter stones and is covered with an elastic membrane. The oxidant flows through the samples from bottom to top. An overpressure of about 80 kPa in the triaxial cell ensures that the membrane is pressed against the specimen so that the gas must flow through the specimen. The triaxial cell and the heating coil device are housed in a heating cabinet at an air temperature of +60 °C (T_{air}). The duration of aging for one specimen is set to three days since former studies, e.g. (Steiner et al. 2016) found that this duration gives an ageing state of the recovered binder that resembles that of an RTFOT+PAV aged bitumen.

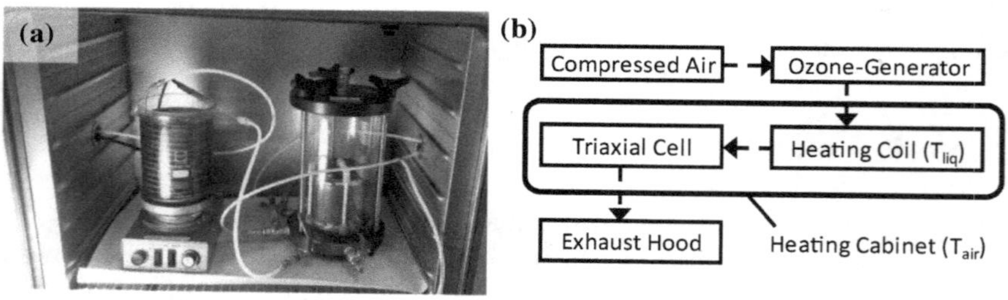

Fig. 1. VAPro setup: (a) Photo and (b) Schematic Diagram of the setup

After aging, the bitumen is extracted with tetrachloroethylene as solvent according to EN 12697-3. The solvent is then separated from the bitumen by vacuum distillation according to EN 12697-3. Dynamic Shear Rheometer (DSR) tests are performed on the bitumen obtained from the VAPro-aged specimens. The DSR-tests are also carried out on the virgin bitumen samples, extracted bitumen samples of slab remnants after drilling (short-term aging condition), RTFOT samples and RTFOT+PAV samples for comparison. The test are carried out at temperatures ranging from −40 °C to +82 °C in 6° steps. The test frequencies are 0.1, 0.3, 1.0, 1.59, 3.0, 5.0 and 10.0 Hz.

3 Results and Discussion

For investigation of the level of aging, the changes in behavior due to VAPro-aging are analyzed. For comparison, the slab remnants, RTFOT and RTFOT+PAV samples are also examined. Figure 2 shows the relative change in dynamic shear modulus. The results presented are for a frequency of 1.59 Hz at 64 °C. Similar trends can be observed for other frequencies and temperatures.

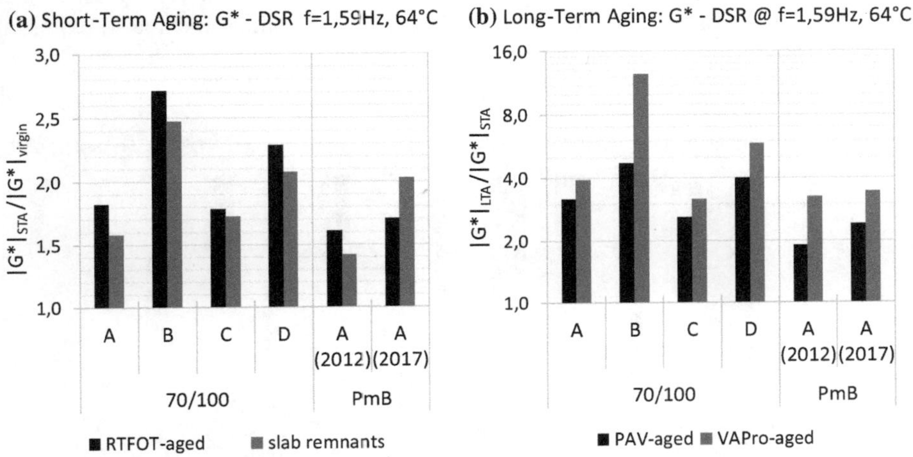

Fig. 2. Dynamic Shear Modulus G*: (a) Comparison of STA (ratio of STA condition to virgin condition) (b) Comparison of LTA (ratio of LTA condition to STA condition)

For STA, it can seen that the achieved level of aging of the slab remnants resembles to that of the RTFOT-samples. The level of aging ranges from 1.5 to 2.5. For each bitumen, with the exception of bitumen from source A(2017), the aging level of the slab remnants is slightly below RTFOT aging. It can be stated that the RTFOT process is a good representation of the aging processes during laboratory production. The differences between the bitumen of different origin can already be seen here: E.g. paving grade bitumen from source B is significantly more aging susceptibly (factor 2.5) than all other paving grade samples.

For LTA, the tendency of the differences is similar. For the paving grade bitumen, the level of aging due to VAPro ranges from 3 to 13. The achieved level of aging corresponds with the dynamic shear modulus of the virgin bitumen. The bitumen with the smallest increase has the biggest virgin shear modulus and vice versa (cf. Table 1). For both polymer-modified bitumen, the achieved level of aging is quite the same, which indicates the constant quality of the supplier. The PAV-aged samples are below the VAPro-samples. A significant difference between PAV and VAPro shows the bitumen from source B. This gives a hint that probably the PAV-procedure does not represent the real aging processes quite well for every bitumen from any source.

Figure 3 shows the changes in phase lag. Similar observations in behavior can be made here. For short-time aging, RTFOT resembles to the slab remnants. For long-term aging the decrease of phase lag of the paving grade bitumen corresponds with the increase of dynamic shear modulus. The bigger the increase in shear modulus, the bigger the decrease in phase lag. The decrease of phase lag for the two polymer-modified bitumen due to VAPro is very similar. This again is a sign for the constant quality of the supplier. It should be noted that the PAV-samples of the polymer-modified bitumen show barely any decrease of the phase angle and the bitumen from source B has again the most significant difference between PAV and VAPro. This, once more indicates that the PAV-procedure might not be the optimal method to represent real aging processes depending on the bitumen source.

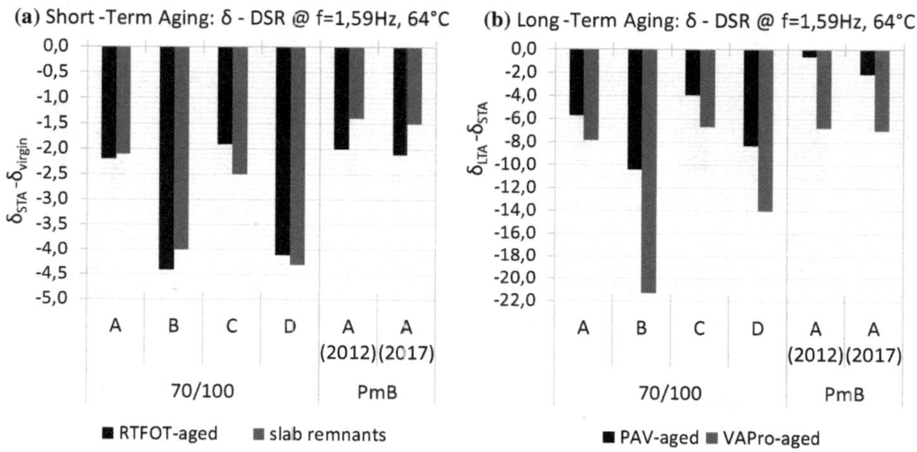

Fig. 3. Phase Lag δ: (a) Comparison of STA (difference of phase lag of STA minus phase lag in virgin condition) (b) Comparison of LTA (difference of phase lag of LTA minus phase lag of STA)

4 Conclusion

The tests carried out in this paper show that the Viennese Aging Procedure (VAPro) is well applicable on a broad basis. VAPro simulates aging processes by perfusing cylindrical specimens with ozone and nitric oxide enriched air with realistic boundary conditions concerning temperature and pressure (+60 °C, ~0.3 bar).

Stiffness and embrittlement of paving grade bitumen from different sources as well as two polymer-modified bitumen from the same source but different production years were investigated using the Dynamic Shear Rheometer (DSR). There are clear differences in the results of the individual bitumen, which indicates that the origin of the bitumen has an influence on aging behavior. The level of aging of some VAPro-samples differs significantly from the RTFOT+PAV-samples, which suggests that the PAV method does not capture the entire aging processes that occur in the field.

Additional results of cyclic indirect tensile test, which are not included in this paper, are available for future discussion.

To investigate the low-temperature behavior of VAPro-aged specimens, the device will be developed further to age prismatic specimens for TSRST and UTST.

References

Baek, C., Underwood, B.S., Kim, Y.R.: Effects of oxidative aging on asphalt mixture properties. Transp. Res. Record **2296**, 77–85 (2012)

Bell, C.A., AbWahab, Y., Cristi, M.E., Sosnovske, D.: Selection of Laboratory Ageing Procedures for Asphalt Aggregate Mixtures (SHRP-A-383), Washington, DC (1994)

da Costa, M.S., Farcas, F., Santos, L., Eusebio, M.I., Diogo, A.C.: Chemical and thermal characterization of road bitumen ageing. Adv. Mater. Forum, **636–637**, 273–279 (2010). V, Pt 1 and 2

Hofko, B., Hospodka, M., Blab, R., Eberhardsteiner, L., Fussl, J., Grothe, H., Handle, F.: Impact of field ageing on low-temperature performance of binder and hot mix asphalt. Asphalt Pavements, vol. **1** and **2**, 381–395 (2014)

Morian, N., Hajj, E.Y., Glover, C.J., Sebaaly, P.E.: Oxidative aging of asphalt binders in hot-mix asphalt mixtures. Transp. Res. Rec. J. Transp. Res. Board **2207**(1), 107–116 (2011)

Steiner, D., Hofko, B., Hospodka, M., Handle, F., Grothe, H., Fussl, J., Eberhardsteiner, L., Blab, R.: Towards an optimised lab procedure for long-term oxidative ageing of asphalt mix specimen. Int. J. Pavement Eng. **17**(6), 471–477 (2016)

Teshale, E.Z., Moon, K.-H., Turos, M., Marasteanu, M.: Pressure aging vessel and low-temperature properties of asphalt binders. Transp. Res. Rec. J Transp. Res. Board **2207**, 117–124 (2011)

Chemo-Mechanical Characterization of Bituminous Materials: Chemo-Mechanical Coupling

Ageing Effect on Chemo-Mechanics of Bitumen

Ruxin Jing(✉), Aikaterini Varveri, Xueyan Liu, Athanasios Scarpas, and Sandra Erkens

Delft University of Technology, Delft, The Netherlands
R.Jing@tudelft.nl

Abstract. Ageing has a significant impact on the chemical and mechanical behavior of bituminous materials. In this study, Fourier Transform Infrared (FTIR) spectrometer and Dynamic Shear Rheometer (DSR) tests were utilized to investigate the effect of ageing on the chemical and mechanical properties of bitumen. Bitumen films with thickness of 2 mm were exposed to laboratory ageing at various conditions. Specifically, different combinations of ageing time, temperature and pressure were applied on the materials. The FTIR results were used to quantify the changes in the chemical functional groups and to calculate the combined ageing index (summation of carbonyl and sulfoxide indices) of bitumen. In addition, the DSR test results were analyzed to determine the evolution of the crossover frequency and crossover modulus with ageing. A linear relationship was found between the combined ageing index and the distance in the crossover map, providing thus a chemo-mechanics framework to describe bitumen ageing.

Keywords: Bitumen · Ageing · FTIR spectroscopy · Rheology Chemo-mechanics

1 Introduction

In the Netherlands, raveling is the most common distress of porous asphalt pavements. Bitumen ageing is believed to be one of the main contributors to ravelling. It is well known that as bitumen ages its ductility and penetration index reduce while the softening point increases. Ultimately, the viscosity of the bitumen increases and bitumen becomes stiffer. This may cause the mixture to become excessively brittle and susceptible to damage (Saoula et al. 2013).

Recently, more and more researchers attempted to correlate the chemical composition of bituminous materials with their performance. Studies have indicated that the ageing mechanism affects the chemical composition of bitumen and it has been made clear that the rheological properties would change as well (Petersen and Glaser 2011). Unfortunately, the specific relationship between the chemical properties and mechanical response of bitumen is still undefined.

The main objective of this study is to determine the changes in the chemical properties of bitumen due to ageing and link them to its mechanical response, by means of FTIR and DSR tests. For this, bitumen films were aged in the laboratory at various times, temperatures and pressures. On the basis of the experimental results, a chemo-mechanics relationship for aged bitumen is established.

L. D. Poulikakos et al. (Eds.): RILEM 252-CMB 2018, RILEM Bookseries 20, pp. 71–76, 2019.
https://doi.org/10.1007/978-3-030-00476-7_12

2 Materials, Ageing Protocols and Test Methods

2.1 Materials

The tests were performed on a penetration bitumen 70/100 with no polymer or other chemical additive, a typical bitumen used in Netherlands.

2.2 Ageing Protocols

In this study, bitumen films with 2 mm thickness were aged using two different ageing methods: oven ageing and PAV (Pressure Ageing Vessel) ageing. Part of the samples were aged in the oven for 20, 40, 80, 160, 320 h at 100 °C and atmospheric pressure. To investigate the effect of temperature, other subsets of the samples were oven aged at 50 °C and 150 °C for 40 h.

Moreover, the samples were subjected to the standard ageing treatments: (i) RTFO at 163 °C for 75 min (EN 12607-1), which simulates bitumen after plant mixing, production, transportation and construction and (ii) PAV at 100 °C and 20 atm for 20 h (EN 14769), which simulates the state of bitumen after the first 5–10 years of pavement service life.

Finally, the standard PAV protocol was modified and bitumen was subjected to ageing for 40 h at 100 °C and at four different pressures of 5, 10, 15, 20 atm.

2.3 Test Methods

Fourier Transform Infrared Spectrometer. The tests were performed using the Spectrum 100 FT-IR spectrometer of Perkin-Elmer. A single-beam configuration was used. The sample was scanned 20 times, with a fixed instrument resolution of 4 cm^{-1}. The wavenumbers was set to vary from 600 to 4000 cm^{-1}.

Dynamic Shear Rheometer. DSR tests were performed using the Anton Paar MCR502. The bitumen samples were tested using the parallel-plates configuration. Initially, the linear viscoelastic (LVE) strain range of bitumen samples was determined using amplitude sweep tests. The frequency sweep tests were performed at five different temperatures (0, 10, 20, 30 and 40 °C). During the tests the frequency varied in a logarithmic manner from 50 Hz to 0.01 Hz.

3 Results and Discussion

3.1 Fourier Transform Infrared Spectrometer

At least three replicate samples were tested in FTIR at each condition. Using the obtained spectra, the ageing indices (carbonyl and sulfoxide index) of each sample were calculated. The calculation methodology is described in a previous study (Van den bergh 2011). Figure 1 shows the carbonyl and sulfoxide indices (average value of three measurements).

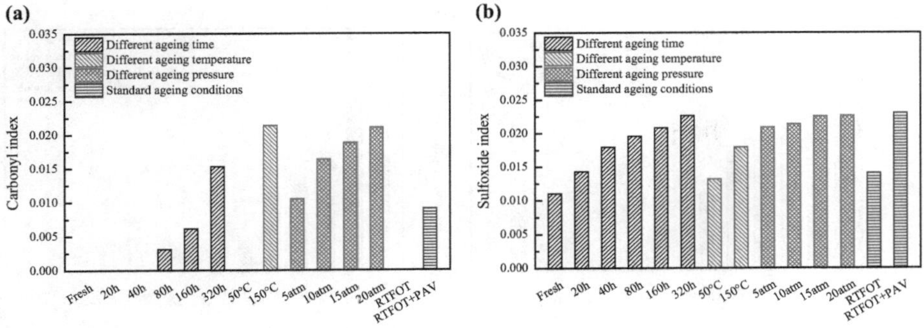

Fig. 1. Ageing indices of PEN 70/100 at different ageing conditions: (a) Carbonyl index, (b) Sulfoxide index

The results show that both indices increased with increasing ageing time, temperature and pressure. It is interesting to note that sulfoxides are formed earlier than carbonyls, because sulfur is more reactive than carbon in bitumen. It can be observed that, under weak ageing conditions (short ageing time, low ageing temperature and pressure), only sulfoxides are formed, and further increase, while no (or few) carbonyls are present in the aged bitumen samples.

On the contrary, the formation of carbonyls starts under strong ageing conditions (long ageing time, high ageing temperature and pressure), whereas the sulfoxide index is stable probably due to the full consumption of sulfur. In order to fully consider the chemical changes of bitumen due to ageing, a combined ageing index (the summation of carbonyl and sulfoxide indices) is used in this study.

Moreover, comparing the results of different ageing protocols in Fig. 1, it can be observed that temperature is the most influential parameters for ageing, probably because of the fact that the ageing rate coefficient increases exponentially with temperature based on the Arrhenius equation (Boysen and Schabron 2011).

3.2 Dynamic Shear Rheometer

Similarly to the FTIR tests, at least three replicate samples at each condition were tested by means of the DSR. The evolution of the complex shear modulus and phase angle master curves of bitumen with increased ageing time, temperature and pressure was discussed in a previous study (Jing et al. 2018).

In this study, the frequency when the storage shear modulus is equal to the loss shear modulus (phase angle is 45°), explicitly the crossover frequency was used to characterize the viscoelastic fluid to solid transitory behavior. The complex shear modulus corresponding to the crossover frequency is named crossover modulus. The crossover frequency and modulus for all samples are represented in Fig. 2(a) and (b), respectively.

Figure 2(a) shows that aged bitumen has lower crossover frequency than fresh bitumen. This suggests that aged bitumen has higher molecular mass (Liu et al. 2006). In the meanwhile, lower crossover frequency denotes longer relaxation time and higher

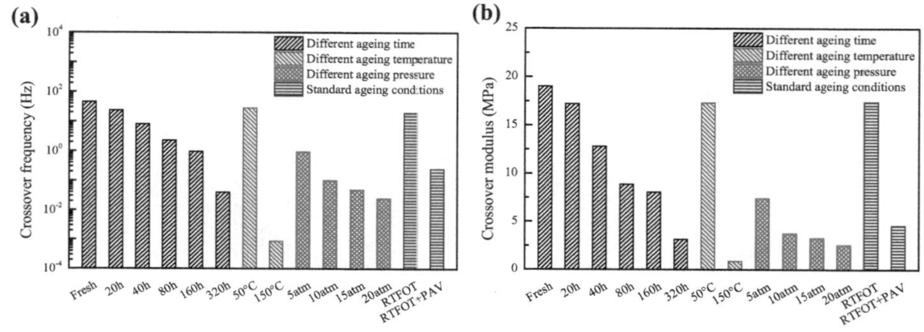

Fig. 2. Crossover frequencies and crossover modulus of bitumen at different ageing condition: (a) Crossover frequency, (b) Crossover modulus

softening point (Nivitha and Krishnan 2016). In Fig. 2(b), the crossover modulus of bitumen reduces due to ageing. This is indicative of wider molecular mass distribution and increased polydispersity of aged bitumen (Scarsella et al. 1999).

3.3 Chemo-Mechanics of Ageing

Chambon and Winter have shown that there was a good logarithmic relationship between crossover frequency and crossover modulus for the same type of polymers (Chambon and Winter 1985). The findings of this research validate this statement for studied bitumen. Figure 3(a) plots the crossover frequency against the crossover modulus of bitumen subjected to the various ageing protocols and, for the convenience of description, it is referred hereinafter as the crossover map.

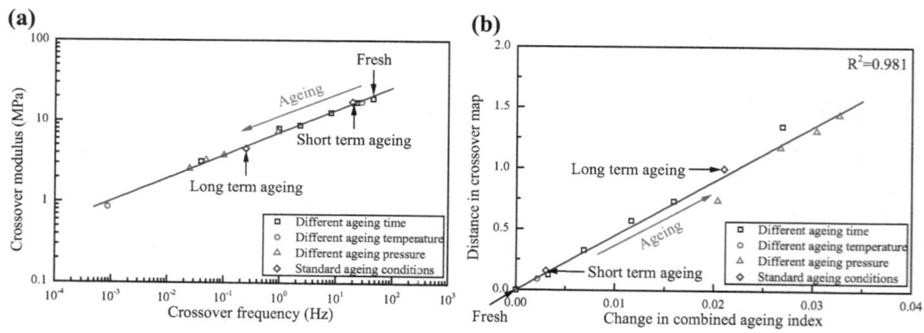

Fig. 3. Chemo-mechanics of ageing: (a) Crossover frequency vs crossover modulus of bitumen at different ageing condition, (b) Chemo-mechanical coupling of aged bitumen

In Fig. 3(a), the points move from the top right corner to the down left corner along the blue line due to ageing. The ageing state of bitumen, which can be described by the changes in the values of the combined ageing index, would determine how far one point can move. Then the relationship between chemical and mechanical properties of aged bitumen (for each ageing condition) can be described by using this distance in the crossover map and the combined ageing index (carbonyl index + sulfoxide index). Figure 3(b) shows that there is a good linear relationship between the two parameters for the specific bitumen type. Interestingly it appears that this relationship does not depend on the ageing methods. In another words, the results indicate that the different ageing conditions can be used interchangeably.

4 Conclusion

Given the strong relationship between the mechanical pavement response and ageing, which is a chemically-induced process, the knowledge of the evolution of the chemical properties in bituminous materials is of uppermost importance. For this reason, a series of ageing experiments were conducted on bitumen films at different times, temperatures, and pressures.

In order to develop a chemo-mechanics model of ageing, a series of FTIR and DSR tests were carried out to determine the changes in chemical properties and rheological response of aged bitumen. A linear relationship was found to exist between the combined ageing index and the distance in crossover map. The results suggest that different ageing conditions can yield the same ageing effect.

As a continuation of this research, the chemo-mechanics relationship will be validated for other bitumen types and a mathematical model will be develop to describe this relationship.

Acknowledgements. The authors gratefully acknowledge the Dutch Ministry of Transport, Public Works and Water Management for funding this project.

References

Boysen, R., Schabron, J.: Laboratory and field asphalt binder ageing: chemical changes and influence on asphalt binder embrittlement. Technical white papers of WRI (2011)

Chambon, F., Winter, H.H.: Stopping of crosslinking reaction in a PDMS polymer at the gel point. Polym. Bull. **13**, 499–503 (1985)

Jing, R., Varveri, A., Liu, X., Scarpas, A., Erkens, S.: Chemo-mechanics of ageing on bitumen materials. In: Transportation Research Board 97th Annual Meeting, Washington, DC, USA, 7–11 January 2018

Liu, C., et al.: Evaluation of different methods for the determination of the plateau modulus and the entanglement molecular weight. Polymer **47**, 4461–4479 (2006)

Nivitha, M.R., Krishnan, J.M.: What is transition temperature for bitumen and how to measure it. Transp. Devel. Econ. **2**, 3 (2016)

Petersen, J.C., Glaser, R.: Asphalt oxidation mechanism and the role of oxidation products on age hardening revisited. Road Mater. Pavement Des. **12**, 795–819 (2011)

Saoula, S., Soudani, K., Haddadi, S., Munoz, M., Santamaria, A.: Analysis of the rheological behavior of ageing bitumen and predicting the risk of permanent deformation of asphalt. Mater. Sci. Appl. **4**, 312–318 (2013)

Scarsella, M., et al.: Petroleum heavy ends stability: evolution of residues macrostructure by ageing. Energy Fuels **13**, 739–747 (1999)

Van den bergh, W.: The effect of ageing on the fatigue and healing properties of bituminous motars, Delft, The Netherlands (2011)

Chemo-mechanical Characterization of Bitumen Binders with the Same Continuous PG–Grade

Jean-Pascal Planche(✉), Michael D. Elwardany, and Jeramie J. Adams

Asphalt and Petroleum Technologies, Western Research Institute, 3474 North 3rd Street, Laramie, WY 82072, USA

Abstract. Chemo-mechanical analysis tools were used to provide plausible reasons behind different binder properties that are not well captured by the conventional Superpave PG-grading system. In this study, six binders were divided in two groups based on their continuous PG-grades. The binder matrix includes: two SBS modified binders, one air blown bitumen, one bitumen with high wax content and two other unique binder blends. Although the binders in each group have the same continuous PG-grades based on AASHTO M320, they exhibit very different low-temperature performance based on the binder relaxation ΔTc index, and they have different upper PG performance according to MSCR testing and AASHTO MP19, which takes into account both traffic load and climate conditions. Based on the results, high apparent molecular weight waxes appear to lead to poor low-temperature cracking and lower molecular weight waxes lead to poor rutting performance. Meanwhile incompatible polymer modification seems to lead to poor rutting and cracking performance relative to unmodified binders or even air blown binder.

Keywords: Binder waxes · Polymer compatibility · SAR-AD Waxphaltene Determinator · 4-mm DSR · MSCR · Glass transition Crystallizable fractions

1 Introduction and Objectives

Bitumen streams from refineries have experienced a decrease in consistency in recent years due to number of reasons linked to the ever evolving crude oil supply that is driven by economics and supply (Planche et al. 2018). Consequently, bitumen quality can change dramatically in a short time. This presents a unique challenge to the bitumen industry, as changes in feedstocks and blend formulations usually go unnoticed, because most refineries are able to tune their processing conditions to meet specifications. However, tuning refinery conditions to meet a specification does not necessarily reflect the performance of the bitumen binders (Planche et al. 2018). For these reasons, a chemo-mechanical characterization approach was used to study six binders, presented in this paper, as part of a much wider Asphalt Industry Research Consortium project led by the Western Research Institute (WRI) in an attempt to study the influence of bitumen chemistry on its physical properties. Results for a third set of

L. D. Poulikakos et al. (Eds.): RILEM 252-CMB 2018, RILEM Bookseries 20, pp. 77–83, 2019.
https://doi.org/10.1007/978-3-030-00476-7_13

binders involving three additional high-asphaltene blends with different performance are not presented in this version for paper due to length constraints.

The objectives of this study were as follows: (1) use a chemo-mechanical approach to study binder performance not captured by conventional PG-grading system, (2) evaluate binder waxes and their effect on binder performance, and (3) investigate polymer compatibility and its effect on binder performance.

2 Experimental Plan, Methods and Analysis

Six binders were selected and splitted in two groups of three based on their continuous PG-grades. Although the binders in each group have very similar continuous PG-grades based on AASHTO M320, they display very different low-temperature performance based on the relaxation related ΔTc index and different upper temperature performance measured by MSCR testing, according to AASHTO MP 19 which accounts for both traffic and climate. Binders in each group were ranked based on their ΔTc index, so that (A) indicates good low-temperature performance, (B) indicates moderate low-temperature performance, and (C) indicates poor low-temperature performance. The more negative the ΔTc, which is the difference between the limiting stiffness temperature (T_cS) and the limiting m-value temperature (T_cm), the higher the cracking tendency. Values below −5 °C difference are assumed to be prone to significant low-temperature cracking (Anderson et al. 2011). Table 1 provides a summary of wax characterization, binders PG-grade, ΔTc and MSCR data, in addition to some data that will be discussed later in the paper.

SAR-AD™. The automated HPLC system that separates saturates, aromatics, and resins (SAR) coupled to the automated Asphaltene Determinator (AD) separation (SAR-AD™), as developed by WRI, was used to study the relative composition of eight chemically different sub-fractions (FHWA 2016).

Differential Scanning Calorimetry (DSC). DSC was used to measure temperature based flow properties and phase transitions such as the glass transition (T_g) and crystallized fraction (wax) (Turner and Branthaver 1997). In the non-modulated mode, samples were cooled to −90 °C at 5 °C per minute and then heated at 10 °C per minute to 165 °C. In the modulated mode, with ±0.5 °C modulation every 80 s, the samples were cooled to −90 °C at 2 °C per minute, and then heated to 165 °C at 2 °C per min.

Size Exclusion Chromatography (SEC). SEC was used to measure molecular weight and molecular associations. 3% wt./vol. solutions were filtered and aliquots were injected onto an HPLC system equipped with a refractive index (RI) detector (Boysen 2009).

Fourier Transform Infrared Spectroscopy (FT-IR). FTIR was used to track oxidation, modification, and chemical functional groups (Petersen 2009). The spectra were obtained from carbon disulfide, carbon tetrachloride and tetrachloroethylene solutions.

Table 1. Binders grouped by PG-grades, ranked by their ΔTc index and a summary of Waxphaltene Determinator data, % crystallizable fraction by DSR, and MSCR test results

Binders grouped by their PG-grades (M 320) and ranked based on their ΔTc index

Binder ID	Continuous Upper PG-Temp. (M 320) (°C)	Continuous Low PG- Temp. (M 320)(°C)	Intermediate PG-Temp. (M 320) (°C)	Low-Temp. Index (ΔTc)(°C)	Upper PG-Grade (MP 19)
C-PG 69-13 (A)	69	-13	24.8	-0.6	PG 64H -10
C-PG 67-14 (B)	67	-14	22.3	-3.7	PG 64H -10
C-PG 68-14 (C)	68	-14	20.6	-11.2	PG 64S -10
C-PG 72-18 (A)	72	-18	21.5	0.4	PG 70V -16
C-PG 72-15 (B)	72	-15	22.4	-5.9	PG 70H -10
C-PG 72-17 (C)	72	-17	20.5	-7.6	PG 70H -10

Waxphaltene Determinator data, % crystallizable fractions, and MSCR results

Binder ID	Waxy A (%)	Waxy B (%)	Waxy A + B	% Crystallizable Fraction	J_{nr} (3.2) at 64°C (1/kPa)	J_{nr} (3.2) at 70°C (1/kPa)
C-PG 69-13 (A)	12.2	1.6	13.8	0.60	1.95	4.69
C-PG 67-14 (B)	9.3	0.9	10.2	1.02	1.84	4.63
C-PG 68-14 (C)	45.3	10.3	55.6	1.65	2.30	5.42
C-PG 72-18 (A)	10.2	1.9	12.1	0.69	0.51	1.49
C-PG 72-15 (B)	7.0	0.2	7.2	0.00	1.05	2.69
C-PG 72-17 (C)	13.7	1.1	14.8	0.25	1.08	2.79

Universal Simple Aging Test (USAT). Binder conditioning to simulate short-term and long-term aging were conducted using the USAT procedure to mimic RTFO and RTFO+PAV conditions with minimum amounts of material (Farrar et al. 2015).

Dynamic Shear Rheometer (DSR). DSR was used to measure binder rheological properties, construct binder mastercurves, and conduct rheological analysis in the black space (Planche et al. 2018; King et al. 2012). DSR was used to find the continuous upper PG-temperature and intermediate PG-temperature, according to AASHTO M 320. Additionally, 4-mm-diameter parallel plate DSR was used to measure binder rheological properties at sub-zero temperatures instead of BBR (FHWA 2017). It was used to determine the rheological index Delta T_c (ΔT_c), which is the difference between the critical temperature based on stiffness limit T_c (S) and the critical temperature based on relaxation rate T_c (m).

Multiple Stress Creep and Recovery Test (MSCR). MSCR test was used to evaluate upper PG-grade considering both climate and traffic according to AASHTO MP-19.

Waxphaltene Determinator (WD). WD is an HPLC based separation developed by WRI that rapidly separates waxes and asphaltenes into different types (Schabron et al. 2012). A low molecular weight wax enriched fraction is eluted with heptane while maintaining the column temperature at −24 °C. This fraction is referred to as Waxy A. After elution of Waxy A, the column is heated to 60 °C to melt the remaining waxes so they elute with the warm heptane. This fraction, defined as Waxy B which contains high molecular weight waxes that are enriched in n-paraffin microcrystalline waxes.

3 Results and Discussion

In addition to PG-grade and ΔTc, Table 1 presents a summary of the Waxphaltene Determinator data and the DSC crystallizable fraction results of the six binders. The MSCR non-recoverable creep compliance (J_{nr}) values at 3.2 kPa stress measured at 64 °C and 70 °C are also presented in Table 1. Waxy A and B was demonstrated to correlate with the rutting performance as measured by J_{nr} at 3.2 kPa for all binders. Higher Waxy A content, in particular, led to poor rutting performance as measured by higher J_{nr} values. Generally, higher Waxy B contents led to higher % crystallizable fraction. It is important to note that Waxy B linearly correlated with waxes extracted by ion exchange chromatography for SHRP asphalts.

3.1 Effect of High Molecular Wax

The first set of binders with an average continuous PG 68-14 consists of two special unmodified blends C-PG 69-13 (A) and C-PG 67-14 (B) and a single binder with high wax content from West Texas C-PG 68-14 (C). Figure 1.a shows black space diagrams of the three binders with same grade. A flat black space curve, as for binder (C), generally corresponds to cracking failure tendency, as it indicates a more elastic behavior with a lower phase angle at a given binder shear modulus. Typically, flat black space curve associates to low relaxation in the high modulus/low phase angle region (Planche et al. 2018; King et al. 2012).

Table 1 shows that binder (C) has much higher waxy A and waxy B contents compared to typical binders. Percent crystallizable fraction of binder (C) is also the highest among all the six binders presented in Table 1. Higher wax content explains the flat curve in black space as well as the feathering associated to phase transitions induced by crystallization or melting, as shown in Fig. 1a, and the low ΔTc (−11.2), as presented in Table 1. Waxes were also observed in the FT-IR data in CS_2 solvent as a strong peak at 721.4 cm^{-1} wavenumber, which corresponds to (CH_2) chains greater than C4.

SEC data, presented in Fig. 1b, shows that binder (C) binder contains significantly more abundant higher apparent molecular weight species than the other two binders. SAR-AD results of this binder showed very little asphaltenes (3.3%) compared to typical binders. Based on these results, binder (C) contains higher molecular weight maltenes and waxes compared to the other two binders, which is consistent with its West Texas Intermediate source. Table 1 shows that high molecular weight waxes can also have a negative effect on the rutting performance as measured by MSCR test.

Binder (C) has the highest Waxy A value which again correlates with higher J_{nr} (3.2 kPa) values compared to the other two binders.

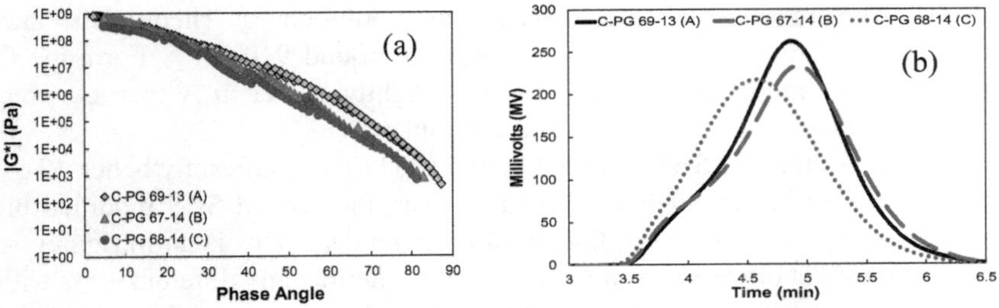

Fig. 1. Binder properties of the first set of binders with average continuous PG 68-14: (a) black space diagram and (b) SEC RI detector chromatograms

3.2 Polymer Compatibility Effect

The second set of binders, having an average PG 72-17, consists of two SBS modified binders C-PG 72-18 (A) and C-PG 72-17 (C) with different base binders and an air blown binder C-PG 72-15 (B). Figure 2a shows the black space diagrams of the binders. Air blown binder (B) have a flatter black space curve compared to a classical bitumen. The air blown sample features ΔTc of −5.9. Figure 2a shows the "polymer plateau" effect of the SBS modifier in black space for binder (A) and a lesser effect for binder (C).

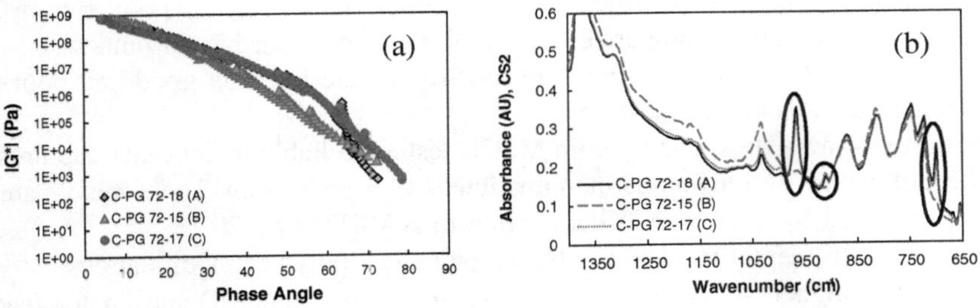

Fig. 2. Binder properties of the second set of binders with average continuous PG 72-17: (a) rheological black space diagram and (b) FTIR spectra in CS_2

SEC data shows the characteristic feature of air blown binders: binder (B) has a significantly more highly associating high apparent molecular weight material visualized by a more pronounced shoulder at shorter elution times. The greater amount of high apparent molecular weight material probably contributes to this binder's high upper PG-grade. It may also explain the higher m-value limited temperature, which causes the ΔT_c to fall below −5 °C. These features are typical of highly air blown bitumen binder. Figure 2b shows the FTIR spectra of the three binders in CS_2. Since, binders (A) and (C) are modified with a SBS elastomer or styrene-butadiene copolymer, two prominent bands were observed in CS_2 solutions which correspond to the C-H of the trans di-substituted R-CH = CH-R butadiene units (965 cm^{-1}) and to the C-H of the aromatic mono-substituted styrene units (699 cm^{-1}). There is also another small butadiene vinyl C-H band, which shows up around 910 cm^{-1}. Form the FTIR data the concentration of the SBS polymers is slightly higher in A than C, which is consistent with what was reported for these samples.

Table 1 shows that the SBS modified binder (A) has a significantly better ΔTc (0.4) than that of the air blown binder (−5.9). However, the second SBS modified binder (C) has a much worse ΔTc (−7.6) than both other binders. This is an intriguing result because it shows that binders (A) and (C) interact in significantly different ways with the SBS. This significant behavior is assumed to be linked to the compatibility between the base binder and the polymer, and the homogeneity of the resulting binder. Although FTIR results indicate similar polymer modification between binders (A) and (C), the black space diagram was able to discriminate between the two binders – thus, Binder (C) shows a minimal "polymer plateau" compared to Binder (A). Data in Table 1 also shows that MSCR test was able to clearly discriminate between (A) and (C) binders: Binder (A) has a smaller J_{nr} at both temperatures compared to binder (C). These results show that incompatible polymer modification may lead to both lower rutting and cracking performance compared to unmodified binders and even air blown binders.

4 Conclusions

- High wax content in bitumen binders may lead to poor low-temperature cracking and rutting performance.
- Incompatible polymer modification of bitumen may lead to poor rutting and cracking performance compared to unmodified or even air blown binders.
- Black space plot is an important tool to discriminate between good and poor performing blends and modifications.
- ΔTc from 4-mm-DSR and J_{nr} from MSCR test are reliable indices that captures the effect of waxes, air blown and incompatible SBS modification. These factors are not well captured by Supepave PG according to AASHTO M320.
- FTIR is instrumental to identify styrene-butadiene polymer modifications.
- Generally, higher Waxy B contents from the Waxphaltene Determinator lead to higher % DSC crystallizable fraction.
- Generally, Waxy A content seem to follow fairly well MSCR J_{nr} results.
- These trends were obtained from a limited number of binders. A study including more than 50 binders is ongoing to validate these findings.

Acknowledgments. The authors would like to acknowledge AIRC-1st Iteration members (Surfax, Repsol, Husky, BRRC, IFSTTAR, Eiffage, Eurovia, and WRI). The authors would also like to acknowledge P. Coles for rheological testing, J. Forney for DSC measurements, J. Loveridge and J. Rovani for SAR-AD analysis, N. Bolton for running FT-IR, and R. Boysen for performing SEC analysis.

References

Anderson, R.M., King, G.N., Hanson, D.I., Blankenship, P.B.: Evaluation of the relationship between Asphalt binder properties and non-load related cracking. J. Assoc. Asphalt Paving Technol. **80**, 615–649 (2011)

Boysen, R.B.: Adaptation of existing analytical scale size exclusion chromatography methods. Technical white paper FP22 (2015)

Farrar, M., Grimes, R.W., Turner, T.F., Planche, J.P.: The universal simple aging test and low temperature performance grading using small plate dynamic shear rheometry. Technical white paper FP14 (2015)

Federal Highway Administration (FHWA) Technical Brief: Automated High-Performance Liquid Chromatography Saturate, Aromatic, Resin, and Asphaltene Separation. US Department of Transportation, No. FHWA–HRT–15–055. (2016)

Federal Highway Administration (FHWA): Technical brief: four-mm dynamic shear rheometer. US Department of Transportation, No. FHWA–HRT–15–053 (2017)

King, G., Anderson, M., Hanson, D., Blankenship, P.: Using black-space diagrams to predict age-induced cracking. In: RILEM Fatigue Cracking Conference, Delft, June 2012

Petersen, J.C.: A review of the fundamentals of asphalt oxidation. Transportation Research Circular, E-C 140 (2009)

Planche, J.P., Adams, J.J., Elwardany, M.D., Boysen, R.B., Rovani J.: Understanding the trends in asphalt binders matters! In: TRB Annual Meeting, Washington D.C. (2018)

Schabron, J., Rovani, J., Sanderson, M., Loveridge, J., Naydong, L., McKenna, A., Marshall, A.: Waxphaltene determinator method for automated precipitation and redissolution of wax and asphaltene components. Energy Fuels **26**, 2256–2268 (2012)

Turner, T.F., Branthaver, J.F.: DSC studies of Asphalts and Asphalt components. In: Usmani, A. M. (ed.) Asphalt Science and Technology, pp. 59–101. Marcel Dekker, New York (1997)

Field Aging Evaluation of Asphalt Binders by Chemical and Rheological Characterization

Marcia Midori Takahashi[1(✉)], Kamilla L. Vasconcelos[1], Margareth Carvalho Coutinho Cravo[2], and Liedi Légi Bariani Bernucci[1]

[1] Department of Transportation Engineering, Polytechnic School, University of São Paulo, São Paulo, Brazil
{marcia.takahashi,kamilla.vasconcelos,liedi}@usp.br
[2] Centro de Pesquisas e Desenvolvimento, "Leopoldo Américo Miguez de Mello" (CENPES), PETROBRAS, Rio de Janeiro, Brazil
cravo@petrobras.com.br

Abstract. Two field aged asphalt binders were evaluated in the study. The binders were extracted and recovered from samples that were collected from two 4 cm HMAs in the field at different times. Chemical and rheological tests were performed to quantify the aging. The binder's chemical properties were evaluated by SARA fractions (saturates, asphaltenes, resins and aromatics), gel permeation chromatography (GPC), nuclear magnetic resonance (NMR) and Fourier transform infrared (FTIR) spectroscopy. The rheological tests were carried out in a dynamic shear rheometer (DSR). Dynamic shear modulus (|G*|) and multiple stress creep and recovery (MSCR) tests were performed to evaluate the change in the binders rheological properties. Both chemical and rheological results confirm the continuous aging of the asphalt binder in the field, mostly up to 24 months after construction, although the greatest aging occurs during mixing. Rheological results indicated that after 36 months of field aging, the two binders were performed similarly even though they aged differently.

Keywords: Field aging binder · Chemical characterization
Rheological characterization

1 Introduction

Asphalt binder is an organic material that ages in the presence of oxygen, especially under heat. Aging occurs by the binder oxidation over time during mixing, transport, compaction and service (Lu and Isacsson 2002; Masson et al. 2006; Dondi et al. 2016) and it has considerable influence on the asphalt binder properties, since the binder is responsible for the adhesion and cohesion of the asphalt mixture.

The oxidation process causes the binder to become stiffer and brittle, resulting in loss of mechanical properties and affecting the pavement durability (Qin et al. 2014). The process of binder aging is complex duo to the influence of the binder source, the mixing conditions and the environmental conditions during the service life of pavement (Lu and Isacsson 2002).

L. D. Poulikakos et al. (Eds.): RILEM 252-CMB 2018, RILEM Bookseries 20, pp. 84–89, 2019.
https://doi.org/10.1007/978-3-030-00476-7_14

In laboratory, the binder aging can be simulated in short and long term by standard aging procedures, as: (i) Thin film oven test (TFOT – ASTM D1754) and Rolling thin film oven test (RTFOT – ASTM D2872) for short term aging, (ii) Pressure aging vessel (PAV – D6521) and Ultraviolet radiation (ASTM D4799) for long term aging.

The aging promotes changes on physical, chemical and rheological binder properties. Several studies have been conducted to understand the aging effect on these properties and how they influence in pavement durability (Lin et al. 1995; Lu and Isacsson 2002; Feng et al. 2013; Qin et al. 2014; Osmari et al. 2017; Zeng et al. 2018).

Qin et al. (2014) studied the field aging effect on binder rheology and structure, binders were extract and recovery from pavement samples varying depths. It was observed that the field aging on the pavement surface is far more severe than the long-term laboratory.

Aging causes the increase of hardening and viscosity of asphalt binder by the increase of asphaltene content, as result of carbonyl formation during oxidative aging (Lin et al. 1995). An increase of |G*| values are favorable changes with respect to rutting performance, but they are unfavorable for thermal cracking performance (Kim 2008).

The evaluation of ultraviolet UV irradiation is important because it is closely related to the asphalt binder degradation during pavement service. Zeng et al. (2018) studied UV aging depth of asphalt using ultraviolet spectrophotometer tests (UST). The authors observed that UV radiation could only age a limited depth of asphalt (4.5 m).

The objective of the present study is to characterize and evaluate the field aging of two different neat asphalt binders applied in the field at different levels of aging.

2 Materials and Methods

Field aged asphalt binders from two test sites located in São Paulo state were evaluated. A dense HMA, applied in the field with 4 cm, constituted by basalt aggregates with nominal maximum aggregate size of 12,5 mm was used for the study with two different neat asphalt binders: (i) one AC 30/45 penetration grade asphalt binder (PG 70S-XX) at 4,85% and (ii) one AC 50/70 penetration grade asphalt binder (PG 64S-XX) at 4,70%. Samples were collected from the field at different times (0, 12, 24 and 36 months after pavement construction) and binders were extracted and recovered in accordance with ASTM D2172 – Test Method B (2017) and ASTM D1856 (2015), respectively. To quantify the field aging, chemical and rheological binder properties were evaluated.

SARA fractions (saturates, asphaltenes, resins and aromatics) of binders were determined by Iatroscan MK-3 (thin-layer chromatography with flame ionization detection, TLC-FID). The asphalt molar mass distribution was determined by gel permeation chromatography (GPC) using an Agilent 1200 chromatographic pump. Nuclear magnetic resonance (NMR) tests were performed on an Agilent INOVA 300 to analyze the structural changes in the binders. The amounts of aromatic carbon (Car), aromatics hydrogen (Har), hydrogen alpha (H-alpha) were measured. Fourier transform infrared (FTIR) spectroscopy test with attenuated total reflection (ATR) were performed using a Nicolet Avatar 360 spectrometer with a diamond cell. All spectra were obtained with 32 scans and 4 cm^{-1} resolution in the region of 4000 to 650 cm^{-1}. The method analyzes the changes of carbonyl index that are calculated based on the ratio between the areas of carbonyl compounds and the areas of CH_2 and CH_3 groups, Eq. 1.

$$I_{c=o} = \frac{A_{c=o}}{A_{CH_2 + CH_3}} \tag{1}$$

Where $A_{C=O}$ is the area of carbonyl compounds (C = O) in the region of 1720 to 1650 cm^{-1} and $A_{CH2+CH3}$ is the area between 1480 to 1410 cm^{-1} of ethylene (CH_2) and the methyl (CH_3) groups.

The rheological properties of the samples were characterized in a dynamic shear rheometer (DSR) with 25 mm parallel plate and gap of 1 mm. Frequency sweep tests (from 1 to 100 rad/s) were performed at different temperatures between 46 to 82 °C (increments of 6 °C) with 0.01% of strain amplitude. Dynamic shear modulus (|G*|) of the samples were determined and master curves were then plotted at a reference temperature of 52 °C, using the shift factor method. The multiple stress creep and recovery (MSCR) test was also performed to evaluate the binder's potential for rutting deformation (ASTM D7405 2010). Tests were conducted at the temperature of 64 °C for the two binders (temperature representative of the region where both binders were applied).

3 Results and Discussions

Figure 1 presents the chemical analyzes by SARA fraction and GPC test. Aging causes chemical changes into the asphalt binder that reflects in an increase of binder viscosity and stiffness. In Fig. 1a, asphaltenes content increases and aromatic content decreases mostly during mixing (t = 0), and the large molecular size (LMS) also increased with aging (Fig. 1b), as has also been observed by Osmari et al. (2017).

The results of FTIR, NMR and MSCR tests are present in Table 1. According to Lin et al. (1995), binder oxidation process results of carbonyl formation. For binders with no aging condition, the carbonyl compounds of the FTIR results were not observed. The field aging increased the carbonyl index ($I_{c=o}$) on both asphalt binders.

ASI (Aromatic-substitution index) increases with aging due to condensation reactions of the aromatic rings (except for AC 30/45 at 24 months and for AC 50/70 at 36 months). H-alpha compound has the tendency to reacts with other molecules and initiate oxidation reactions (Osmari et al. 2017). H-alpha decreases with binder oxidation, however the result presented the opposite trend after 24 months, for both asphalt binders, what may have occurred due to other reactions that took place at the same time, interfering with these results.

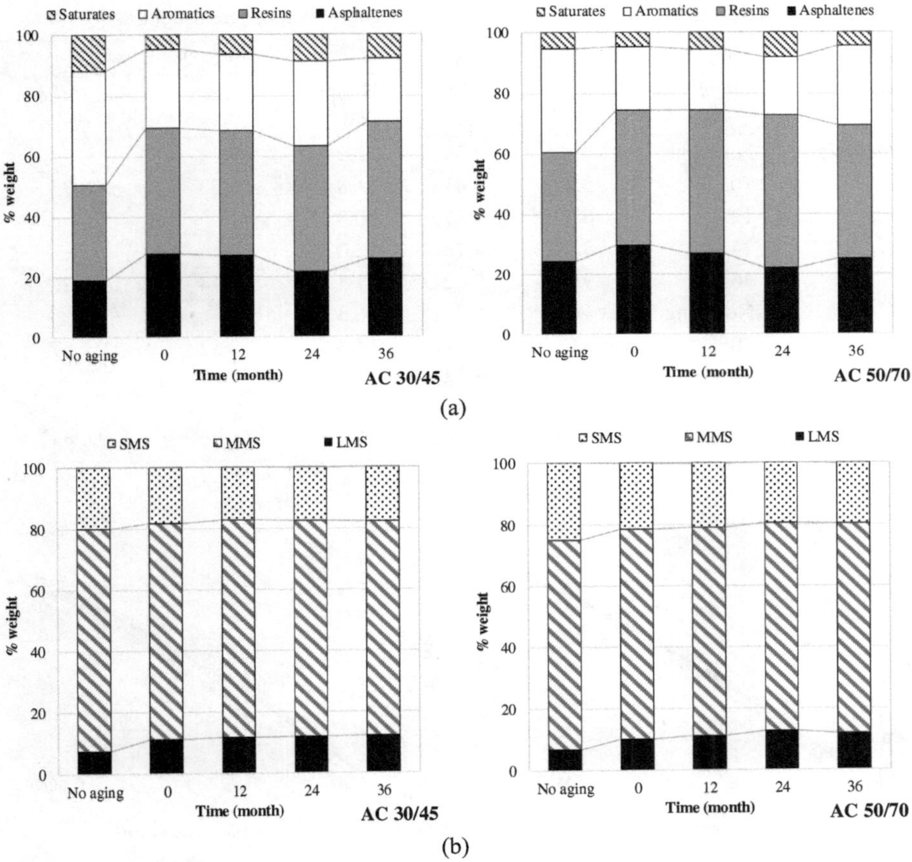

Fig. 1. Results of (a) SARA fractions and (b) GPC tests

The permanent deformation was evaluated with MSCR test, in two different aging conditions: (i) t = 0 (after mixing and compacted), and (ii) t = 36 months. The results show that AC 30/45 has the lower value of Jnr (non-recoverable compliance) for t = 0, which indicate that this binder tends to have better permanent deformation resistance than AC 50/70, but after 36 months of field aging, they performed similarly.

Figure 2 presents the |G*| master curves of both asphalt binder in four aging conditions. The dynamic shear modulus increased with aging, specially AC 50/70 that got stiffer after 12 months in the field. For both asphalt binder, the master curves at t = 24 and t = 36 are practically superposed, what might indicate that the most severe aging process occurred up to 24 months after construction.

Table 1. FTIR, NMR and MSCR test results of binders with different field aging time

Asphalt binder	Time (months)	FTIR	NMR		MSCR	
		$I_{C=O}$ (%)	ASI (%)	H-alpha (%)	$Jnr_{3,2}$ (kPa^{-1})	Jnr_{diff} (%)
AC 30/45	No aging	–	4,891	12,2	–	–
	0	5,8	5,411	11,8	0,23	8,75
	12	6,5	5,579	10,8	–	–
	24	8,5	4,781	12,0	–	–
	36	9,6	5,622	12,3	0,14	3,22
AC 50/70	No aging	–	4,731	13,3	–	–
	0	6,6	5,448	13,4	0,52	7,67
	12	6,5	5,648	11,8	–	–
	24	9,8	5,768	13,0	–	–
	36	10,4	5,553	11,6	0,17	2,84

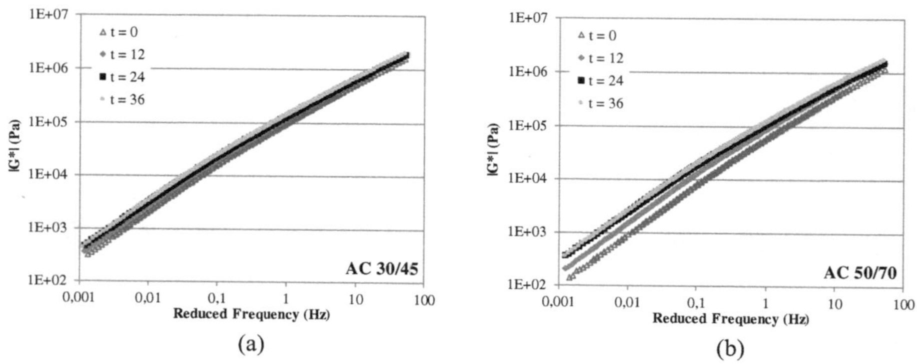

Fig. 2. Dynamic shear modulus at 52 °C for (a) AC 30/45 and (b) AC 50/70

4 Conclusions

This paper presents the results of the evaluation of the field aging of two neat asphalt binders at different aging levels (in time). The chemical and rheological results confirm the continuous aging of the asphalt binder in the field, however the most severe aging process occurred up to 24 months after construction. The increase of asphaltenes and LMS content shows that the process of oxidation occurs mainly during mixing where the binder is exposed at high temperatures. Although the aging of the two binders is different, rheological results indicated that they presented similar chemical and rheological characteristics after 36 months in the field.

References

Dondi, G., Mazzotta, F., Simone, A., Vignali, V., Sangiorgi, C., Lantieri, C.: Evaluation of different short term aging procedures with neat, warm and modified binders. Constr. Build. Mater. **106**, 282–289 (2016). https://doi.org/10.1016/j.conbuildmat.2015.12.122

Feng, Z.G., Yu, J.Y., Zhang, H.L., Kuang, D.L.: Effect of ultraviolet aging on rheology, chemistry and morphology of ultraviolet absorber modified Bitumen. Mater. Struct. **46**, 1123–1132 (2013). https://doi.org/10.1617/s11527-012-9958-3

Kim, Y.R.: Modeling of Asphalt Concrete. McGraw-Hill, New York (2008)

Lin, M.S., Davison, R.R., Glover, C.J., Bullin, J.A.: The effects of Asphaltenes on Asphalt recycling and aging. Trans. Res. Rec. **1507**, 86–95 (1995)

Lu, X., Isacsson, U.: Effect of ageing on bitumen chemistry and rheology. Constr. Build. Mater. **16**(1), 15–22 (2002). https://doi.org/10.1016/S0950-0618(01)00033-2

Masson, J., Leblond, V., Margeson, J.: Bitumen morphologies by phase-detection atomic force microscopy. J. Microsc. **221**, 17–29 (2006). https://doi.org/10.1111/j.1365-2818.2006.01540.x

Osmari, P.H., Aragão, F.T.S., Leite, L.F.M., Simão, R.A., Motta, L.M.G., Kim, Y.R.: Chemical, microstructural, and rheological characterizations of binders to evaluate aging and rejuvenation. Trans. Res. Rec. **2632**, 14–24 (2017). https://doi.org/10.3141/2632-02

Qin, Q., Schabron, J.F., Boysen, R.B., Farrar, M.J.: Field aging effect on chemistry and rheology of asphalt binders and rheological predictions for field aging. Fuel **121**, 86–94 (2014). https://doi.org/10.1016/j.fuel.2013.12.040

Zeng, W., Wu, S., Pang, L., Chen, H., Hu, J., Sun, Y., Chen, Z.: Research on Ultra Violet (UV) aging depth of asphalts. Constr. Build. Mater. **160**, 620–627 (2018). https://doi.org/10.1016/j.conbuildmat.2017.11.047

Modifying Surface Properties of Model and Pavement Aggregates with Silanes

Gabriel Orozco[1(✉)], Cédric Sauzéat[1], Jules Galipaud[2], and Hervé Di Benedetto[1]

[1] Université de Lyon, ENTPE, LTDS (CNRS UMR 5513), Lyon, France
{gabriel.orozco, cedric.sauzeat, herve.dibenedetto}@entpe.fr
[2] Université de Lyon, Ecole Centrale de Lyon, LTDS (CNRS UMR 5513), Lyon, France

Abstract. Adhesion between binder and aggregate is a major concern when studying bituminous mixtures' mechanical performances. Many parameters are to be taken into account, for instance chemical composition of binder and aggregate, interface roughness, etc. Isolating each parameter for a comprehensive approach of adhesion phenomenon can be very challenging but necessary. This paper is a first step of a larger study to isolate the role of surface chemical properties of aggregates in binder/aggregate interface. In order to alter only the very surface of aggregate and its chemical composition, a simple 3-step-procedure for chemical surface modification at molecular scale using two different hydrophobic organosilanes is given. Two substrates proposed here are made of classical rock (microdiorite) for pavement construction from a French quarry, and of glass as model material. The change of physicochemical properties of modified substrates is then assessed through contact angle method on smooth plates, while the nano-coating's composition is assessed with X-Ray Photoelectron Spectroscopy (XPS) on aggregates. Contact angle method significantly revealed hydrophobic properties for both modified substrates, up to a 30 degrees' increase. XPS analysis show that the coating is a success on both substrates, and probably with a strong siloxane bond.

Keywords: Surface modification · Organosilane · XPS · Contact angle Adhesion

1 Introduction

Adhesion between bitumen and aggregate has an important role in bituminous mixtures thermomechanical properties. Hefer and Little (2005) noticed that the first studies on adhesion in bituminous materials focused on moisture damage, which is a major cause of pavement deterioration. More recent works showed that adhesive failure between bitumen and mineral substrates can occur during mechanical stripping tests (Packham 1992, Canestari et al. 2010). Adams (2005) summed up the complexity when studying adhesion: although fundamental interactions at microscopic scale are well-known, adhesion between two media can be the combination of several possible mechanisms (physical interlocking, chemical adsorption…), which depend on several factors, such

L. D. Poulikakos et al. (Eds.): RILEM 252-CMB 2018, RILEM Bookseries 20, pp. 90–95, 2019.
https://doi.org/10.1007/978-3-030-00476-7_15

as surface chemical composition, surface roughness or even porosity (Corté and Di Benedetto 2004, Vassaux et al. 2017, Baldi-Sevilla et al. 2017).

In this paper, a procedure to modify only surface chemical composition of model (silica glass) and construction (microdiorite) substrates with organosilanes was tested. Silane coupling agents have been well studied (Arkles 1977, Plueddemann 1991) in textile and medical fields and can ensure nano-coating on many types of substrate, preserving surface topology and providing new surface properties (e.g. hydrophobicity). Recent silane treatments on bituminous materials helped reducing moisture sensibility (Min et al. 2015). In this paper, physicochemical properties were investigated with contact angle method, whereas surface atomic composition was obtained with X-Ray Photoelectron Spectroscopy (XPS) for modified and reference substrates.

2 Organosilane Coating Procedure

Considering possible industrial implementation, a simplified 3-step-silanization procedure was performed. The first step, referred as “A”, is a washing of the substrate. It is washed in a recipient full of distilled water with a mechanical stirrer, at least for 30 min. The substrate is then rinsed with water and washed again for 30 min. After filtering water, the substrate is dried at 70 °C during 4 h. The main purpose of “A” is to drain dusts from substrates, especially aggregates which come from quarry crusher. The second step, referred as “B”, is the silanization itself. A solution of 95% ethanol (Chimie-Plus Laboratoires, purity >99.9%) + 5% distilled water is mixed, with the desired concentration of silanes. Shortly after adding the silanes, the substrate is immerged and stirred in the solution for 1 h. Whereas ethanol is a good solvent for most silanes, a small quantity of water is required to “activate” the silanes and substrate’s surface, i.e. creating respectively Si–OH and –OH groups that ultimately leads to a strong covalent bond (e.g. siloxane bond on silica) between the coupling agent and the substrate’s surface (Arkles 1977). If not subjected to the last step, the modified substrate is dried at 70 °C during 4 h. The last step, referred as “C”, is a post-washing. The modified substrate is washed several times with ethanol and distilled water to drain the remaining adsorbed silanes solution, and eventually dried at 70 °C during 4 h.

3 Materials

Two types of substrate have been studied: Microdiorite rock (“R”) from the quarry of Corbigny, France, used in pavement construction and silica glass (“G”) from Silibeads as a model material. Each substrate was investigated in aggregate form (“Agg”) or plate form (“P”).

Two organosilanes were used: n-Octadecyltrimethoxysilane, Gelest Inc., referred as Carbon chain (“+”) and Nonafluorohexyltrimethoxysilane, Gelest Inc., referred as Fluroniated chain (“–”). While both hydrophobic coupling agents, carbon chains are lipophilic and fluorinated chains are lipophobic (Arkles 2011), which could lead to different behaviors regarding adhesion between modified substrates and bitumen, in further studies.

All tested materials are presented in Table 1. They are subjected to different combinations of steps (A, B, C) presented in Sect. 2, in order to have a better understanding of the coating procedure and ensure proper silanization. Reference materials, referred as "O", did not receive any treatment.

Table 1. Tested materials

	Microdiorite for road construction (R)		Glass model material (G)		Carbon chain (+)	Fluorinated chain (−)	Surface modification steps (see Sect. 2)			
	Aggregate (Agg)	Cored plate (P)	Glass bead (Agg)	Glass plate (P)	c = 1%	c = 0.2 %	A	B	C	Ref. (O)
R_Agg_A	X						X			
R_Agg_ABC +_1%	X				X		X	X	X	
R_Agg_AB-_0.2%	X					X	X	X		
G_Agg_O			X							X
G_Agg_B +_1%			X		X			X		
G_Agg_B-_0.2%			X			X		X		
R_P_O		X								X
R_P_ABC +_1%		X			X		X	X	X	
R_P_ABC-_0.2%		X				X	X	X	X	
G_P_O				X						X
G_P_C				X			X		X	
G_P_AB-_0.2%				X		X	X	X		
G_P_ABC-_0.2%				X		X	X	X	X	

4 Effect of Surface Modification on Contact Angle

Contact angle of sessile drop of water (3 μl) was measured at 23.9 °C with a Krüss DSA30 goniometer. The contact angle θ (°) is the average of left and right contact angles. θ was measured at several instants, during at least 6 seconds. 5 repetitions were performed for each material. The evolution of mean contact angle and its standard deviation is given for each tested plates in Fig. 1.

Firstly, it has been observed that the drop requires 5s to approximately stabilize, as expected on microdiorite plates which are porous (Boulangé et al. 2013). After 5s, θ keeps decreasing due to physical absorption and evaporation of water, but at a significantly lower pace. The contact angle value at 5s was used to compare substrates. All

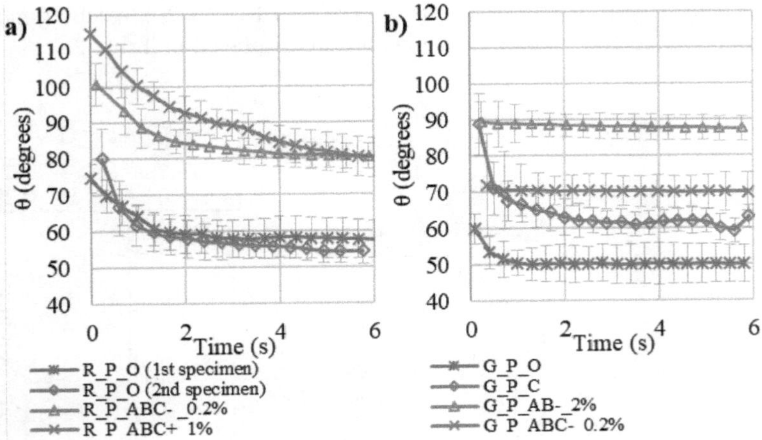

Fig. 1. Evolution of mean contact angle θ and its standard deviation of water sessile drop during time on (**a**) cored rock plates and (**b**) glass plates

modified rock plates, R_P_ABC-_0.2% / R_P_ABC+_1%, presented a neat increase of contact angle compared with the reference rock plates R_P_O, approximately 25°. For glass specimens, G_P_AB-_0.2% / G_P_ABC-_0.2%, the increase was respectively 25.6° / 19.8°. Every modified material showed a more hydrophobic behavior. Post-washing ("C") for a given silane concentration seemed to lower significantly the increase of θ on glass plates. However, θ was still higher than the contact angle of only post-washed reference glass plate G_P_C. These differences could be explained by the fact that post-washing may remove or replace adsorbed chemicals during silanization. This phenomenon should be further investigated. Regarding repeatability, R_P_O was tested twice showing close results of θ, within the standard deviation range, but requiring further confirmation for other materials.

5 Modified Surface Analysis with XPS

XPS was performed with a VersaProbe 2 spectrometer and SmartSoft VP acquisition software, all from ULVAC-PHI. Examples of spectrum obtained for glass beads G_Agg_O / G_Agg_B+_1% / G_Agg_B-_0.2% are shown in Fig. 2.

Intense F1s peaks for G_Agg_B-_0.2% and R_Agg_AB-_0.2% were observed, as well as a double peak near C1s peak (285.0 eV) which indicates the presence of carbon atoms bonded to fluorine (287.8 eV), brought by the fluorinated chains.

For a quantitative analysis of relative atomic composition of a given spectrum, a "Shirley"-type background ("BG") curve is calculated for each selected peak with Multipak software (example given in Fig. 1). Each local area between the peak and BG curves, using a specific sensitivity factor, can be compared to give an approximate atomic composition. The relative surface composition of tested aggregates was calculated with only 4 main components (C1s / O1s / F1s / Si2p), reaching more than 90% of total atomic concentration. The results are displayed in Table 2.

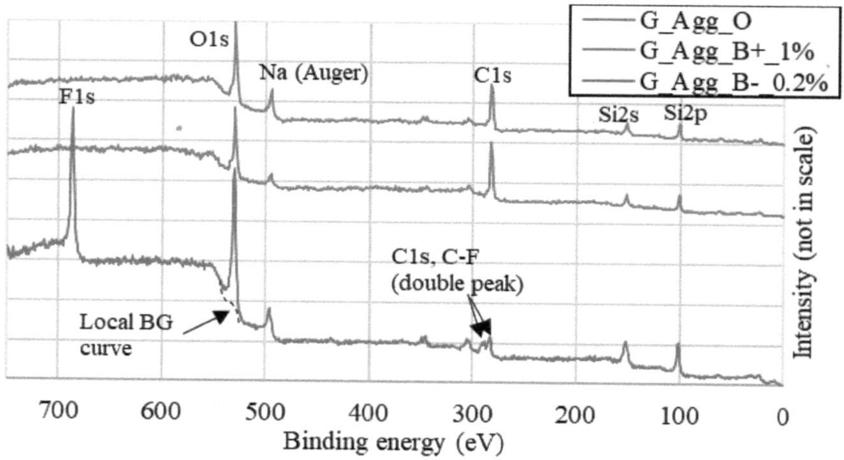

Fig. 2. Surface spectra with Al source of reference and modified glass beads

Table 2. Relative atomic concentration of substrates' main components calculated from XPS

	C1s	O1s	F1s	Si2p
R_Agg_A	13%	72%	0%	14%
R_Agg_ABC-_0,2%	6%	66%	13%	15%
R_Agg_ABC+_1%	14%	69%	0%	16%
G_Agg_O	44%	41%	0%	15%
G_Agg_B-_0.2%	20%	38%	29%	13%
G_Agg_B+_1%	53%	35%	0%	13%

Strong fluorine presence was confirmed for R_Agg_ABC-_0.2% and G_Agg_B-_0.2%, up to 29%. Neither post-washing on R_Agg_ABC-_0.2%, nor desorption during XPS (performed at 10^{-6} Pa) did remove fluorinated silanes, which is a good indication for a strong siloxane bond. Reference substrates presented no peak for fluorine (0%), making this element an excellent marker of surface modification.

The initial presence of carbon biased analysis for carbon chain. Only a slight increase from 44 to 53% for glass beads was noticed. A stronger washing step "A" could be developed for better understanding.

6 Conclusion

Silane surface modification procedure has been proposed and tested on microdiorite and model materials. Both contact angle method and XPS showed a clear change in surface properties and chemical composition that suggests a proper silanization. The expected hydrophobic behavior of modified substrates has been observed with contact angle method. The investigation of carbon chain silanes coating through XPS was biased by initial carbon presence, whereas fluorinated chain silanes appeared to be an excellent marker for surface modification on all substrates. Post-washing after silanization seemed to alter surface properties without removing silanes. This study is a first step to further investigate the influence of adhesion between bitumen and aggregate on thermomechanical properties of mixtures.

Acknowledgments. The authors thank Eiffage Route company and the Eiffage/ENTPE industrial chair for providing support and resources that helped to carry out this research.

References

Adams, R.D.: Adhesive Bonding. W.P. Limited, Ed. (2005)

Arkles, B.: Tailoring surfaces with silanes. In Chemtech (Ed.) (1977)

Arkles, B.: Hydrophobicity, hydrophilicity and silane surface modification (Gelest, Inc.) (2011)

Baldi-Sevilla, A., et al.: Influence of bitumen and aggregate polarity on interfacial adhesion. Road Mater. Pavement Des. **18**, 304–317 (2017)

Boulangé, L., Bonin, E., Saubot, M.: Physicochemical characterisations of the bitumen–aggregate interface to get a better understanding of stripping phenomena. Road Mater. Pavement Des. **14**, 384–403 (2013)

Canestari, F., Cardone, F., Graziani, A., Santagata, F.A., Bahia, H.U.: Adhesive and cohesive properties of asphalt-aggregate systems subjected to moisture damage. Road Mater. Pavement Des. **1**(Special Issue 1), 11–32 (2010)

Corté, J.-F., Di Benedetto, H.: Matériaux Routiers Bitumineux (Lavoisier – Hermès) (2004)

Hefer, A., Little, D.: Adhesion in bitumen-aggregate systems and quantification of the effects of water on the adhesive bond. Technical report, International Center for Aggregate Research (2005)

Min, Y., et al.: Surface modification of basalt with silane coupling agent on asphalt mixture moisture damage. Appl. Surf. Sci. **346**, 497–502 (2015)

Packham, D.E.: Handbook of adhesion. Wiley (ed.) (1992)

Plueddemann, E.P.: Silane Coupling Agents. Springer, New York (1991)

Vassaux, S., et al.: Towards a better understanding of wetting regimes at the interface asphalt/aggregate during warm-mix process of asphalt mixtures. Constr. Build. Mater. **133**, 182–195 (2017)

Chemo-Mechanical Characterization of Bituminous Materials: Low, Intermediate and High Temperature Behavior

Effect of Morphology on High-Temperature Rheological Properties of Polymer-Modified Bitumen

Jiqing Zhu[1(✉)] and Xiaohu Lu[2]

[1] Swedish National Road and Transport Research Institute (VTI), Olaus Magnus väg 35, 581 95 Linköping, Sweden
jiqing.zhu@vti.se
[2] Nynas AB, 149 82 Nynäshamn, Sweden
xiaohu.lu@nynas.com

Abstract. This paper employs isothermal annealing at different temperatures to obtain different microstructures of the same modified bitumen with styrene-butadiene-styrene (SBS) copolymer. The effect of polymer-modified bitumen (PMB) morphology on high-temperature rheological properties is investigated based on the same material. The characteristic wavelength ξ of PMB microstructure was calculated from the fluorescence microscopy images. Dynamic shear rheometer (DSR) measurements and multiple stress creep and recovery (MSCR) tests were performed. The PMB morphology-rheology correlation was discussed within the studied temperature and frequency range. Comparing with a binary droplet-in-matrix microstructure, a homogenous PMB morphology tends to store more energy during shear cycles and reach higher recovery after the loading.

Keywords: Polymer-modified bitumen · Morphology · Rheology Creep

1 Introduction

The morphology of polymer-modified bitumen (PMB) has important influences on its rheological properties (Lu et al. 2010). Some previous studies (Liang et al. 2017, Wang et al. 2017) used additives to obtain different PMB microstructures and investigated their effects on PMB rheology. However, the use of additives may bring unknown impacts into PMB. It is still preferred to study various PMB microstructures based on the same material composition.

Soenen et al. (2006) reported that thermal history can largely affect the PMB morphology and thus the rheology, especially at low frequencies. This might have provided a feasible and practical way to control PMB morphology, i.e. by controlling its thermal history. This paper employs isothermal annealing at different temperatures to obtain different microstructures of the same modified bitumen with styrene-butadiene-styrene (SBS) copolymer. The effect of PMB morphology on high-temperature rheological properties is investigated.

L. D. Poulikakos et al. (Eds.): RILEM 252-CMB 2018, RILEM Bookseries 20, pp. 99–104, 2019.
https://doi.org/10.1007/978-3-030-00476-7_16

2 Materials and Method

2.1 Materials

A base bitumen of penetration grade 70/100 was used to prepare PMB. Its SARA fractions were 8% of saturates, 55% of aromatics, 22% of resins and 15% of asphaltenes according to thin-layer chromatography with flame ionisation detection (TLC-FID). A linear triblock SBS copolymer was mixed with the base bitumen at 180 °C. The polymer content was 5% by weight of the blend. The related properties were tested and are presented in Table 1.

Table 1. Properties of the base bitumen and polymer-modified bitumen

Property	Base bitumen	Polymer-modified bitumen
Penetration, 25 °C (0.1 mm)	86	56
Softening point, ring & ball (°C)	43.4	77.8
Penetration index	−1.8	4.3

2.2 Method

To obtain different microstructures, the prepared PMB samples were conditioned by isothermal annealing for 1 h at 160 °C and 120 °C respectively before each test. Fluorescence microscopy was employed to capture the PMB morphology and the two-dimensional fast Fourier transform (2D-FFT) method was applied to analyse the captured images (Zhu et al. 2018). As for rheology, dynamic shear rheometer (DSR) was used to measure the complex modulus and phase angle at different test temperatures (64 °C, 70 °C, 76 °C and 82 °C) and frequencies (10 rad/s, 1 rad/s and 0.1 rad/s) according to ASTM D7175. In addition, multiple stress creep and recovery (MSCR) tests were performed at different temperatures (76 °C, 70 °C and 64 °C) according to ASTM D7405. The relation between PMB morphology and rheological properties is discussed.

3 Results and Discussion

3.1 Morphology

The captured PMB morphology images are shown in Fig. 1. It can be observed that the PMB presents a homogenous microstructure after isothermal annealing for 1 h at 160 °C, but a two-phase pattern with SBS-rich droplets in the bitumen-rich matrix after conditioning at 120 °C. For further morphological analysis, please refer to Lu et al. (2010). In this paper, values of the characteristic wavelength ξ of the observed PMB microstructures are calculated, as defined by:

$$\xi = 2\pi / k_m \tag{1}$$

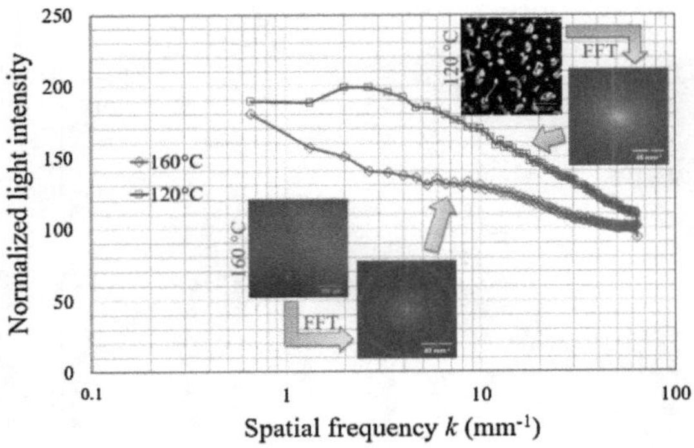

Fig. 1. Microscopy images and the analysis by two-dimensional fast Fourier transform (2D-FFT)

where k_m is the characteristic spatial frequency corresponding to a peak in the 2D-FFT power spectrum (Fig. 1). For the sample after isothermal annealing for 1 h at 160 °C, the curve does not present a peak. Its k_m is read as infinite. Thus, the calculated ξ value is 0 mm. But the calculated ξ value is 2.356 mm for the sample after conditioning at 120 °C. These results indicate that the isothermal annealing at different temperatures has resulted in significantly different PMB microstructures.

3.2 Complex Modulus and Phase Angle

The DSR measurement results of complex modulus G*, phase angle δ and other derivative parameters (G*/sinδ and elastic modulus G') at different test temperatures and frequencies are presented in Fig. 2. It is indicated that the two samples after isothermal annealing at different temperatures show only very limited difference between each other at 10 rad/s, both fulfilling the 1000 Pa criterion of G*/sinδ at 76 °C (ASTM D6373).

As the frequency decreases, however, their difference starts to enlarge, especially in phase angle δ and elastic modulus G'. This is attributed to the different microstructures of the two samples. Comparing with a binary droplet-in-matrix microstructure, the homogenous PMB morphology after isothermal annealing for 1 h at 160 °C can assist in storing more energy during the shear cycles at 1 rad/s and 0.1 rad/s. This means that, assuming a slow traffic, PMB with homogenous morphology would provide higher resistance to the loading than the one with binary droplet-in-matrix microstructure.

3.3 Multiple Stress Creep and Recovery (MSCR)

Figure 3 shows the MSCR test results at different temperatures. The results indicate that, at all the tested temperatures, the sample after isothermal annealing for 1 h at 120 °C has a higher accumulated strain level than that at 160 °C. For both two samples, the recovery is very limited at 76 °C, resulting in a very slight difference between them. As the temperature decreases, however, the recovery starts to increase. Their difference also starts to enlarge, especially at the higher stress level of 3200 Pa.

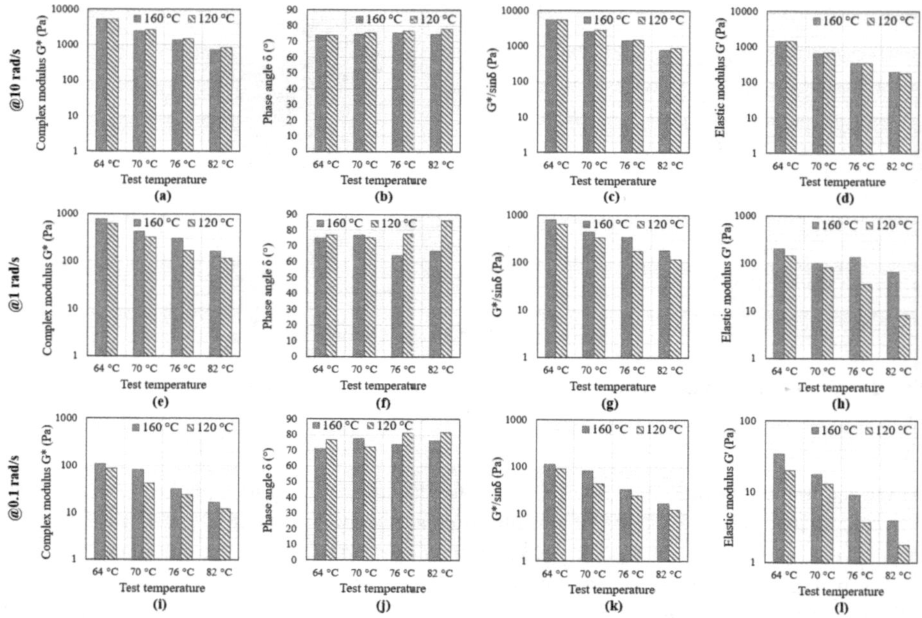

Fig. 2. Complex modulus, phase angle and other derivative parameters

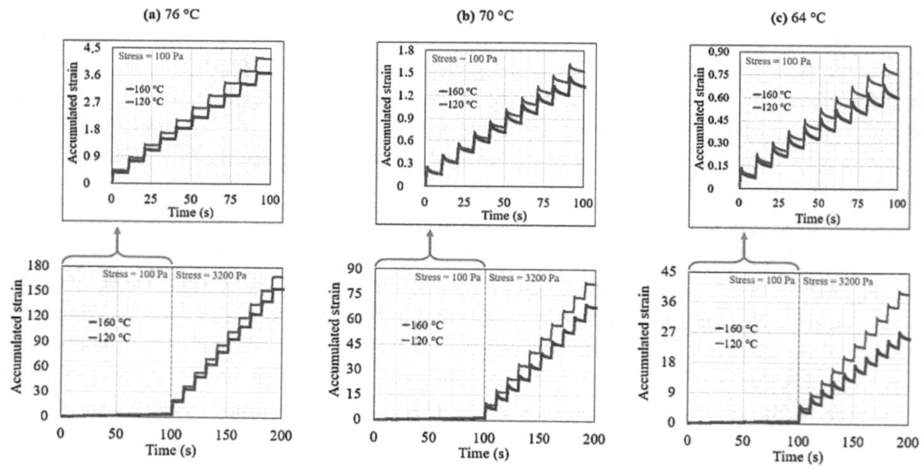

Fig. 3. Multiple stress creep and recovery (MSCR) test results at different temperatures. (In all cases, the upper curve represents the sample of 120 °C; and the lower for 160 °C). (Color figure online)

The related parameters, including the average total strain ε_1, percent recovery R, non-recoverable creep compliance J_{nr} at both two stress levels and the percent difference between stress levels, are calculated according to ASTM D7405. Calculation results are listed in Table 2.

Table 2. Calculation of multiple stress creep and recovery (MSCR) test results

Tests	@ 76 °C		@ 70 °C		@ 64 °C	
	160 °C	120 °C	160 °C	120 °C	160 °C	120 °C
$\varepsilon_1$100	0.413	0.454	0.250	0.244	0.136	0.139
$\varepsilon_1$3200	15.8	16.8	9.32	9.32	4.69	4.99
R100	10.7%	8.4%	46.9%	37.5%	55.5%	45.4%
R3200	4.6%	2.2%	28.8%	13.9%	47.1%	24.3%
R_{diff}	56.7%	74.2%	38.7%	62.9%	15.2%	46.6%
J_{nr}100	3.69 kPa^{-1}	4.16 kPa^{-1}	1.33 kPa^{-1}	1.53 kPa^{-1}	0.605 kPa^{-1}	0.757 kPa^{-1}
J_{nr}3200	4.72 kPa^{-1}	5.15 kPa^{-1}	2.08 kPa^{-1}	2.51 kPa^{-1}	0.775 kPa^{-1}	1.18 kPa^{-1}
$J_{nr\text{-}diff}$	27.9%	23.8%	56.8%	64.3%	28.0%	56.2%

The calculation shows that the two samples mostly reach similar levels of total strain ε_1. The higher conditioning temperature (160 °C) has leaded to higher percent recovery. This confirms that the difference between samples is not caused by different levels of aging at high temperatures but does result from the different microstructures. Comparing with a binary droplet-in-matrix microstructure, the homogenous PMB morphology after isothermal annealing for 1 h at 160 °C can provide higher resistance to permanent deformation in the tested temperature range. The percent difference between stress levels is mostly lower for the homogenous PMB morphology, indicating a lower stress sensitivity. These results mean that, to some extent, PMB with homogenous morphology would be resistant to heavier traffic loading than one with binary droplet-in-matrix microstructure.

4 Summary: Towards PMB Morphology-Rheology Relation

In the previous sections, it has been discussed that PMB morphology has influences on the rheological properties, although not necessarily with the same significance in the full temperature and frequency range. In certain range, however, it is possible to build the quantitative correlation between the characteristic wavelength ξ of the PMB microstructure and rheological parameters, such as the elastic modulus G' and MSCR percent recovery at 3200 Pa (Table 3). Comparing with a binary droplet-in-matrix microstructure, a homogenous PMB morphology tends to store more energy during shear cycles and reach higher recovery after the loading.

Table 3. Correlation between morphological and rheological parameters of PMB

Sample	ξ	G' @76 °C & 1 rad/s	R3200 @70 °C	R3200 @64 °C
160 °C	0 mm	134 Pa	28.8%	47.1%
120 °C	2.356 mm	36.2 Pa	13.9%	24.3%

Due to the limited number of samples and tests presented in this paper, the PMB morphology-rheology correlation is only preliminarily discussed within the studied temperature and frequency range. Towards a complete relation, however, this paper provides a quantitative morphological parameter for future studies, i.e. the characteristic wavelength ξ of the PMB microstructure by 2D-FFT method. In addition, it is demonstrated in this paper that controlling thermal history is a feasible and practical way to control PMB morphology based on the same material. As a recommendation, it would be interesting to further investigate the effect of PMB morphology on the linear viscoelastic region and zero-shear viscosity.

References

Liang, M., Xin, X., Fan, W., Wang, H., Ren, S., Shi, J.: Effects of polymerized sulfur on rheological properties, morphology and stability of SBS modified asphalt. Constr. Build. Mater. **150**, 860–871 (2017). https://doi.org/10.1016/j.conbuildmat.2017.06.069

Lu, X., Soenen, H., Redelius, P.: SBS modified bitumens: does their morphology and storage stability influence asphalt mix performance? In: Paper presented at the 11th International Conference on Asphalt Pavements, Nagoya, Japan, 1–6 August 2010

Soenen, H., De Visscher, J., Vanelstraete, A., Redelius, P.: Influence of thermal history on rheological properties of various bitumen. Rheol. Acta **45**(5), 729–739 (2006). https://doi.org/10.1007/s00397-005-0032-8

Wang, P., Dong, Z., Tan, Y., Liu, Z.: Identifying the rheological properties of polymer-modified bitumen based on its morphology. Road Mater. Pavement Des. **18**(S3), 249–258 (2017). https://doi.org/10.1080/14680629.2017.1329879

Zhu, J., Balieu, R., Lu, X., Kringos, N.: Microstructure evaluation of polymer-modified bitumen by image analysis using two-dimensional fast Fourier transform. Mater. Des. **137**, 164–175 (2018). https://doi.org/10.1016/j.matdes.2017.10.023

Experimental Investigation of Rutting in the Different Phases of Asphalt Mixtures

Chiara Riccardi(✉), Augusto Cannone Falchetto, and Michael P. Wistuba

Technische Universität Braunschweig, Brunswick, Germany
{chiara.riccardi,a.cannone-falchetto, m.wistuba}@tu-bs.de

Abstract. Rutting is one of the most severe failure mechanisms for asphalt pavements. This phenomenon is due to the accumulation of permanent deformation in consequence of traffic loading. The behavior of asphalt mixture is highly affected by the properties of the asphalt binder used in the mix design. For this reason, the Multiple Stress Creep and Recovery (MSCR) test procedure was recently introduced with the objective of better evaluating the rutting resistance while replacing the conventional Superpave parameter, $G^*/\sin\delta$. Good understanding of the rutting mechanism within the asphalt binder component is essential for correctly studying the mutual interactions of the asphalt mixture components: binder, fine aggregate and large particles. This paper presents the results of an experimental campaign consisting of MSCR tests performed on asphalt binder, mastic and fine aggregate mixture which compose a typical mixture for asphalt binder layer. All the tests were conducted using a Dynamic Shear Rheometer (DSR). The classical plate-plate configuration having 25 mm diameter and 1 mm gap was selected for asphalt binder and mastic tests. The cylindrical geometry was used for torsional tests on fine aggregate mixture presenting aggregate as large as 1.16 mm. A single testing temperature of 60 °C and three different stress levels, 100, 1600, 3200 Pa, were imposed. The results indicate that creep and recovery are functions of filler concentration and stress level.

Keywords: Binder · Mastic · Fine aggregate mixture · Rutting

1 Introduction

Rutting is one of the major distresses of asphalt pavements. This phenomenon is induced by repeated traffic loading and manifests as an accumulated permanent deformation within the asphalt pavement layers. This is particularly true during the early pavement service life when residual void content is high and the asphalt binder component is less aged, presenting a softer consistency especially at high temperature and under slow traffic (low loading frequency) resulting in an asphalt mixture with a considerable viscous behaviour. Therefore, the binder represents a key constituent governing the high temperature behaviour of asphalt mixtures. Nevertheless, asphalt mastic (consisting of filler and binder) and fine aggregate mixture (FAM) (composed of

L. D. Poulikakos et al. (Eds.): RILEM 252-CMB 2018, RILEM Bookseries 20, pp. 105–110, 2019.
https://doi.org/10.1007/978-3-030-00476-7_17

filler, fine aggregate and binder) are more representative of the mixture response in comparison to asphalt binder since they also take into account the binder-filler interaction.

The present study aims to experimentally investigate the rutting behaviour in different mixture phases in order to better understand the effect of adding filler and fine particles to the binder phase on the rheological properties of the material. The correlation to the mixture performance is not addressed in this research study.

2 Materials and Testing

In this study, a Penetration grade 50/70 binder with the following characteristics was used: penetration equals to 55 dmm (EN 1426 2007), softening point equals to 49.8 °C (EN 1427 2007) and the Performance Grade (PG) 64-22 (AASHTO M320 2017). Together with the binder, filler particles with a density of 2.605 g/cm^3 were mixed in order to produce asphalt mastic. In addition, fine aggregate mixture (FAM) samples were prepared using fine particles with a diameter smaller than 1.18 mm and having a density of 2.609 g/cm^3.

Mastics were produced mixing the preheated binder and filler in three different volume proportions of aggregate particles (*Vp* 20, 35 and 50%) (Riccardi et al. 2017b) at 160 °C, continuously stirring in order to avoid any lump. Mastic was then tested in the Dynamic Shear Rheomether (DSR) performing Multiple Stress Creep and Recovery tests (MSCRT) at 60 °C using the parallel plate geometry with a diameter of 25 mm and 1 mm gap. Three levels of stress were applied: 100 Pa, 1600 Pa and 3200 Pa. The test consists in ten cycles of creep (1s) and recovery (9s).

FAM was produced in order to recreate this specific material phase as in a binder layer asphalt mixture AC22 to be tested in the progress of this research. The maximum diameter of the fines and the correct proportions of fine particles and filler in the FAM were determined in accordance to the Bailey method (Zollinger 2005). FAM was then compacted with a gyratory compactor producing samples of 150 mm in diameter and having a voids content of 2.5%. From these samples, small cylinders with a diameter of 1.5 cm and a height of 4.7 cm were cored and tests in the DSR using the torsion bar represented in Fig. 1 performing MSCRT. In order to prevent specimen failure at the two ends, an aluminium C-shaped reinforcement was glued to the specimens as previously proposed in a different study on asphalt mixture (Moon et al. 2013) The same stresses and the same loading and rest periods were applied to the cylindrical samples.

3 Results

Figure 2 reports the results of MSCRT on the different material phases. As shown, the total shear strain decreases and the recoverable strain increases as the volume of the fine particles increases. In addition, a decreasing of the accumulated strain is observed moving from binder to FAM. This is due to the presence of filler and aggregate particles. As the amount of the aggregate component increases as structure lithic skeleton develops in the material, leading to interaction phenomena among the aggregates components resulting in a more robust elastic component which mitigate the viscous effect of the binders.

Fig. 1. FAM tests in DSR using the torsion bar

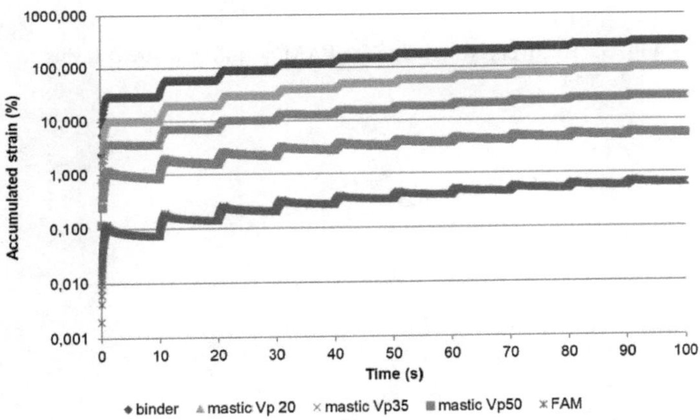

Fig. 2. MSCRT results on the different material phases

From MSCRT results the following parameters were calculated for the different material phases:

- the non-recoverable creep compliance *Jnr* reported in Eq. 1:

$$Jnr = \gamma_{nr}/\sigma \tag{1}$$

Where γ_{nr} is the unrecovered shear strain and σ is the applied shear stress.

- The Recovery (R) for each cycle given by Eq. 2:

$$R = 100\% \frac{\gamma_{peak} - \gamma_{nr}}{\gamma_{peak}} \tag{2}$$

In the case of FAM, the recoverable strain is larger than for mastic. In particular, plotting the recovery percentage versus *Jnr*, a power trend was found for all the different stress levels applied, as shown in Fig. 3.

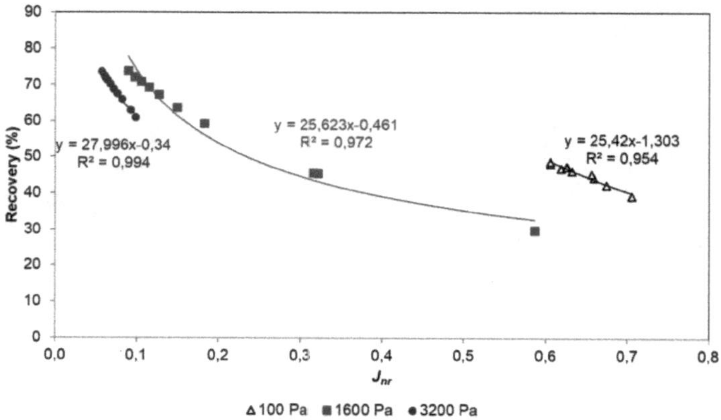

Fig. 3. Recovery versus Jnr for FAM at different stress levels

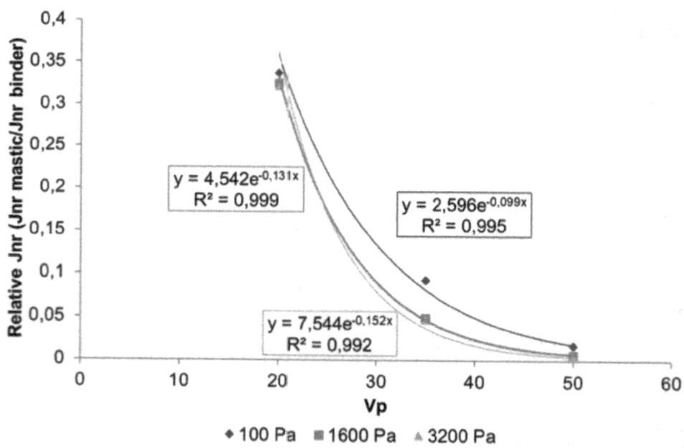

Fig. 4. Relative *Jnr* versus the volume fraction of filler

Figure 4 shows the relative *Jnr* vs. filler volume fraction for different stress levels. An exponential trend was observed, as confirmed by the equations in the plots, which can be potentially used to predict the relative non-recoverable creep compliance at any volume proportion *Vp* for the mastics studied in this research.

During MSCRT, viscosity was also measured to investigate the effect of filler content for the different mastic prepared. As known, the viscosity of a suspension increases with flocculation, in part due to the relative immobilisation of a fraction of the suspended particles trapped in the agglomeration. This increase is clearly shown in

Fig. 5, where the relative viscosity at 60 °C is plotted against the volume filler fractions. An exponential trend is exhibited for all three stress levels. This is in accordance with the semi empirical formula of Krieger–Dougherty relation that is commonly used to interpret the relative viscosity of suspension incorporating hard sphere (Adamczyk et al. 2004).

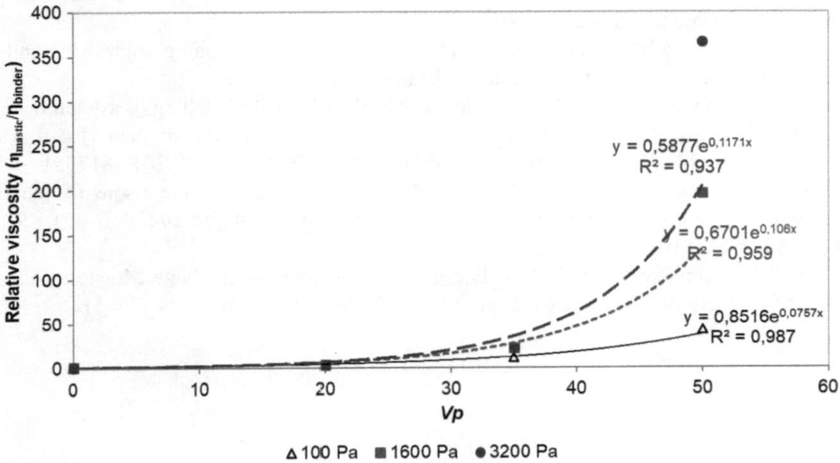

Fig. 5. Relative viscosity versus volume fraction of filler

4 Conclusions

In the present work a laboratory investigation was conducted on different asphalt mixture phases (binder, mastics at different volume concentration, FAM) based on the Multiple Stress Creep and Recovery test (MSCRT). Based on the experimental results the following conclusions can be drawn:

- The total shear strain decreases as the volume of the fine particles increases, while the recoverable strain increases when the fine particles increases.
- For FAM, the recoverable strain is larger than that of mastic. A decrease in recovery following a power trend as function Jnr was observed for the different stress levels applied.
- Plotting the accumulated strain versus time for the different phases, a decrease of the accumulated strain is exhibited across the material phases from binder to FAM.
- The relative viscosity shows an increasing exponential trend with the volume percentage of the fine particles.
- The relative Jnr shows a decreasing exponential trend with the volume percentage of the fine particles.

In future activities, tests on asphalt mixture will be performed in order to determine correlations with the parameters of the MSCRT on other material phases.

References

AASHTO M320: Standard Specification for Performance-Graded Asphalt Binder. American Association of State Highway and Transportation Officials (2017)

Adamczyk, Z., Jachimska, B., Kolasinska, M.: Structure of colloid silica determined by viscosity measurements. J. Colloid Interface Sci. **273**, 668–674 (2004)

EN 1426: Bitumen and bituminous binders. Determination of needle penetration. European Committee for Standardization (2007)

EN 1427: Bitumen and bituminous binders. Determination of the softening point. Ring and Ball method. European Committee for Standardization (2007)

Moon, K.H., Cannone Falchetto, A., Marasteanu, M.: Rheological modelling of asphalt materials properties at low temperatures: from time domain to frequency domain. Road Mater. Pavement Des. **14**, 810–830 (2013). https://doi.org/10.1080/14680629.2013.817351

Riccardi, C., Cannone Falchetto, A., Wistuba, M., Losa, M.: Fatigue comparisons of mortars at different volume concentration of aggregate particles. Int. J. Fatigue **104**, 416–421 (2017b). https://doi.org/10.1016/j.ijfatigue.2017.08.005

Zollinger, C.R.: Application of Surface Energy Measurements to Evaluate Moisture Susceptibility of Asphalt and Aggregates. Master thesis Texas A&M University (2005)

Investigation on the Effect of Physical Hardening and Aging Condition on Low-Temperature Properties of Asphalt Binder Based on BBR

Di Wang, Augusto Cannone Falchetto(✉), Chiara Riccardi, and Michael P. Wistuba

Department of Civil Engineering – ISBS, Technische Universität Braunschweig, 38106 Brunswick, Germany
{di.wang, a.cannone-falchetto, chiara.riccardi, m.wistuba}@tu-bs.de

Abstract. Low-temperature properties of asphalt binders are fundamental for designing asphalt mixture in cold regions. This is especially true for alternative technologies such as Warm Mix Asphalt (WMA), for which a temperature reduction during production may potentially lead to substantial benefits in terms of long-term aging conditions. At low temperature, asphalt binder is conventionally characterized based on creep tests conducted with the Bending Beam Rheometer (BBR) in ethanol at a single conditioning time of 1 h. However, asphalt binders undergo significant time-dependent stiffening, often referred to physical hardening, when stored at such low temperatures. In this paper, the effect of aging temperatures and physical hardening on the low temperature rheological properties is experimentally investigated and modeled. First, BBR tests are performed on four long-term aged asphalt binders, which were previously short-term aged at three different temperatures (123 °C, 143 °C, and 163 °C), after three different conditioning times: 1 h, 24 h and 72 h. Next, the creep stiffness, $S(t)$, relaxation parameter, m-value, and difference in critical temperature, ΔT_c are calculated and compared. Finally, the Huet model is fitted to the experimental data with the goal of comparing the effect of aging temperatures and physical hardening on the rheological parameters. Results indicate that physical hardening causes a significant increase in creep stiffness in the first 24 h while only moderate contribution is obtained when condition time is extended. In addition, the reduced production temperature of 40 °C can significantly improve the aging properties of asphalt binders at low temperatures while mitigating the effect of physical hardening.

Keywords: Asphalt binder · Bending Beam Rheometer (BBR) Huet model · Aging temperatures · Physical hardening

L. D. Poulikakos et al. (Eds.): RILEM 252-CMB 2018, RILEM Bookseries 20, pp. 111–116, 2019.
https://doi.org/10.1007/978-3-030-00476-7_18

1 Introduction

Low temperature properties of asphalt binders play an important role in designing asphalt mixtures especially in cold regions. The Performance Grade (PG) specifications (Anderson and Kennedy 1993; AASHTO M320 2016) were proposed with the goal to characterize asphalt binder over the entire service temperature, where the Bending Beam Rheometer (BBR) (AASHTO T313 2012) was developed to determine the low PG.

Currently, the standard BBR procedure consists of low temperature creep tests on small asphalt binder beams in an ethanol bath with a conditioning time of 1 h (AASHTO T313 2012). However, there is still ongoing scientific debates on this method with respect to the conditioning time. When asphalt binders undergo isothermal storage at low temperatures, higher stiffness can be observed for longer conditioning times (Anderson *et al.* 1994); this phenomenon was reported for amorphous polymers with the name of physical hardening (Anderson *et al.* 1994). Several studies (Anderson *et al.* 1994; Tabatabaee *et al.* 2012) indicate that the physical hardening significantly affect the creep response of asphalt binders when isothermally stored. Hence, the physical hardening of asphalt binders plays a major role in thermal cracking of asphalt pavements. More recently, the use of Warm-Mix Asphalt (WMA) has found increasing application both in Europe and US (Chowdhury and Button 2008; EAPA 2015). In spite of significant research addressing WMA, limited studies (Bonaquist 2011; Rashwan 2012) addressed the effect of different short-term aging temperatures on the low temperature properties of asphalt binders while most of the efforts focused on asphalt mixture (Gandhi 2008).

Although significant attention was devoted to study both physical hardening and aging temperatures, little is available on their combination. In this paper, the combined effect of different short-term aging temperatures and conditioning times on the low temperature properties of asphalt binder after long term aging is investigated. First, BBR measurements with different conditioning times are performed on a set of four long-term aged asphalt binders subjected to an initial stage of short-term aging at different temperatures. Then, creep stiffness, $S(t)$, m-value and ΔT_c are calculated and evaluated. Finally, the rheological Huet model (Huet 1963) is fitted to the creep stiffness and the model parameters are used to evaluate the combined effect of short term aging temperatures and conditioning time.

2 Materials and Testing

Four different 70/100 pen-graded asphalt binders (B501, B502, B503 and B504), which were provided by the RILEM Technical Committee 252-CMB project, were used in this study. The virgin binders were identified as B501_virgin to B504_virgin. The binders were then short-term aged with the RTFOT at three different temperatures: 123 °C, 143 °C, and the standard 163 °C and, next, long-term aged according to the standard PAV procedure. Based on previous characterization, the PG of the four asphalt binders was determined to be PG 70-22, PG 70-22, PG 70-22 and PG 64-22, respectively.

BBR creep tests (AASHTO T313 2012) were performed to obtain the low temperature properties of the entire asphalt binders. Creep stiffness, $S(t)$, relaxation parameter, m-value, and difference in critical temperature $\Delta T_c = Tc_S - Tc_m$, were computed according to the current standard (AASHTO T313 2012). In order to evaluate the effect of different aging temperatures and physical hardening on the low temperature properties of asphalt binders, BBR tests were performed (AASHTO T313 2012) with different conditioning time for all materials. Three replicates were tested at two temperatures: −12 °C and −18 °C for each binder.

3 Results and Analyses

3.1 Creep Stiffness Results and Analysis

Figure 1 summarizes the results of creep stiffness, $S(t)$, and m-value at −12 °C, while Table 1 lists ΔT_c for the entire asphalt binders. It can be observed that there is a substantial increase in S(60 s) and a decrease in m(60 s) for longer conditioning times. The maximum difference between 24 h and 1 h are 38% and 16% for S(60 s) and m (60 s), respectively. These values reach 55% and 20% when the condition time is extended to 72 h, suggesting that the condition time leads to a significant increase in creep stiffness in the first 24 h. The creep stiffness decreases while the relaxation parameter increases with a reduction in aging temperatures; therefore, it may be hypothesized that the reduced production temperature might result in a lower stiffness and higher relaxation capability. However, only a reduction of 40 °C appears to provide significant variation in stiffness.

The parameter ΔT_c represents the difference of T_c between S(60 s) and m(60 s); this provides an indication on the loss of relaxation properties as asphalt binders age. According to Table 1, the ΔT_c value becomes smaller (toward negative values) when conditioning time and aging temperature are increased. This trend is more remarkable when conditioning time is extended from 1 h to 24 h. In addition, when 163 °C and 143 °C (20 °C reduction in temperature) are used for short term aging, a significant reduction in ΔT_c is obtained when condition time gets longer. However, when a 40 °C reduction in aging temperature is imposed, only a moderate difference can be observed with different conditioning time, suggesting that the effect of physical hardening is actually less robust.

3.2 Huet Modeling and Analysis

The rheological Huet model (Huet 1963) which consists of an assembly of two parabolic elements and one spring combined in series, was used to better understand the effect of cooling media and aging temperatures. The creep compliance, $D(t)$, is expressed according to the following equation:

$$D(t) = \frac{1}{S(t)} = \frac{1}{E_\infty}\left(1 + \delta\frac{(t/\tau)^k}{\Gamma(k+1)} + \frac{(t/\tau)^h}{\Gamma(h+1)}\right) \quad (1)$$

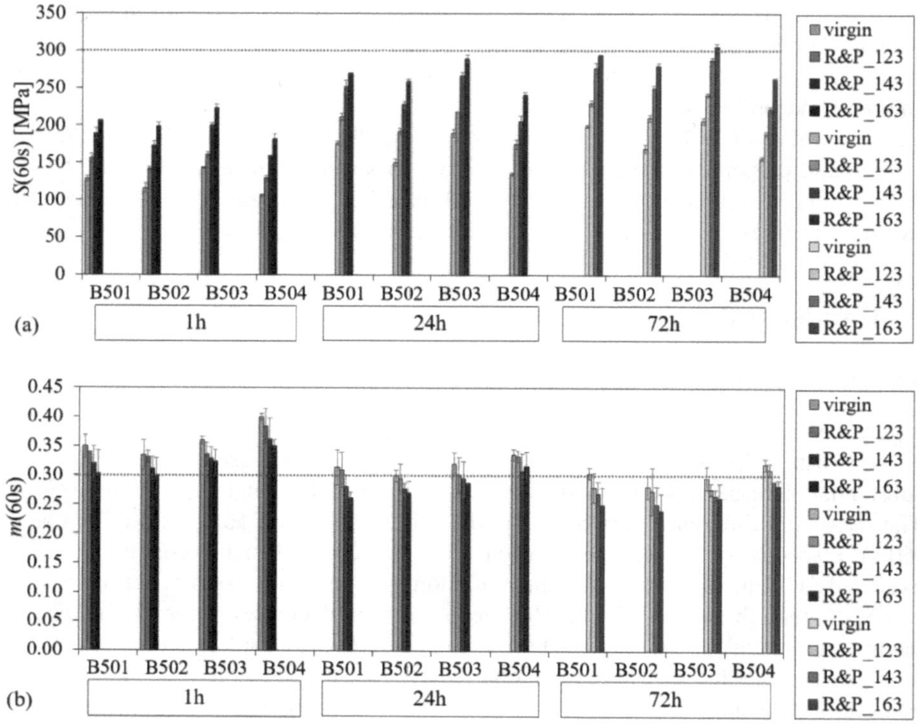

Fig. 1. (a) Creep stiffness $S(t)$ and, (b) m-value at −12 °C

Table 1. ΔT_c (°C) for the entire set of asphalt binders

Asphalt binder	1 h	24 h	72 h	Asphalt binder	1 h	24 h	72 h
B501_R&P_123	1.32	1.04	0.86	B503_R&P_123	1.58	1.22	0.93
B501_R&P_143	0.81	0.22	0.03	B503_R&P_143	0.71	0.12	−0.66
B501_R&P_163	0.07	−0.68	−1.32	B503_R&P_163	−0.03	−0.52	−1.21
B502_R&P_123	−1.13	−1.66	−2.14	B504_R&P_123	0.03	−0.61	−0.83
B502_R&P_143	−2.81	−3.52	−3.99	B504_R&P_143	−1.05	−1.66	−1.91
B502_R&P_163	−4.02	−4.98	−5.62	B504_R&P_163	−1.93	−2.73	−3.25

where, $S(t)$ is creep stiffness; E_∞ is glassy modulus; h and k are exponents such that $0 < k < h < 1$; δ is dimensionless constant; Γ is the gamma function and τ is characteristic time, associated with the relaxation time of the material. Parameters, k and h are commonly in the range of 0.08 ~ 0.3 and 0.3 ~ 0.8, respectively, although lower values were found as reported in a different research (Cannone Falchetto and Moon 2015); stiffer materials are associated with lower values of these parameters. The fitting of Huet model was performed on the low temperature creep stiffness obtained at −12 °C. The corresponding model parameters for binder B504 are listed in Table 2.

Table 2. Huet model parameters of $S(t)$ at T = low PG + 10 °C for binder B504

Asphalt binder	Condition time	E_∞ (MPa)	δ	k	h	$\log(\tau_0)$	R^2
B504_virgin	1 h	3000	2.37	0.23	0.80	2.01	0.998
	24 h	3000	1.54	0.21	0.56	2.09	0.996
	72 h	3000	1.34	0.18	0.49	2.21	0.995
B504_R&P_123	1 h	3000	4.86	0.24	0.80	1.67	0.998
	24 h	3000	4.31	0.21	0.62	2.43	0.999
	72 h	3000	4.16	0.20	0.54	2.68	0.997
B504_R&P_143	1 h	3000	5.26	0.24	0.80	1.83	0.996
	24 h	3000	4.80	0.22	0.58	2.46	0.994
	72 h	3000	4.36	0.21	0.6	2.72	0.996
B504_R&P_163	1 h	3000	6.70	0.27	0.80	1.97	0.994
	24 h	3000	5.30	0.23	0.64	2.51	0.999
	72 h	3000	4.83	0.21	0.58	2.79	0.997

The parameters of the Huet model provide additional understanding of the combined effect of aging temperatures and physical hardening. An overall larger characteristic time, τ, and smaller shape parameter, δ, are obtained for the measurements with longer condition time and higher aging temperatures. As previously mentioned, this suggests that materials may have poorer relaxation properties and may be more brittle under this condition further confirming the trend observed from the result of creep stiffness, m-value and ΔT_c.

4 Summary and Conclusion

In this paper, the effects of different aging temperatures and conditioning time on the low temperature properties of asphalt binder were experimentally investigated. It can be stated that both physical hardening and reduced aging temperature have a significant impact on the low temperature properties of asphalt binders. Longer conditioning times cause a substantial increase in creep stiffness in the first 24 h. In addition, a reduced aging temperature of 40 °C can result in better the low temperatures properties and while mitigating the effect of physical hardening.

References

AASHTO M320: Standard Method of Test for Performance Graded Asphalt Binder, American Association of State Highway and Transportation Officials (2016)

AASHTO T313: Standard Method of Test for Determining the Flexural Creep Stiffness of Asphalt Binder Using the Bending Beam Rheometer (BBR), American Association of State Highway and Transportation Officials (2012)

Anderson, D.A., Christensen, D.W., Bahia, H.U., Dongre, R., Sharma, M.G., Antle, C.E., Button, J.: Binder Characterization and Evaluation (SHRP A-369), vol. 3 in Physical Characterization. Strategic Highway Research Program, National Research Council (1994)
Anderson, D.A., Kennedy, T.W.: Development of SHRP binder specification (with discussion). J. Assoc. Asphalt Paving Technol. **62** (1993)
Bonaquist, R.F.: Mix design practices for warm mix asphalt, vol. 691. National Cooperative Highway Research Program-NCHRP Report 691. Washington, DC: Transportation Research Board (2011)
Cannone Falchetto, A., Moon, K.H.: Micromechanical-analogical modelling of asphalt binder and asphalt mixture creep stiffness properties at low temperature. Road Mater Pavement **16** (S1), 111–137 (2015). https://doi.org/10.1080/14680629.2015.1029708
Chowdhury, A., Button, J.W.: A review of warm mix asphalt (No. SWUTC/08/473700-00080-1). Texas Transportation Institute, the Texas A & M University System (2008)
European Asphalt Pavement Association (EAPA): The Use of Warm Mix Asphalt. EAPA position paper 2010 (2015). http://www.eapa.org/userfiles/2/Asphalt%20in%20Figures/2016/AIF_2015_v6.pdf. Accessed 26 Feb 2018
Gandhi, T.: Effects of warm asphalt additives on asphalt binder and mixture properties, Doctoral dissertation, Clemson University (2008)
Huet, C.: Etude par une Méthode d'Impédance du Comportement Viscoélastique des Matériaux Hydrocarbonés. Thèse de Doctorat d'Ingénieur. Faculté des Sciences de l'Université de Paris. (in French) (1963)
Rashwan, M.H.: Characterization of Warm Mix Asphalt (WMA) performance in different asphalt applications, Doctoral dissertation, Iowa State University (2012)
Tabatabaee, H.A., Velasquez, R., Bahia, H.U.: Predicting low temperature physical hardening in asphalt binders. Constr. Build. Mater. **34**, 162–169 (2012). https://doi.org/10.1016/j.conbuildmat.2012.02.039

Laboratory and Field Experience with PMMA/ATH Composite in Asphalt Mixtures

Marjan Tušar[1(✉)] and Mojca Ravnikar Turk[2]

[1] National Institute of Chemistry,
Hajdrihova 19, 1001 Ljubljana, Slovenia
marjan.tusar@ki.si

[2] Slovenian National Building and Civil Engineering Institute,
Dimičeva Ulica 12, 1000 Ljubljana, Slovenia

Abstract. To guarantee the requested efficiency and safety standards of the roads high quality materials must be used. Nowadays higher sensibility towards environmental issues and increasing difficulty in finding quality materials force us in search of alternative resources, able to substitute the traditional ones. Many different alternative materials were used in asphalt mixes. Variety of different materials can be used not only to substitute traditional materials but also to improve performance of built in asphalt mixtures. With alternative materials we can partly or fully substitute binder or particular stone fractions. In this work Polymethylmethacylate/Aluminium hydroxide composite dust (PMMA/ATH composite) was introduced in asphalt mixture as an additive to bitumen and filler. With laboratory tests and field trials we proved that PMMA/ATH composite as an additive to asphalt increases adhesion between stone aggregates and bitumen binders and improve the workability of asphalt mixture.

Keywords: Asphalt additive · Poly-methyl methacrylate
Aluminium trihydrate

1 Introduction

Reuse of existing alternative materials as aggregate in asphalt mixes implies an environmental and economic advantage by limiting the extraction of natural aggregates. The mining activity is reduced and, as a result, there is a reduced environmental impact belonging to. If waste material is used, then performance of end product- asphalt mix must be almost as good for the mixes made from traditional materials (Agostinacchio et al. 2009). When waste material improves the final product, it can be considered as alterative material. In our study we intend to show how a by-product, which was in former times considered as waste, can become an additive for improving the performance of asphalt mixtures.

Polymethylmethacylate/Aluminium hydroxide (PMMA/ATH) composite plates are used mainly as desks and counters. Due to high hardness, resistance to most chemical substances, mechanical and volume stability at low and high temperatures of such plates, they can be used also outdoor as house front elements. Surface of plates must be

L. D. Poulikakos et al. (Eds.): RILEM 252-CMB 2018, RILEM Bookseries 20, pp. 117–122, 2019.
https://doi.org/10.1007/978-3-030-00476-7_19

completely smooth, so grinding of the surface is an important phase of production. As by-product at the grinding phase large quantities of PMMA/ATH composite dust are produced. In this work PMMA/ATH composite dust was introduced in asphalt mixture as an additive to filler.

The main component of the dust is Aluminium hydroxide (Al $(OH)_3$). This compound acts analogous as Hydrated lime (Ca $(OH)_2$) (Tušar 2016). Hydrated lime is commonly known additive to asphalt mixtures. It is known that hydrated lime mitigates moisture sensitivity. Aluminium hydroxide is also an effective flame retardant (Jianying et al. 2009). This property can be useful when asphalt is used in the tunnels.

The second component of PMMA/ATH composite dust Polymethylmethacrylate (PMMA) is a thermoplastic. For asphalt application it is important that PMMA act as glue. Epoxy mortar and PMMA-based mortar can be used to repair concrete cracks (Kan et al. 2008).

There are two ways to introduce PMMA/ATH as additive in asphalt mixture. First it was treated as additive to filler in asphalt mixtures (dry process) and second PMMA/ATH was treated as additive to bitumen (wet process). For dry process we had to determine the optimal ratio between ordinary filler and PMMA/ATH to achieved good mechanical properties of asphalt layer. Also for wet process optimal ratio between bitumen and PMMA/ATH had to be determined.

For 5 different types of asphalt mixtures industrial production was carried out, which means that we produced around 1000 tons of asphalt mixtures, and laid them in a test fields test and normal roads.

2 Laboratory Tests on PMMA/ATH Composite as Filler Additive

We had to solve several problems before PMMA/ATH composite could be used as filler in asphalt mixture. First we had to check, if adhesion between PMMA/ATH composite and bitumen is as good as expected. Than we had to fulfill the requirements for filler, so we had to check sieving curve, hardening bitumen with Delta ring and ball test and sieving curve.

Only in case when PMMA/ATH composite fulfills requirements for filler, we were enabled to make laboratory trial with asphalt mixture.

2.1 Adhesion and Sieving Analyses of PMMA/ATH Composite

Adhesion was tested according standard EN 12697-11, procedure A. First with laboratory milling from the PMMA/ATH composite plates fraction 5/8 mm was prepared (Fig. 1a). PMMA/ATH composite fractions demonstrated excellent adhesion for both test methods; from 95% to 100% of grains was coated with bitumen (Fig. 1b). All results were over required limit 80%.

To determine sieving curve of PMMA/ATH composite dust air jet sieving was performed. Due to the fact that sieving curve of PMMA/ATH composite did not meet the requirement, we prepared and analyzed additionally two mixtures of ordinary filler and PMMA/ATH composite in weight ratios 5:1 and 8:1. Both mixtures of fillers met the requirements.

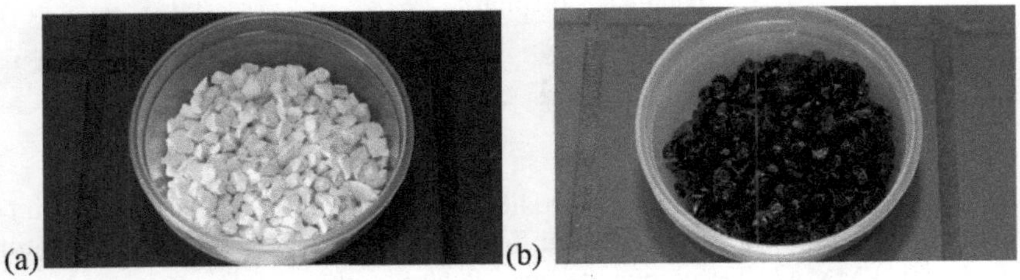

Fig. 1. PMMA/ATH composite fraction 5/8 mm (a). and grains of PMMA/ATH composite fraction 5/8 mm coated with bitumen B 50/70 after test according standard EN 12697-11 (b).

2.2 Delta Ring and Ball Test and Density of PMMA/ATH Composite

To determine ageing of bitumen due to effect of PMMA/ATH composite test was performed according to method EN 13179-1:2002 - Tests for filler aggregate used in bituminous mixtures - Part 1: Delta ring and ball test. The softening point of original bitumen according to ring and ball method (RB) was 46.6 °C. For the mixture of bitumen and PMMA/ATH composite we determined RB 161 °C. The difference in softening points 114.4 °C was highly above the requirement according to EN 13043 (less than 25 °C). We noticed that bitumen and PMMA/ATH composite did not only mix, but they were glued in the lump. We heated the samples again and found out that at elevated temperatures (over 150 °C) PMMA/ATH composite is sticky and is acting like glue.

We prepared and analyzed also two mixtures of ordinary filler and PMMA/ATH composite in ratio 5:1 and 8:1. Even at elevated temperatures both mixtures of fillers were acting like normal filler. Both mixtures of fillers met the requirement (less than 25 °C): for ratio between ordinary filler and PMMA/ATH composite = 8:1 we got difference in RB 14.2 °C and for ratio between ordinary filler and PMMA/ATH composite = 5:1 we got difference in RB 17.2 °C.

The maximal density of PMMA/ATH composite determined according to EN 1097-7:2008 was 1.74 Mg/m^3.

3 Laboratory Tests on PMMA/ATH Composite as Bitumen Additive

To mix PMMA/ATH composite in bitumen high shear mixer was needed. Silverson L5 M homogenizer was used for mixing different quantities of PMMA/ATH composite in paving grade bitumen B 70/100 (Šušteršič et al. 2013a). To ensure a good dispersion dust was mixed in bitumen for 1.5 h at 170 °C. Several test methods were used to evaluate quality of produced modified bitumen such as needle penetration at 25 °C, softening point, Fraass breaking point and rut resistance potential (G*/sin(δ) (Table 1). With RTFOT ageing procedure also ageing potential of modified bitumen was evaluated.

Table 1. Properties of PMMA/ATH composite modified bitumen.

PMMA/ATH composite cont. wt.%	Softening point (°C)	Fraass breaking point (°C)	Pen. at 25 °C, 0.1 mm	G*/sin(δ) before ageing at 60 °C	G*/sin(δ) after RTFOT ageing at 60 °C
0	46.2	−11	78	1330	4790
25	53.6	−15	55	3650	14500

With simple test methods such as needle penetration at 25 °C, softening point and Fraass breaking point only insignificant differences between base bitumen and bitumen modified with PMMA/ATH composite were determined.

From G*/sin(δ) measurements with dynamic shear rheometer we found significant differences between asphalts containing base bitumen and asphalts containing bitumen modified with PMMA/ATH composite dust. From these results increased resistance to permanent deformations was expected.

4 Laboratory Tests on PMMA/ATH Composite as Filler Additive

For laboratory testing AC 8 asphalt mixtures were prepared. First mixture contained PMMA/ATH composite as additive to filler in mass ratio 1:5, second contained PMMA/ATH composite in paving grade bitumen B 70/100 in mass ratio 1:3, third was similar to second with additionally 3 wt% paraffin wax and fourth reference was without PMMA/ATH composite (Šušteršič et al. 2013b).

Wheel tracking tests were performed at 50 °C. The results of wheel tracking test were in good agreement with G*/sin(δ) values determined with binder test.

Increased water resistance of samples containing PMMA/ATH composite (ITS ratio) implies that waste PMMA/ATH particles in asphalt binder improve the adhesion performance between aggregate and bitumen (Table 2) (Tušar 2014).

Table 2. Higher temperature properties of PMMA/ATH composite modified asphalt.

Samples of AC 8 surf	ITS at 25 °C (kPa)	ITS ratio at 25 °C (%)	Proportional rut depth at 50 °C (%)	WTS_{AIR} at 50 °C
Reference mixture	907	93.1	18.3	0.46
PMMA/ATH composite added in filler	895	94.4	9.0	0.16
PMMA/ATH composite added in bitumen	1102	99.2	6.3	0.09

For all asphalt mixtures compactability and low temperature properties were determined. Samples with PMMA/ATH composite added in filler give similar or even better results than reference mixture at these two tests (Table 3).

Table 3. Compactability and low temperature properties of PMMA/ATH composite modified asphalt.

Samples of AC 8 surf	Compactability at 160 °C (kPa)	TSRST failure temperature (°C)	TSRST failure stress σ cry (MPa)	Tensile strength reserve T Δβ, max. (°C)	Tensile strength reserve, Δβ, max. (MPa)
Reference mixture	26.4	−24.7	4.0	−8.4	3.8
PMMA/ATH composite added in filler	26.5	−28.1	4.4	−10.0	4.3
PMMA/ATH composite added in bitumen	30.4	−26.3	4.5	−6.3	4.4

5 Conclusions

With this study we found out that PMMA/ATH composite can be used as additive to asphalt. With laboratory tests we found out that PMMA/ATH composite can be added to the filler and bitumen. With high hardness, resistance to most chemical substances, mechanical and volume stability PMMA/ATH composite can improve some properties of asphalt mixture. With laboratory tests and field trials we proved that PMMA/ATH composite as an additive to filler increases adhesion between stone aggregates and bitumen binders and improve the workability of asphalt mixture. The quality of asphalt pavement at the test field is still good and in future we will perform observation also in future.

Acknowledgments. We are grateful to Marko Oven, Dušanka Bohinc, Miloš Kmet, Ema Šušteršič and Andreja Zupančič-Valant for laboratory work. The research is partly funded by Slovenian Research Agency (grant P1-0017).

References

Agostinacchio, M., Diomedi, M., Olita, S.: The use of marginal materials in road constructions: proposal of an eco-compatible section. In: Advanced Testing and Characterization of Bituminous Materials –Loizos, Partl, Scarpas, Al-Qadi (eds.), Taylor & Francis Group, London (2009). ISBN 978-0-415-55854-9

Jianying, Y., Peilianga, C., Shaopenga, W.: Investigation of the properties of asphalt and its mixtures containing flame retardant modifier. Constr. Build. Mater. **23**(6), 2277–2282 (2009)

Šušteršič, E., Tušar, M., Zupančič-Valant, A.: Rheological and mechanical characterization of waste PMMA/ATH modified bitumen. Constr. Build. Mater. **38**, 119–125 (2013)

Šušteršič, E., Tušar, M., Zupančič-Valant, A.: Effective concept of asphalt concrete modification with waste PMMA/ATH. Materials and Structures (2013b)

Tušar, M.: Laboratory and field experience with PMMA/ATH composite dust in asphalt mixtures. In: Lakušić, S. (Ed.). Road and Rail Infrastructure III: Proceedings of the 3rd International Conference CETRA 28–30 April 2014, Split, Croatia, ISSN 1848-9842). Zagreb: Department of Transportation, Faculty of Civil Engineering, University of Zagreb, pp. 345–350 (2014)
Tušar, M.: Asphalt for heavily loaded pavement: proceedings of the 6th Eurasphalt & Eurobitume Congress, Prague, 1–3 June 2016. Prague (2016)
Kan, Y.C., Yen, T., Lee, M.G.: Restored strength of cracked concrete beam repaired by epoxy and polymethyl methacrylate. ACI Materials J. (2008)

On the Use of a Novel Binder-Fast-Characterization-Test

Johannes Schrader(✉) and Michael P. Wistuba

ISBS Braunschweig Pavement Engineering Centre, Technische Universität Braunschweig, Beethovenstr. 51b, 38106 Braunschweig, Germany
{jo.schrader,m.wistuba}@tu-braunschweig.de

Abstract. The Binder-Fast-Characterization-Test in the DSR can be used instead of conventional test methods for characterizing plain and modified asphalt binders in the high temperature range. The test is called "Bitumen-Typisierungs-Schnell-Verfahren (BTSV)" and was recently introduced to the German standards. With two rheological key parameters (T_{BTSV} and δ_{BTSV}), asphalt binders can be differentiated in regard to hardness and degree of modification. This study presents possible applications of the procedure. The test method can be used to clearly characterizing virgin unknown asphalt binders and to differentiate between modified and unmodified binders. In addition, the two key parameters can be used to evaluate aged asphalt binders and the ageing susceptibility. A linear change of the two key parameters was detected for ageing and blending processes of asphalt binders. Thus, the effectiveness of different rejuvenating materials can be evaluated and necessary amounts of rejuvenator identified to produce target binders with specific rheological properties in the high temperature range.

Keywords: DSR · Softening point · Ring & ball · High temperature properties Binder-Fast-Characterization-Test · BTSV

1 Introduction

The asphalt binder classification system in Europe is based on the empirical values of needle penetration (EN 1426 2015) and Ring-and-Ball softening point ($SP_{R\&B}$) (EN 1427 2015). Often, the $SP_{R\&B}$ is used during the asphalt mix design for evaluating Reclaimed Asphalt Pavement (RAP) binder and for identifying suitable amounts of rejuvenator. Thus, different limitations for the $SP_{R\&B}$ of asphalt binders are found in national technical standards when using RAP and rejuvenators in the asphalt mix design. However, a number of studies have reported significant problems and misleading results using this traditional test method for evaluating asphalt binders with increasing complexity due to the high degree of modification and rejuvenating additives (Alisov et al. 2018). As a replacement for the conventional $SP_{R\&B}$ a new test method was recently introduced under the name "BTSV" in the German standards (AL DSR-Prüfung (BTSV) 2017; E DIN 52050 2018) for rheological binder characterization in the high temperature range.

L. D. Poulikakos et al. (Eds.): RILEM 252-CMB 2018, RILEM Bookseries 20, pp. 123–128, 2019.
https://doi.org/10.1007/978-3-030-00476-7_20

2 Background and Objective

Properties of asphalts binders are highly temperature dependent. The transition between the solid state at low temperatures and the liquid state at high temperatures is determined by the $SP_{R\&B}$, which has been used for approximately 100 years to characterize asphalt binders (Mallison 1928). The $SP_{R\&B}$ represents the state of the softening and is therefore considered as the upper limit of the service temperature (EN 12591 2007; EN 14023 2010). However, with increasing complexity of asphalt binders the $SP_{R\&B}$ can not sufficiently determine the state of the binder softening (Alisov 2017). For modified asphalt binders the results of the $SP_{R\&B}$ are inconsistent and do not represent an equivalent rheological property. Thus, the novel Binder-Fast-Characterization-Test is used instead to determine an equivalent rheological material condition for characterization of asphalt binders.

3 Methodology

A Dynamic Shear Rheometer (DSR) is used for the Binder-Fast-Characterization-Test (in German: BTSV) with a plate-plate setup. The diameter of the geometry is 25 mm and the gap between the two plates is 1 mm (AL DSR-Prüfung (BTSV) 2017). Following the $SP_{R\&B}$ test procedure, the test temperature is continuously increased from 20 °C to a maximum of 90 °C with a temperature rate of ΔT = 1.2 °C/min, resulting in a testing time of approximately 60 min. During the temperature ramp, an oscillatory shear stress is applied to the binder sample. A constant shear stress of 500 Pa and a constant frequency of 10 rad/s are chosen for all binders to ensure testing within the linear viscoelastic (LVE) range (Alisov and Wistuba 2016). Three rheological parameters are recorded at least every 2.5 s: temperature T, complex shear modulus G^*, and phase angle δ. If the complex shear modulus falls below the default value of 1 kPa the test can be aborted. As the outcome of the test, the temperature T_{BTSV} and the phase angle δ_{BTSV} which are corresponding to G^* = 15 kPa are obtained. Whereas G^* = 15 kPa is the rheological equivalent to the $SP_{R\&B}$ (Alisov 2017).

4 Applications

4.1 Virgin Binder Characterization

So far, the BTSV has been performed on 205 different virgin asphalt binders, which are typically used in Europe and classified according to their penetration grade. All the binder properties are within the specifications of the European Standards (EN 12591 2007; EN 14023 2010). Figure 1 illustrates the range of the two key parameters, temperature T_{BTSV} and phase angle δ_{BTSV}, for different binder grades in form of box-domains.

As shown in Fig. 1, plain binders only differ in their temperature T_{BTSV}, while having an approximate constant phase angle δ_{BTSV} of 80 °C, whereas polymer modified binders appear with much smaller values of δ_{BTSV}. Thus, T_{BTSV} is an indicator for

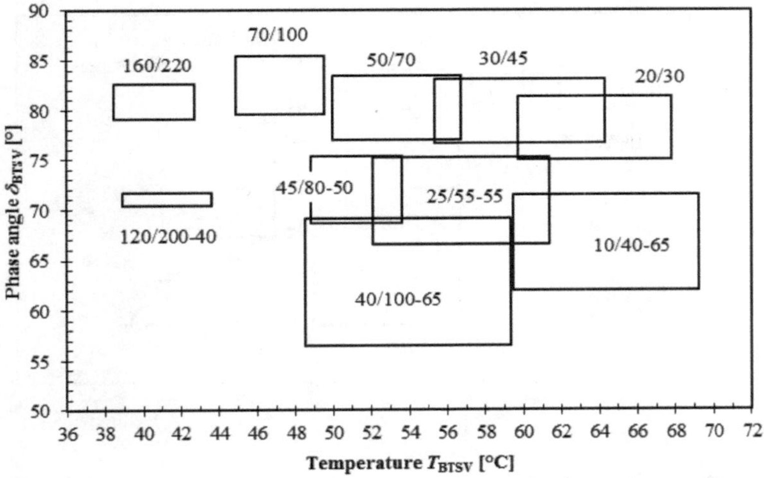

Fig. 1. Binder characterization based on the Binder-Fast-Characterization-Test (in German: BTSV) for a set of asphalt binders typically used in Europe, and using key parameters: temperature T_{BTSV} and the corresponding phase angle δ_{BTSV}

the binder hardness and δ_{BTSV} an indicator for the degree of modification, allowing a simple differentiation between different binder grades. Based on these two key parameter the BTSV presents a distinct rheological characterization system for plain and polymer-modified binders in the high temperature range.

4.2 Aged Binder Characterization

Both plain binder 50/70 and polymer modified binder 25/55–55 were exposed to laboratory ageing. Using Rolling Thin Film Oven Test (RTFOT) (EN 12607-1, 2014), aging time was varied stepwise from 75 min, which is standard RTFOT ageing time, to 8 times 75 min. After every step, BTSV key parameters were determined. As seen in Fig. 2, the change in T_{BTSV} and δ_{BTSV} follows a linear trend. Interestingly, this is true for any binder: The change in rheological properties due to aging can be expressed by a linear function, which can advantageously be used to rank different binders with respect to their aging susceptibility (see Alisov et al. 2018).

4.3 Rejuvenated Binder Characterization

In addition, BTSV parameters can successfully assist for rejuvenating binder from reclaimed asphalt pavements (RAP) during the mix design process. Exemplarily, Fig. 3 depicts T_{BTSV} and δ_{BTSV} of a bitumen extracted from RAP (see 'red dot'), and of corresponding T_{BTSV} and δ_{BTSV} after having mixed with virgin plain binder 160/220 (see 'blue rectangle'), and with a chemical rejuvenator in different amounts (see 'green rectangles'). Depending on the amount of added bitumen, and of the added rejuvenator respectively, the positions of T_{BTSV} and δ_{BTSV} for the resulting blend will always end up on a linear "blending line" (see 'hatched lines').

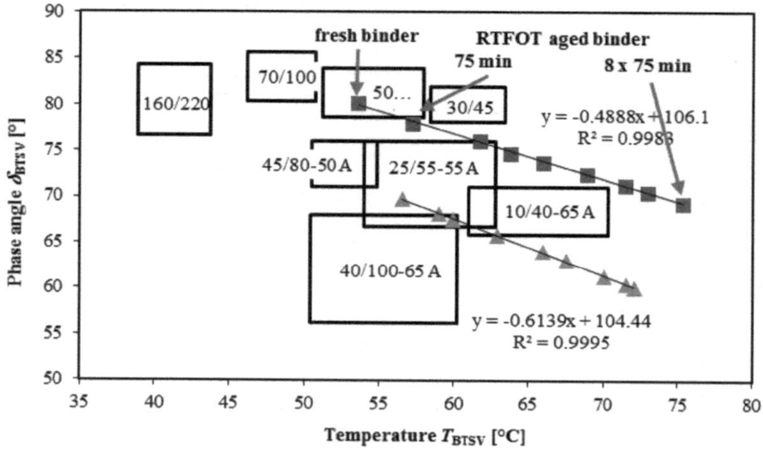

Fig. 2. Effects of RTFOT laboratory aging on T_{BTSV} and δ_{BTSV} (Alisov et al. 2018)

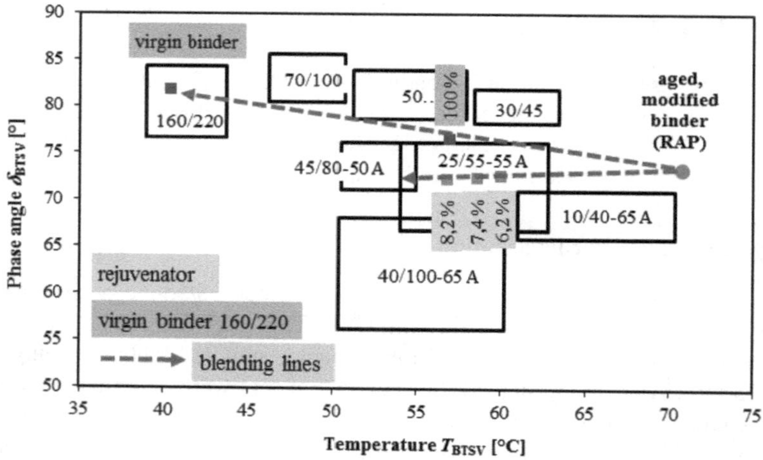

Fig. 3. Example on using the blending line based on BTSV parameters (Alisov et al. 2018)

Figure 3 makes obvious, that BTSV can also be used for studying the effect of different rejuvenating materials on the resulting rheological properties at high temperature, and for finding the optimum amount of any rejuvenating additive. In Fig. 3, the effects of virgin binder and chemical rejuvenator are quite different: For the chemical rejuvenator, a horizontal shift of T_{BTSV} is observed mainly, which means that binder hardness is reduced only. When adding virgin binder, a reset of both parameters is observed, T_{BTSV} and δ_{BTSV}, which means that real rejuvenation is achieved. However, a big amount of virgin binder is needed, which will be hard to accomplish in the asphalt mix. Contrary, the needed amount of the chemical rejuvenator is rather limited (see 6.2, 7.4, 8.2 mass%).

5 Conclusions

With increasing complexity of asphalt binders, the traditional softening point Ring and Ball method is insufficient and inconsistent. The Binder-Fast-Characterization-Test (BTSV) (Bitumen-Typisierungs-Schnell-Verfahren, in German) can instead be used to differentiate binder characteristics in the high temperature range based on rheological properties. It consists in performing a DSR test with a constant oscillating shear stress in the LVE range over a temperature ramp. As soon as the decreasing shear modulus G^* reaches a pre-defined value of 15.0 kPa, the corresponding temperature, called T_{BTSV} and the phase angle δ_{BTSV} are obtained. While T_{BTSV} provides information on material hardness, δ_{BTSV} indicates the degree of modification. Both key parameters form a solid basis for further rheological asphalt binder evaluation.

The procedure can be used to clearly characterize unknown plain and unmodified binders and to show the influence of binder modification. The BTSV can also be applied to investigate laboratory aging of virgin binder and blending of extracted RAP binder. As a consequence to aging and blending the two key parameters was found are following linear trends. Hence, these linear trends can be used for evaluating aging state and aging susceptibility of asphalt binders and for evaluating the effectiveness of rejuvenators (chemical additives of soft virgin binder) in the high temperature range.

Acknowledgement. The authors would like to acknowledge Malvern Panalytical Ltd. for providing a part of the instruments used to determine rheological material properties.

References

AL DSR-Prüfung (BTSV): Arbeitsanleitung zur Bestimmung des Verformungsverhaltens von Bitumen und bitumenhaltigen Bindemitteln im Dynamischen Scher-Rheometer (DSR) - Teil 4: Durchführung des BTSV (Bitumen-Typisierungs-Schnell-Verfahren). Forschungsgesellschaft für Straßen- und Verkehrswesen (FGSV), Köln (2017)

Alisov, A.: Typisierung von Bitumen mittels instationärer Oszillationsrheometrie, Dissertation, Institut für Straßenwesen, Technische Universität Braunschweig (2017)

Alisov, A., Wistuba, M.P.: Von der Differenzierung komplexer Bitumen. Asph. Bitum. **4**, 58–63 (2016)

Alisov, A., Riccardi, C., Schrader, J., Cannone Falchetto, A., Wistuba, M.P.: A Novel Method to Characterize Asphalt Binder at High Temperature. Road Materials and Pavement Design (2018)

E DIN 52050: Bitumen und bitumenhaltige Bindemittel - BTSV-Prüfung. Norm-Entwurf, Erscheinungsdatum: 2018-05-25. Deutsches Institut für Normung e. V., Berlin (2018)

EN 12591: Bitumen and bituminous binders – Specifications for paving grade bitumen. European Committee for Standardization (CEN), Brussels (2007)

EN 14023: Bitumen and bituminous binders – Specification framework for polymer modified bitumens. European Committee for Standardization (CEN), Brussels (2010)

EN 1426: Bitumen and bituminous binders – Determination of needle penetration. European Committee for Standardization (CEN), Brussels (2015)

EN 1427: Bitumen and bituminous binders - Determination of the softening point - Ring and Ball method. European Committee for Standardization (CEN), Brussels (2015)

Mallison, H.: Die Ring und Kugel-Methode zur Bestimmung des Erweichungspunktes von Pechen und Asphalten. Industrie-Zeitung Asphalt und Teer (1928)

Use of Microencapsulated Phase Change Materials in Bitumen to Mitigate the Thermal Distresses in Asphalt Pavements

Muhammad Rafiq Kakar[1(✉)], Zakariaa Refaa[1,2], Jörg Worlitschek[2], Anastasia Stamatiou[2], Manfred N. Partl[1], and Moises Bueno[1]

[1] Empa, Swiss Federal Laboratories for Material Science and Technology, 8600 Dübendorf, Switzerland
muhammad.kakar@empa.ch

[2] Lucerne University of Applied Sciences and Arts, 6048 Horw, Switzerland

Abstract. In asphalt pavements, the temperature variations due to extreme weather conditions are mainly responsible for thermal distresses. Cracking and rutting are the main types of damages in asphalt mixtures from extreme low and high temperatures. Recently, in building construction, phase change materials (PCM) are efficiently used as a source of thermal energy preservation by storing thermal energy in a latent form. The use of microencapsulated phase change material (μPCM) enables thermal energy storage and release inside bitumen. The PCM's crystallization/melting temperature define the temperature at which thermal energy can be stored/released. In this study, a low temperature μPCM was incorporated in bitumen and its rheological and thermal properties were evaluated using dynamic shear rheometer (DSR) and differential scanning calorimeter (DSC) respectively. It was observed that upon cooling, the addition of 25% μPCM by mass of bitumen crystallizes and releases the stored thermal energy. The surface temperature variations captured by means of thermal infra-red camera illustrate that compared to controlled specimen, in modified bitumen a delaying effect of cooling was found during the crystallization temperature of μPCM.

Keywords: Thermal energy · Phase change material · Crystallization
Infra-red image

1 Introduction

Temperature is a fundamental parameter when analyzing the behavior of road surfaces based on asphalt or bituminous materials. The thermal balance of road surfaces involves both environmental impact and in service performance. It is known that at high temperature, the asphalt pavements are more prone to suffer permanent deformation (rutting) and, upon cooling, the material becomes stiffer until reaching a brittle state at which thermal cracking usually occurs (Du et al. 2018). The performance of asphalt materials is directly influenced by their temperature. Hence, ageing behavior and distresses are related not only to traffic conditions but also to the environmental climate and weather situation. At low temperature, when the pavement cools, the

L. D. Poulikakos et al. (Eds.): RILEM 252-CMB 2018, RILEM Bookseries 20, pp. 129–135, 2019.
https://doi.org/10.1007/978-3-030-00476-7_21

bituminous binder slowly transforms from a viscoelastic ductile into an elastic brittle material. Initially, at warm temperatures, the binder is soft enough for any shrinkage stress to relax through viscous flow, but later, at colder temperature, it stiffens losing its thermal stresses relaxation property. Asphalt pavements start to experience cracks when binders reach the brittle state during cooling (low temperature cracking) (Desseaux et al. 2018).

As additives with desirable properties of storing and releasing thermal energy, phase change materials (PCM) have been widely used in the field of concrete buildings for energy saving purposes (Song et al. 2018; Laaouatni et al. 2017). Pielichowska and Pielichowski (2014) summarized a number of thermal, physical, kinetic and chemical criteria which a suitable PCM should meet to be used as storage material for thermal energy. Nevertheless, the applications of PCM in thermoplastic materials, such as asphalt, have only been explored recently (Ma et al. 2014; Kong et al. 2017). Chen et al. 2012 found that the use of PCM in asphalt concrete was promising in solving problems of asphalt pavement at high temperatures. PCM have a narrow characteristic temperature range where melting and solidification (crystallization) occurs that can be selected to influence operating temperatures in several engineering applications. The application of PCM in roads may reduce failures due to extreme temperatures variations (Cocu et al. 2010).

In this study, the incorporation of low temperature μPCM in bitumen is assessed in terms of its ability to store energy in the form of latent heat and protect the direct interaction of bitumen and PCM. In addition, the rheological and thermal performances of μPCM modified bitumen were evaluated for further studies and future recommendations. For comparison also lime filler modified bitumen is considered.

2 Materials and Methods

Bitumen 160/220 samples were provided by commercial suppliers to study the impact of PCM at low temperature. Tetradecane n-alkane (n-C_{14}) was selected as core material inside the polymeric shell (formaldehydes) with melting point near 6 °C which makes it suitable for low temperature energy storage applications. The μPCM was supplied by MikroCaps (Slovenia), and mineral filler $CaCO_3$ (nekafill® 15) was supplied by KFN (Switzerland). Depending on the average particle size of supplied μPCM which is approximately 7 μm, hereafter in this paper the μPCM is referred as μPCM-7. The conventional mineral filler with an average particle size of less than 63 μm makes it convenient to be replaced with μPCM-7. In order to compare the equal volume modification, density measurements of μPCM-7 (2.780 g/cm^3) and lime filler (0.899 g/cm^3) were used. The blending of 25% μPCM-7 by mass of bitumen and lime filler was performed using a speed mixer at 2000 rpm during 2 min. The temperature at the time of blending was kept 120 °C in order to ensure the necessary bitumen viscosity for blending. Dynamic shear properties were measured with a Physica MCR 301 DSR in a parallel plate configuration on 2 mm high cylindrical specimens with 8 mm in diameter. The upper and bottom plate faces were roughened to ensure good adhesion between plate and binder at low temperature (Bueno et al. 2014). A temperature ramp from 20 °C to −10 °C corresponds to a cooling rate of −0.44 °C/min was used while applying oscillatory shear strain with constant strain amplitude (0.1%) and frequency (1 Hz) in order to avoid damaging the sample. The repeatability of results was ensured with one random specimen tested twice before proceeding to analyze all the samples. The thermal analysis of PCM modified and unmodified bitumen was determined with differential scanning calorimetry (DSC). Heat flow versus temperature was measured during cycle at a cooling rate of 10 °C/min. The infra-red thermal images were captured using testo 885- professional infra-red camera with full videometric resolution of 320 × 240 pixels. The specimens were placed on a cooling stage "Peltier-Thermoelectric Cold Plate CP-200HT-TT" with thermoelectric temperature controller (TC-48-20 OEM). The infra-red thermal scanning on μPCM-7/lime filler modified bitumen specimens of 1 mm thickness was performed with a cooling rate of 10 °C/min. However, due to the optimum infra-red image resolution, a diameter of 25 mm and 18 mm was fixed for both μPCM-7 and lime filler modified bitumen 160/220 respectively.

3 Results and Discussion

The thermal analysis techniques enable the characterization and quantification of thermal transition of bitumen. Both unmodified and PCM modified bitumen were analyzed using DSC and heat flow was recorded during bitumen cooling from 70 °C. In Fig. 1 the heat flow versus temperature is given for unmodified and modified bitumen with μPCM-7. The μPCM-7 modified bitumen shows an exothermic peak below 0 °C, due to the crystallization of microencapsulated Tetradecane. The integration of the peak area resulted in a crystallization enthalpy of 28.9 J/g. The stored thermal energy in bitumen depends on the content of μPCM as well as on its melting enthalpy.

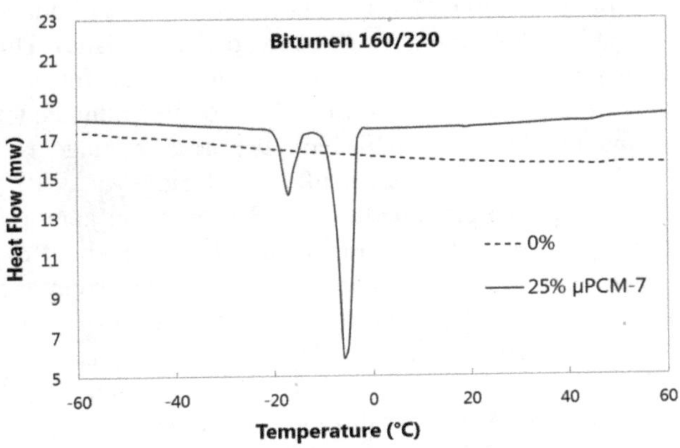

Fig. 1. Heat flow vs. temperature (cooling rate 10 °C/min)

The rheological measurements using DSR temperature sweep test are illustrated in Fig. 2. The results reveal that upon cooling, the complex modulus (G^*) of bitumen 160/220 increases with the decrease in temperature and the lime filler modified bitumen registered the highest values of G^*.

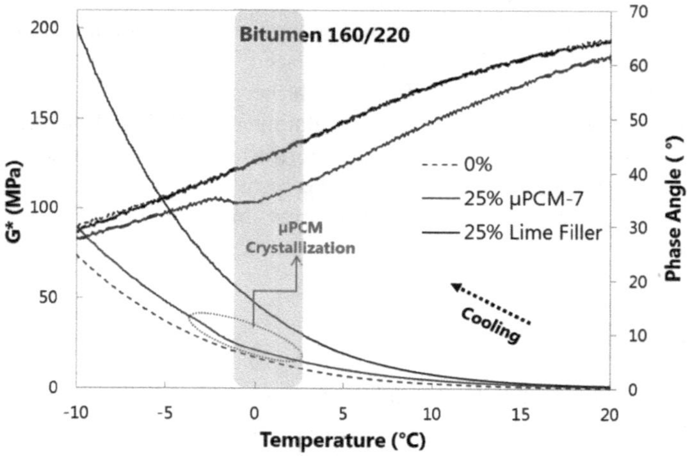

Fig. 2. DSR, G*, phase angle low temperature sweep test

Furthermore, the unmodified bitumen has the lowest G^* compared to both μPCM-7 and lime filler modified bitumen. However, at low temperature (−10 °C) the stiffness of μPCM-7 modified bitumen were closer to the unmodified bitumen compared to the lime filler modified bitumen. According to the DSR low temperature sweep test results the crystallization effect of μPCM-7 can be observed through the change of G^* in the temperatures ranging from 1 °C to −3 °C (cf. Fig. 2).

Indeed, upon cooling, the μPCM-7 slowers the stiffening of bitumen by releasing thermal energy stored in μPCM (i.e. crystallization of Tetradecane). This effect can be seen more clearly by plotting the phase angle values against the temperature (X-axis) as shown in Fig. 2, where the phase angle of μPCM-7 modified bitumen tends to increase upon cooling during the PCM crystallization temperature range (1 °C to −3 °C) compared to unmodified and lime filler modified bitumen.

The Black space diagrams presented in Fig. 3 provide an appropriate means for comparing the rheological properties of modified bitumens without requiring mathematical shifting, like, e.g. in case of modulus-frequency mastercurves. The effect of PCM is more significant in this type of plot, where the temperature range of μPCM-7 crystallization becomes evident through a clear increase in phase angle and slower increase in G^*. Therefore, the delaying of change in phase angle vs G^* indicates the ability of microencapsulated PCM crystallization in bitumen.

The experimental infra-red thermal imaging setup is depicted and schematically shown in Fig. 4. Two specimens of μPCM-7 and lime filler modified bitumen 160/220 were placed on an aluminum plate and simultaneously cooled by means of a cooling

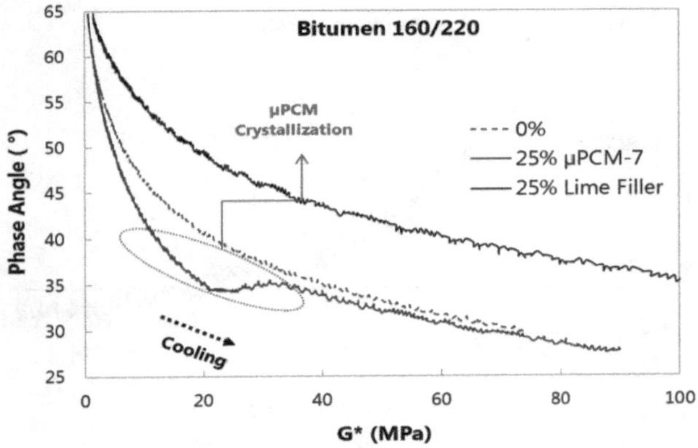

Fig. 3. DSR, Black space diagram

stage at the bottom. The thermal images were captured at a cooling rate of 10 °C/min. The inverted grayscale infra-red thermal image of both specimens is shown on the top (left) of Fig. 4. The results presented in Fig. 5 elucidate a decrease in cooling of μPCM-7 modified bitumen during the crystallization temperature. These results reveal the same effect noticed in the rheological testing of μPCM-7 modified bitumen. Therefore, the use of thermal infra-red image analysis is an appropriate method to assess the thermal behavior of μPCM-7 in bitumen.

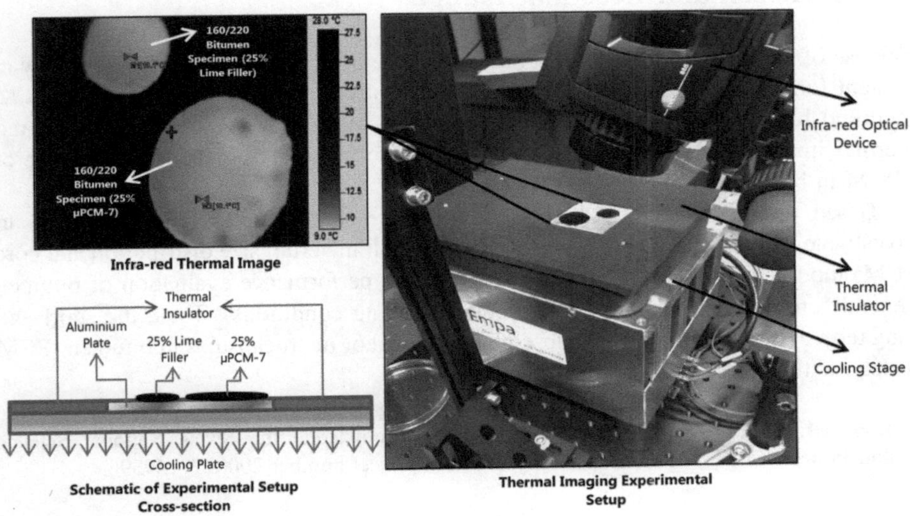

Fig. 4. Infra-red thermal image and schematic experimental setup

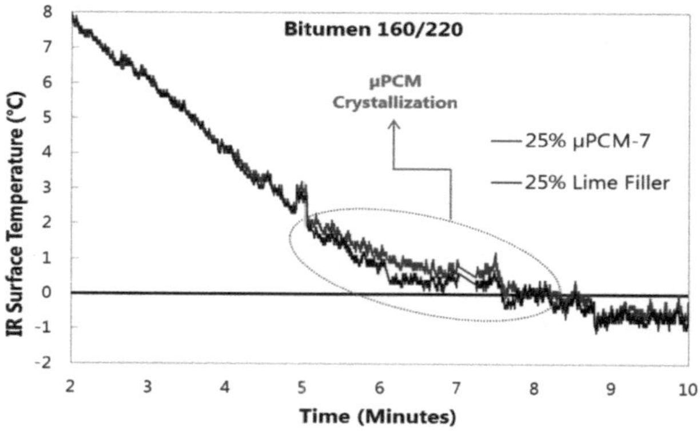

Fig. 5. Average specimen surface temperature of the μPCM-7 and lime filler modified bitumen 160/220 samples measured by infra-red camera during cooling

The rheological effects of μPCM-7 observed in Figs. 2 and 3 are desirable properties of bitumen in terms of the thermal susceptibility and extreme low temperature distresses in asphalt mixtures. Besides the advantages of PCM's thermal energy storage in latent form, the PCM modified bitumen has improved mechanical properties as compared with lime filler modified bitumen. Therefore, μPCM-7 has greater potential to improve the performance against extreme weather distresses in asphalt mixtures.

4 Conclusion and Recommendations

The use of 25% μPCM by mass of bitumen has the potential to store thermal energy in a latent form inside bitumen. The rheological results from DSR show that the μPCM modified bitumen may improve low temperature mechanical performance. Infra-red thermal images analysis was successfully used for quantifying thermal behavior of μPCM in bitumen.

Based on the results of this work, future research studies are encouraged to investigate the use of different types of μPCM (shell material, size distribution and core PCM) modified bitumen. However, the long term performance evaluation of bitumen depends on the environmental conditions (i.e. aging conditions). Thus, the short and long term aging of bitumen and subsequent survival of microcapsules in bitumen (PCM leakage) still need to be analyzed in future.

Acknowledgement. The authors would like to acknowledge the Swiss National Science Foundation (SNSF) for the financial support of the project number 200021_169396.

References

Bueno, M., Hugener, M., Partl, M.N.: Low temperature characterization of bituminous binders with a new cyclic shear cooling (CSC) failure test. Constr. Build. Mater. **58**, 16–24 (2014)

Cocu, X., Nicaise, D., Rachidi, S.: The use of phase change materials to delay pavement freezing. In: Transport Research Arena (TRA) 2010-Thematic Session 11: New Materials (2010)

Desseaux, S., dos Santos, S., Geiger, T., Tingaut, P., Zimmermann, T., Partl, M.N., Poulikakos, L.D.: Improved mechanical properties of bitumen modified with acetylated cellulose fibers. Compos. B Eng. **140**, 139–144 (2018)

Du, Y., Chen, J., Han, Z., Liu, W.: A review on solutions for improving rutting resistance of asphalt pavement and test methods. Constr. Build. Mater. **168**, 893–905 (2018)

Pielichowska, K., Pielichowski, K.: Phase change materials for thermal energy storage. Prog. Mater. Sci. **65**, 67–123 (2014)

Kong, W., Liu, Z., Yang, Y., Zhou, C., Lei, J.: Preparation and characterizations of asphalt/lauric acid blends phase change materials for potential building materials. Constr. Build. Mater. **152**, 568–575 (2017)

Laaouatni, A., Martaj, N., Bennacer, R., El Omari, M., El Ganaoui, M.: Phase change materials for improving the building thermal inertia. Energy Procedia **139**, 744–749 (2017)

Chen, M., Wan, L., Lin, J.: Effect of phase-change materials on thermal and mechanical properties of asphalt mixtures. J. Test. Eval. **40**, 1–8 (2012)

Ma, B., Si, W., Ren, J., Wang, H., Liu, F., Li, J.: Exploration of road temperature-adjustment material in asphalt mixture. Road Mater. Pavement Des. **15**(3), 659–673 (2014)

Song, M., Niu, F., Mao, N., Hu, Y., Deng, S.: Review on building energy performance improvement using phase change materials. Energ. Build. **158**, 776–793 (2018)

Chemo-Mechanical Characterization of Bituminous Materials: Microstructure and Micro-Mechanics

Analysis of Bitumen and PmB Using Fluorescence Spectroscopy and Microscopy

Johannes Mirwald[1], Hinrich Grothe[1(✉)], Bernhard Hofko[2], Daniel Maschauer[2], and Daniel Steiner[2]

[1] Institute of Materials Chemistry, TU Wien, Vienna, Austria
grothe@tuwien.ac.at
[2] Institute of Transportation, TU Wien, Vienna, Austria

Abstract. Physicochemical characterization of bitumen and polymer modified bitumen has been a research subject for the last decades. Various microscopic and spectroscopic techniques have been used to unravel the bitumen microstructure and to establish the link to its chemical composition. Over the years, the usage of bitumen has risen. This is particularly true for polymer modified bitumen (PmB) which has primarily been blended with styrene-butadiene-styrene (SBS) polymers. Since there is a demand for homogeneous blending and an overall quality control, suitable methods need to be developed. Fluorescence spectroscopy and fluorescence microscopy were found useful tools to achieve these goals. This paper focuses on the analysis of a 160/220 pen-graded base bitumen and SBS modified bitumen. Complementary analysis of specific surfaces enables a better understanding of chemical composition (spectroscopic information) and microstructure (microscopy).

Keywords: Bitumen · Polymer modified bitumen · Fluorescence Spectroscopy · Microscopy

1 Introduction

The characterization and chemical analysis of bitumen, a product of the crude oil refinery, is challenging due to its complex composition and microstructure. Polarity chromatography divides the hydrocarbon composition into four fractions (Corbett 1969). These so-called SARA fractions (saturates, aromatics, resins, and asphaltenes) are usually separated according to the ASTM D-4124 standard (Lesueur 2009). The characterization of bitumen and its polymer modified products has involved various microscopic and spectroscopic techniques. Beside FTIR spectroscopy (Hofko et al. 2017, Weigel and Stephan 2017, Hofko et al. 2018), also fluorescence spectroscopy provides molecular information. Previous results from our group show characteristic fluorescence spectra of pure bitumen and those of its four SARA fractions (Handle et al. 2016, Grossegger et al. 2018). The origin of the fluorescence signals mainly lies within the aromatic and the resin fractions. While fluorescence microscopy is a common technique in bitumen and PmB analysis, fluorescence spectroscopy has not been used to the same extent. Concerning spectroscopy, the following questions arises: How can we obtain consistent results and good reproducibility? What differentiates pure

L. D. Poulikakos et al. (Eds.): RILEM 252-CMB 2018, RILEM Bookseries 20, pp. 139–144, 2019.
https://doi.org/10.1007/978-3-030-00476-7_22

bitumen from polymer modified samples? Which spectroscopic and morphological changes are caused by the addition of the polymer? Will microscopy help to understand these spectroscopic results?

For these aims conventional incident light fluorescence microscopy was chosen. Even at low lateral resolution valuable microstructural features can be resolved which help to interpret the spectra. Previous studies on SBS modified bitumen have already shown promising results regarding the microstructure (Sengoz and Isikyakar 2008).

Here we suggest a preparation and measurement routine for fluorescence spectroscopic analysis on pure bitumen and an approach to analyze the surface of SBS modified bitumen by combining spectroscopy and microscopy.

2 Materials and Methods

2.1 Materials

This study was performed using a 160/220 pen-graded base bitumen, the pure SBS polymer and the resulting SBS modified bitumen (blended with base bitumen and 11% SBS). These components are commonly used in roofing membranes production.

2.2 Methods and Procedures

The 160/220 bitumen samples were prepared using a specific preparation procedure that leaves the sample surfaces at a minimum exposure time to air and therefore at a low level of oxidation (s. Fig. 1). A small quantity of bitumen (0.55–0.65 g) was heated up to 100 °C in a silicone mold (volume: 0,51 cm^3). After heating for a maximum of 10 min, the liquid sample was covered with a glass slide and allowed to cool down to room temperature for 20 min. Once cooled down, the silicone mold was removed, and the fresh surface was stored under ambient atmosphere for 3 h before it was used for fluorescence spectroscopic analysis.

The pure SBS pellets were ground using a commercial grinder and liquid nitrogen. The resulting powder was filled in a quartz cuvette and measured directly with the spectrometer.

The SBS modified bitumen samples were heated up to 180 °C, homogenized and poured in a siliconized foil template that ran through a rolling machine, creating a 5 mm thick PmB sample. Removing one side of the siliconized foil allowed the application onto a glass slide, removing the second side made analysis possible.

For the fluorescence spectroscopic measurements, an Edinburgh Instruments FPS920 photoluminescence spectroscopy setup was used. This setup contains a XE900 Xenon Arc Lamp (500 W) as an illumination source, double Czerny-Turner monochromators (type TMS300) at both excitation and emission arms as well as a S900 single-photon photomultiplier (type R928) as the detector. The setup was used in excitation mode (ex. wavelength 240–500 nm, em. wavelength 525 nm).

The fluorescence microscope setup consists of a Nikon Eclipse 50i, a 30 W halogen light source, a color-digital camera (DS-Fl1c) and an epifluorescence unit with a

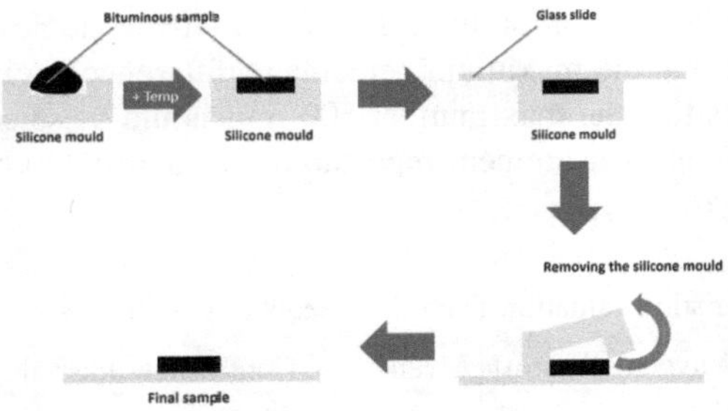

Fig. 1. Scheme of the bituminous sample preparation

excitation filter at 400–440 nm, an dichroic mirror at 455 nm and a emission filter at 470 nm. Two different objectives with magnification of 10x and 100x were used.

3 Results and Discussion

3.1 Reproducibility of Fluorescence Spectroscopy

The fluorescence spectroscopic analysis on the 160/220 base bitumen showed good reproducibility when using the preparation procedure and parameters presented. To check the repeatability of the device itself, a 160/220 base bitumen was measured consecutively 25 times. The resulting fluorescence excitation scan on the left side of Fig. 2 shows a sufficient repeatability, as no significant changes in the spectra are observable. This indicates that the device (e.g. light beam) causes almost no chemical changes on the sample surface when being measured for 7–8 consecutive hours (25 repeats).

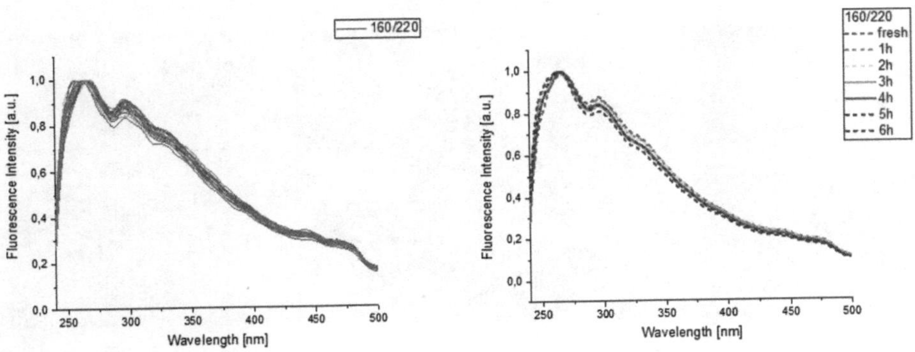

Fig. 2. Fluorescence excitation spectra: Repeatability (left) and time dependency (right) of 160/220 bitumen

Applying statistic evaluation on three significant characteristics of the spectrum (s. Table 1) one can see that the maximum percentage difference is given at 300 nm with ±3.9%. This suggests that the maximum at 300 nm should be selected for verification when looking at general measurement repeatability of a 160/220 bitumen.

Table 1. Statistic evaluation from 25 repeats at three different wavelengths

Wavelength [nm]	Mean		Confidence interval
264	0,9956	±	0,0090
300	0,8809	±	0,0388
480	0,2531	±	0,0146

Due to low viscosity of the 160/220 bitumen the problem regarding sample stability arises. After a short period, the sample starts to flow in all directions, which changes its surface. Therefore, additional time dependent studies were conducted that revealed that the best reproducibility is obtained when conducting measurements within 3–4 h after preparation (s. right side of Fig. 2).

3.2 SBS Modified Bitumen

Figure 3 shows the fluorescence microscopic observations from the SBS modified bitumen samples. The two different magnifications outline a trend in homogeneity. While sample 1 has a rather fine microstructure, sample 2 and 3 show a coarse-grained microstructure, which can only be observed at higher magnification (100x). Sample 4 shows a significant lower degree of mixing, resulting in a bright continuous area and a dark part that has not been observed in samples 1, 2 and 3. We assume that the bright continuous area is an agglomerated polymer phase.

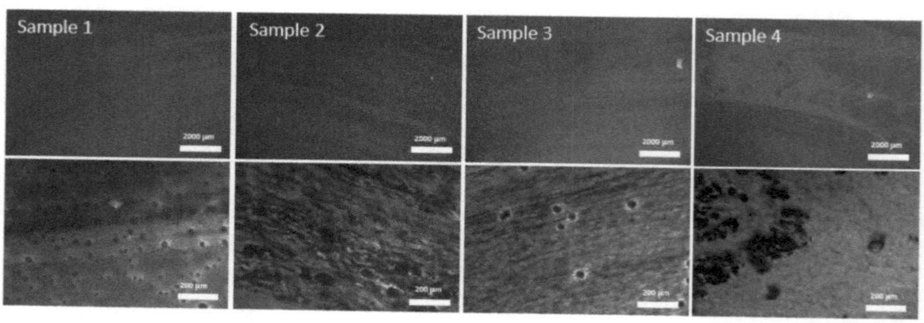

Fig. 3. Surface of the SBS modified bitumen in 10x (upper) and 100x (lower) magnification

These microscopic observations can be correlated with the spectroscopic features on the left side in Fig. 4. The surface of sample 4 contains partially unmixed SBS, which shows a higher fluorescence intensity compared to pure bitumen or a well-blended PmB. This results in a strong signal around 350 nm. The spectrum of sample 1 indicates that once a fine microstructure is obtained, the signals from bitumen and SBS contribute almost equally to the final spectrum. As a result of spectroscopy and microscopy from sample 2 and 3, the coarse-grained microstructure has a stronger SBS signal. Hence, we can conclude that the extent of visible microstructures in microscopy correlates with the respective SBS signal in the fluorescence spectra.

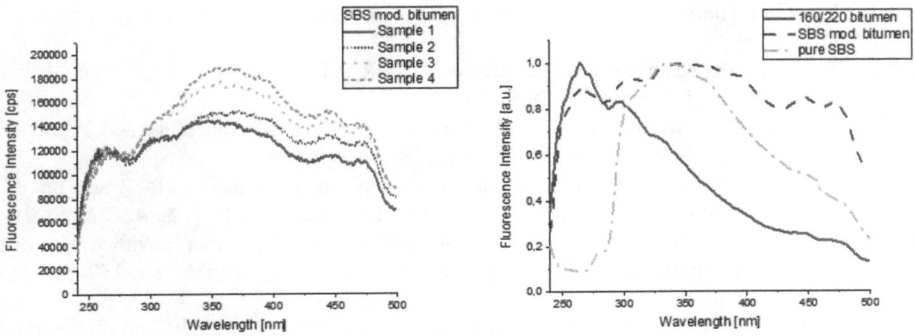

Fig. 4. Fluorescence excitation scans of the SBS modified bitumen samples (left) and the comparison to its 160/220 base bitumen and pure SBS (right)

The right side in Fig. 4 gives evidence that the bitumen exhibits local maxima at 264 and 300 nm and three shoulders at 350, 400 and 480 nm. The pure SBS excels its local maximum at 350 nm. The resulting spectrum of the SBS modified bitumen (sample 1) shows good accordance with the spectra of its pure components.

Hence, the characteristic signals from the two components add up nicely to the resulting PmB spectrum. This gives valuable information about the homogeneity of the sample and the involved blending procedure parameters.

4 Conclusion

The results lead to the conclusion that fluorescence spectroscopy is a viable method for analyzing base bitumen. Sufficient reproducibility is given by using the presented sample preparation technique and the suggested storage time of 3–4 h before conducting the measurement. For the PmB samples, spectroscopic reproducibility is far more complex, since the resulting information is linked to the SBS concentration on the surface and the homogeneity of the sample. Hence, microscopic monitoring is necessary. The gathered spectroscopic results show good accordance from the comparison to its pure components and will certainly contribute to the task of characterizing polymer modified bitumen.

References

Corbett, L.W.: Composition of asphalt based on generic fractionation using solvent deasphaltening elution-adsorption chromatography and densimetric characterization. Anal. Chem. **4**, 41 (1969)

Grossegger, D., et al.: Fluorescence spectroscopic investigation of bitumen aged by field exposure respectively modified rolling thin film oven test. Road Mater. Pavement Des. **19**(4), 992–1000 (2018)

Handle, F., et al.: The bitumen microstructure: a fluorescent approach. Mater. Struct. **49**(1–2), 167–180 (2016)

Hofko, B., et al.: Repeatability and sensitivity of FTIR ATR spectral analysis methods for bituminous binders. Mater. Struct. **50**(3), 187 (2017)

Hofko, B., et al.: FTIR spectral analysis of bituminous binders: reproducibility and impact of ageing temperature. Mater. Struct. **51**(2), 45 (2018)

Lesueur, D.: The colloidal structure of bitumen: consequences on the rheology and on the mechanisms of bitumen modification. Adv. Colloid Interface Sci. **145**(1–2), 42–82 (2009)

Sengoz, B., Isikyakar, G.: Analysis of styrene-butadiene-styrene polymer modified bitumen using fluorescent microscopy and conventional test methods. J. Hazard. Mater. **150**(2), 424–432 (2008)

Weigel, S., Stephan, D.: The prediction of bitumen properties based on FTIR and multivariate analysis methods. Fuel **208**, 655–661 (2017)

Chemical Composition and Microstructure of Bitumen – a Matter of Terminology?

Bernhard Hofko[1(✉)], Daniel Maschauer[1], Daniel Steiner[1], Hinrich Grothe[2], and Johannes Mirwald[2]

[1] Institute of Transportation, TU Wien, Vienna, Austria
bernhard.hofko@tuwien.ac.at
[2] Institute of Materials Chemistry, TU Wien, Vienna, Austria

Abstract. Bitumen is an organic material derived from crude oil refinery. It exhibits a highly complex chemical composition, time- and temperature dependent viscoelastic material behavior and distinct microstructural features. Ongoing efforts try to link these three material characteristics by combining physico-chemical with mechanical analysis. Two different theses that explain which chemical constituents trigger the materials' microstructure: one group of researchers is in favor of the idea that waxes are responsible, the other group favors asphaltenes as the crucial factor. Both groups base their assumptions on experimental evidence. However, no final proof for either theses exists. A recent study on high mass resolution analysis of bitumen and its constituents shows that in fact both groups might be referring to the same thing and simply using different terminology. This paper suggests a temporary solution for overcoming the differences by substituting the controversial terms "wax" and "asphaltene" by the more general but still correct term "n-heptane insolubles" for the time being.

Keywords: Bitumen · Microstructure · Composition · Chemo-mechanics Asphaltenes · Waxes

1 Introduction

Bitumen or asphalt binder as a crude oil derivative is an organic material with a highly complex chemical composition and a temperature- and time-dependent viscoelastic mechanical behavior. In addition, it is prone to aging by thermal processes, as well as oxidation. It is by now common knowledge that bitumen exhibits microstructural features that can be visualized with various microscopic techniques, such as atomic force microscopy (AFM) (Jaeger et al. 2004; Loeber et al. 1996; Soenen et al. 2013), confocal laser scanning microscopy (CLSM) (Handle et al. 2014; Mikula and Munoz 2000) or environmental scanning electron microscopy (ESEM) (Mikhailenko et al. 2017; Stangl et al. 2006). Recent studies have shown that these microstructural features correlate with binder source (Nahar et al. 2013a), thermal history (Nahar et al. 2013b) and chemical composition (Hofko et al. 2016). Links between microstructure and mechanical behavior, as well as between microstructure and chemical composition have been established (Eberhardsteiner et al. 2015b). It has been validated that the microstructural features visible in AFM and in CLSM as fluorescing centers have a

L. D. Poulikakos et al. (Eds.): RILEM 252-CMB 2018, RILEM Bookseries 20, pp. 145–149, 2019.
https://doi.org/10.1007/978-3-030-00476-7_23

strong impact on the stiffness and elasticity of the material. In addition, these features are linked to the concentration of asphaltenes in the material (Eberhardsteiner et al. 2015a). A microstructural model explains how the visible features and the chemical composition can be linked and describe the mechanical behavior: the higher the concentration of asphaltenes, the more microstructural features are visible and the interaction between these features increase with increasing asphaltene concentration. This leads to an increasing ratio of elasticity in the mechanical behavior, as well as increasing stiffness and brittleness.

However, there is an on-going scientific debate whether the crucial factor in affecting microstructural features and mechanical behavior are asphaltenes or waxes, sometimes also referred to as asphaltenic waxes. At this point, no final proof for either of the theses can be given.

2 Motivation and Objectives

The field of linking physico-chemical to mechanical and microstructural information of bituminous binders (chemo-mechanics) is very active and has seen a strong gain in interest the last couple of years. To give an example, two RILEM technical committees, namely" Nanotechnology-based bituminous materials (NBM)" and "Chemo-Mechanical Characterization of Bituminous Materials (CMB)" have dedicated their work to this field. While many open questions have been solved in the last decades, some are still up to debate.

This paper is an attempt in providing a possible solution to overcome some differences in terminology, at least for the time being. It is also thought to be an invitation for continued discussion and exchange of experience in this highly interesting and rapidly developing area of research. What this paper does not try to be, and cannot be at the given length, is a state-of-the-are literature review on this topic, for which others (Lesueur 2009) have given excellent examples.

In detail, this paper gives some input on the terminology used to link microstructural features of bitumen to its chemical composition.

3 Asphaltenes or Waxes, Waxes or Asphaltenes

While it is proven beyond doubt that bitumen is not a homogeneous material on the micro-scale but rather exhibits distinguished features, it is not clear up until today what triggers these features and how the chemical composition is linked to the formation of the core-shell particles, as shown exemplary in Fig. 1.

Two seemingly opposing ideas regarding this matter can be found: One group of researchers is in favor of the thesis that the microstructure is closely linked to the existence of waxes or asphaltenic waxes in the bitumen. These waxes are supposed to crystalize on the surface and form the well-known features. The argumentation is mainly based on the correlation between the wax content and the frequency of features found by e.g. AFM (Lesueur 2009; Lu et al. 2005; Redelius and Soenen 2015; Soenen et al. 2014).

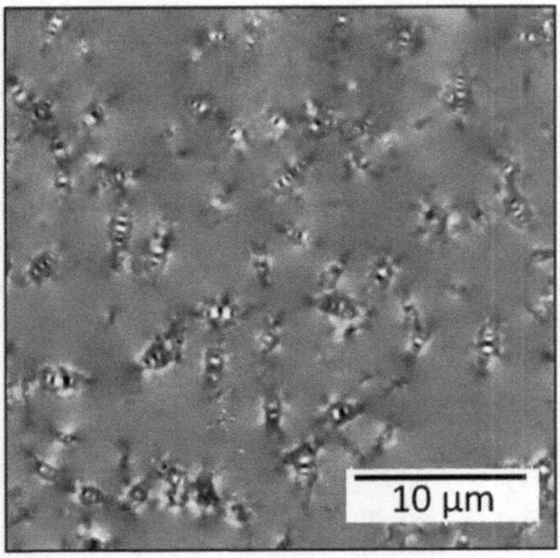

Fig. 1. Example of AFM topology of a bituminous binder

Another set of researchers is in favor of the idea that asphaltenes agglomerate within the maltene phase and a polarity balancing shell surrounds these agglomerates. Together, agglomerate and shell form core-shell structures (Ortega-Rodriguez et al. 2003). This model is supported by the distinctive agglomeration behavior of asphaltenes (Yarranton et al. 2013) and studies on precipitated blends with varying concentrations of asphaltenes in the maltene phase. These studies (Yarranton et al. 2013) showed a strong correlation between asphaltene concentration and occurrence of microstructures features.

However, neither group has been able to finally prove their concepts and thus, dismiss the other concept as wrong. One main reason for this situation is that changes in the composition of bitumen by adding or removing any polarity based fraction will inevitably change the polarity regime within the bitumen. These changes in overall polarity can be the reason for appearance or disappearance of microstructural features. Thus, there is evidence that waxes and asphaltenes play a crucial role in the microstructural composition of bitumen but no final proof that either of them is the actual player that forms these features.

A recently completed study worked with a Negative Electrospray Ionization Fourier Transform Ion Cyclotron Resonance Mass Spectrometry (FT-ICR-MS) (Handle et al. 2017). Unaged and aged bitumen samples, as well as their SARA fraction were analyzed with high mass resolution. One of the most interesting findings of this study is that the asphaltene fraction does not mainly consist of aromatic structures. Only about 25% of the structures found are aromatic by definition. The rest of the structures are either aliphatic or alicyclic. Asphaltenes are by definition n-heptane insolubles. However, long-chain waxes are insoluble in n-heptane as well. Therefore, a certain fraction of asphaltenes is made of waxes.

Looking at these facts, is there not a possibility that a significant part of the commotion in the field might be due to just different terminology used by different groups? One group might speak about waxes or asphaltenic waxes, which the other group puts to the unpolar end of the spectrum and dismisses the idea since their experience tells them that the more polar asphaltene fraction plays a strong role in microstructure. The term "asphaltene" is commonly linked to the higher polar fraction of bitumen. Thus, the waxy party dismisses any involvement of this constituent in the microstructure of bitumen based on their experimental results.

A bridge could be constructing to connect the two thought islands. A small change in wording, i.e. substituting both terms "waxes" and "asphaltenes" with the more general and still correct term "n-heptane insolubles" for the time being, could overcome the differences – at least until final proof is given what exactly triggers bitumen to developing the microstructural features that intrigues so many of us.

References

Eberhardsteiner, L., Fussl, J., Hofko, B., Handle, F., Hospodka, M., Blab, R., Grothe, H.: Influence of asphaltene content on mechanical bitumen behavior: experimental investigation and micromechanical modeling. Mater. Struct. **48**(10), 3099–3112 (2015a)

Eberhardsteiner, L., Fussl, J., Hofko, B., Handle, F., Hospodka, M., Blab, R., Grothe, H.: Towards a microstructural model of bitumen ageing behaviour. Int. J. Pavement Eng. **16**(10), 939–949 (2015b)

Handle, F., Fussl, J., Neudl, S., Grossegger, D., Eberhardsteiner, L., Hofko, B., Hospodka, M., Blab, R., Grothe, H.: Understanding the microstructure of bitumen: a CLSM and fluorescence approach to model bitumen ageing behavior. In: Asphalt Pavements, vol. 1, no. 2, pp. 521–530 (2014)

Handle, F., Harir, M., Fussl, J., Kosyun, A.N., Grossegger, D., Hertkorn, N., Eberhardsteiner, L., Hofko, B., Hospodka, M., Blab, R., Schmitt-Kopplin, P., Grothe, H.: Tracking aging of bitumen and its saturate, aromatic, resin, and asphaltene fractions using high-field fourier transform ion cyclotron resonance mass spectrometry. Energy Fuels **31**(5), 4771–4779 (2017)

Hofko, B., Eberhardsteiner, L., Fussl, J., Grothe, H., Handle, F., Hospodka, M., Grossegger, D., Nahar, S.N., Schmets, A.J.M., Scarpas, A.: Impact of maltene and asphaltene fraction on mechanical behavior and microstructure of bitumen. Mater. Struct. **49**(3), 829–841 (2016)

Jaeger, A., Lackner, R., Eisenmenger-Sittner, C., Blab, R.: Identification of four material phases in bitumen by atomic force microscopy. Road Mater. Pavement Des. **5**, 9–24 (2004)

Lesueur, D.: The colloidal structure of bitumen: consequences on the rheology and on the mechanisms of bitumen modification. Adv. Colloid Interface Sci. **145**(1–2), 42–82 (2009)

Loeber, L., Sutton, O., Morel, J., Valleton, J.M., Muller, G.: New direct observations of asphalts and asphalt binders by scanning electron microscopy and atomic force microscopy. J. Microsc.-Oxford **182**, 32–39 (1996)

Lu, X.H., Langton, M., Olofsson, P., Redelius, P.: Wax morphology in bitumen. J. Mater. Sci. **40**(8), 1893–1900 (2005)

Mikhailenko, P., Kadhim, H., Baaj, H.: Observation of bitumen microstructure oxidation and blending with ESEM. Road Mater. Pavement Des. **18**(sup2), 216–225 (2017)

Mikula, R.J., Munoz, V.A.: Characterization of emulsions and suspensions in the petroleum industry using cryo-SEM and CLSM. Coll. Surf. A: Physicochem. Eng. Asp. **174**, 14 (2000)

Nahar, S.N., Mohajeri, M., Schmets, A.J.M., Scarpas, A., van de Ven, M.F.C., Schitter, G.: First observation of blending-zone morphology at interface of reclaimed asphalt binder and virgin bitumen. Transp. Res. Rec. **2370**, 1–9 (2013a)

Nahar, S.N., Schmets, A.J.M., Scarpas, A., Schitter, G.: Temperature and thermal history dependence of the microstructure in bituminous materials. Eur. Polym. J. **49**(8), 1964–1974 (2013b)

Ortega-Rodriguez, A., Cruz, S.A., Gil-Villegas, A., Guevara-Rodriguez, F., Lira-Galeana, C.: Molecular view of the asphaltene aggregation behavior in asphaltene-resin mixtures. Energy Fuels **17**(4), 1100–1108 (2003)

Redelius, P., Soenen, H.: Relation between bitumen chemistry and performance. Fuel **140**, 34–43 (2015)

Soenen, H., Besamusca, J., Fischer, H.R., Poulikakos, L.D., Planche, J.P., Das, P.K., Kringos, N., Grenfell, J., Lu, X., Chailleux, E.: Laboratory investigation of bitumen based on round robin DSC and AFM tests. Mater. Struct. **47**, 1–16 (2013)

Soenen, H., Besamusca, J., Fischer, H.R., Poulikakos, L.D., Planche, J.P., Das, P.K., Kringos, N., Grenfell, J.R.A., Lu, X.H., Chailleux, E.: Laboratory investigation of bitumen based on round robin DSC and AFM tests. Mater. Struct. **47**(7), 1205–1220 (2014)

Stangl, K., Jager, A., Lackner, R.: Microstructure-based identification of bitumen performance. Road Mater. Pavement Des. **7**, 111–142 (2006)

Yarranton, H.W., Ortiz, D.P., Barrera, D.M., Baydak, E.N., Barre, L., Frot, D., Eyssautier, J., Zeng, H., Xu, Z., Dechaine, G., Becerra, M., Shaw, J.M., McKenna, A.M., Mapolelo, M.M., Bohne, C., Yang, Z., Oake, J.: On the size distribution of self-associated asphaltenes. Energy Fuels **27**(9), 5083–5106 (2013)

ESEM Microstructural and Physical Properties of Virgin and Laboratory Aged Bitumen

Peter Mikhailenko[1(✉)], Hassan Baaj[1], Changjiang Kou[1,2], Lily D. Poulikakos[3], Augusto Cannone Falchetto[4], Jeroen Besamusca[5], and Bernhard Hofko[6]

[1] University of Waterloo, Waterloo, Canada
{p2mikhai,hbaaj}@uwaterloo.ca, kcj_mail@163.com
[2] Yangzhou University, Yangzhou, China
[3] EMPA, Dübendorf, Switzerland
Lily.Poulikakos@empa.ch
[4] Technische Universität Braunschweig, Braunschweig, Germany
augcanno@tu-braunschweig.de
[5] Kuwait Petroleum Research and Technology, Rotterdam, Netherlands
j.besamusca@Q8.com
[6] Technical University of Vienna, Vienna, Austria
bernhard.hofko@tuwien.ac.at

Abstract. The physical and microstructural properties of four straight run asphalt binders were examined and compared in virgin state and in after short term aging (RTFOT) and long-term (PAV) laboratory aging. RTFOT aging was conducted at 123, 143 and 163 °C. Physical testing parameters included penetration and softening point. Selected binders came from four different sources with same penetration grade. They all showed an increase in stiffness with aging, and RTFOT temperatures. The microstructural evolution of the binder was examined by Environmental Scanning Electron Microscopy (ESEM) on aged binders at 123 and 163 °C. The physical transformation corresponded to an evolution in the binders' 'fibril' microstructure under ESEM as a result of electron beam exposure, with the microstructure getting denser with PAV aging. The asphalt binders showed varied ESEM 'fingerprints' and aged in different ways. The ESEM properties generally showed to evolve with the physical properties, although this was not the case for all of the binders due to their unique aging characteristics.

Keywords: Asphalt binder · Aging
Environmental Scanning Electron Microscopy (ESEM) · Microstructure
Penetration · Softening point

L. D. Poulikakos et al. (Eds.): RILEM 252-CMB 2018, RILEM Bookseries 20, pp. 150–155, 2019.
https://doi.org/10.1007/978-3-030-00476-7_24

1 Introduction

A promising but not yet well understood technique for observing the nature of asphalt binder is Environmental Scanning Electron Microscopy (ESEM). There have been a few studies on asphalt binder with this technique that have confirmed the formation of a fibril microstructure (Rozeveld et al. 1997; Stangl et al. 2006). Earlier researchers assumed that the structures developed as a result of electron exposure was related to the heaviest molecules in the binder (i.e. asphaltenes) (Stangl et al. 2006). It has since been proposed that they correspond to a part of the maltenes (intermediate to light molecular weight) fraction and possibly a part of the asphaltenes (Gaskin 2013). The microstructure has been found to evolve with both binder aging (Mikhailenko et al. 2017), tensile forces (Rozeveld et al. 1997) and polymer modification (Mikhailenko et al. 2017). Furthermore, mixes of virgin and oxidized binders have been shown to produce hybrid fibril microstructures in ESEM, sharing the properties of both parent binders (Mikhailenko et al. 2017).

Despite some very interesting results, the relation of the ESEM findings to asphalt binder physical performance and aging needs to be further understood. This study is part of a larger inter-laboratory study of the RILEM Technical Committee 252 CMB (Hofko et al. 2017). The objective of the current paper is to analyze these same four straight run asphalt binders that having same penetration grade, and the evolution of their microstructures after short term (RTFOT) and subsequent long-term (PAV) laboratory aging. The ESEM analysis was performed by the University of Waterloo. This study enable the authors to: (i) understand the differences of plain binders from different sources under ESEM and (ii) compare the physical properties of the binder (penetration, softening point) and the findings of the ESEM analysis. Overall, this would allow for the validation of the ESEM findings and a better understanding of how they can be interpreted to achieve further insight into the nature of the material using a relatively large set of binders.

2 Materials and Experimental Methods

Four 70/100 penetration graded bituminous binders (EN 12591) from different crude sources were used and identified as B501, B502, B503 and B504. Table 1 provides the properties of the four binder samples according to European Standards (EN 1426, 1427) conducted at TU Wien and the US Performance Grade Specifications (AASHTO M 320-10) obtained at TU Braunschweig.

Table 1. Properties of asphalt binder samples tested at various laboratories

Sample	Penetration at 25 °C [1/10 mm]	Softening Point [°C]	PG
B501	77	46.2	70–22
B502	77	47.1	64–22
B503	79	47.5	64–22
B504	84	47.3	64–22

The entire set of binders was short- and long-term aged. Short-term ageing was performed according to the RTFOT method (EN 12607-1) In addition to the standard RTFOT temperature of 163 °C, two additional temperatures, 143 and 123 °C, were used to evaluate the effect of temperature on the ageing process. Long-term ageing was conducted using the PAV device (EN 14769). It was carried out after RTFOT at T = 100 °C and an air pressure of 2.1 MPa for 20 h. The aging on the set of binders used for the ESEM study was carried out at 123 °C and 163 °C at Empa, Switzerland that was one of the participating laboratories.

Softening point test was measured according to EN 1427, while penetration was tested based on EN 1426. Together with conventional tests, asphalt binders were prepared for ESEM observation by a protocol previously developed (Mikhailenko et al. 2017). The observations were conducted at room temperature immediately after being removed from the cooler with a FEI Quanta 250 FEG ESEM. The observation parameters were an acceleration voltage of 20 keV, a spot size of 3.5, a chamber pressure of 0.8 mbar in low vacuum mode, and a magnification of 1000x in secondary electron (SE) mode.

3 Results

3.1 Physical Properties

The physical properties results of all asphalt binders with the three different ageing regimes are a combined effort of the different participants in a RILEM Round Robin. Most participants analysed two different RTFOT ageing temperatures. The results for penetration are shown in Fig. 1 and softening point in Fig. 2. The diagrams show mean values and error bars indicating the standard deviation.

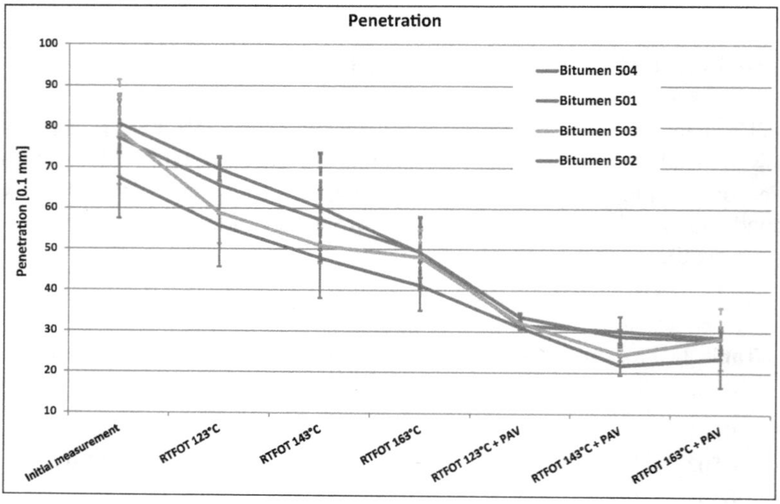

Fig. 1. Penetration values of different asphalt binders at different aging regiments

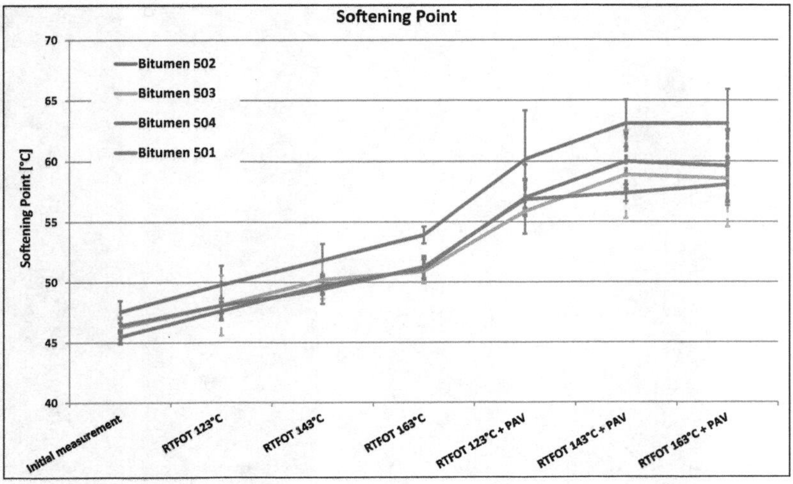

Fig. 2. Softening point values of different asphalt binders at different aging regiments

The influence of temperature is shown in a decrease of penetration and an increase of softening point. The change in penetration and softening point is almost linear with temperature increase in short term conditioning (RTFOT) while the influence of long term conditioning (RTFOT 123 °C/PAV) shows a sharper variation. Combining RTFOT at 143 °C and PAV results are almost similar with combining RTFOT at 163 °C and PAV. This might indicate a higher degree of chemical reaction, possibly with oxygen, that occurs from 143 °C. Asphalt binder 502 is more sensitive in penetration and softening point under laboratory ageing.

3.2 ESEM Microstructure

The irradiation of the binder samples by the ESEM electron beam over a period of time produces a fibril structure. The images of the resulting fibril structure for all of the virgin and aged binders are shown in Fig. 3. All of the unaged binders have a relatively sparse fibril structure. The RTFOT aging at 123 °C does not seem to induce much fibril evolution to the virgin binder, while the structure does appear to become denser with RTFOT aging at 163 °C. The structures for all of the binders evolve much more significantly with PAV aging, with the structure getting much denser, especially with binder 502. Binders 501 and 503 show a similar evolution with PAV, with the fibrils also getting smaller in diameter. The PAV evolution of binder 504 was unique in that the fibrils did not decrease in diameter significantly, while the structure became denser. The evolution time of the fibril structure with ESEM irradiation has been shown to correlate with binder aging (Mikhailenko et al. 2017) and stiffness (Hofko et al. 2017). It is calculated from observing video of the image evolution in the ESEM and taking the time for the fibril structure to stabilize, with the results presented in Fig. 4.

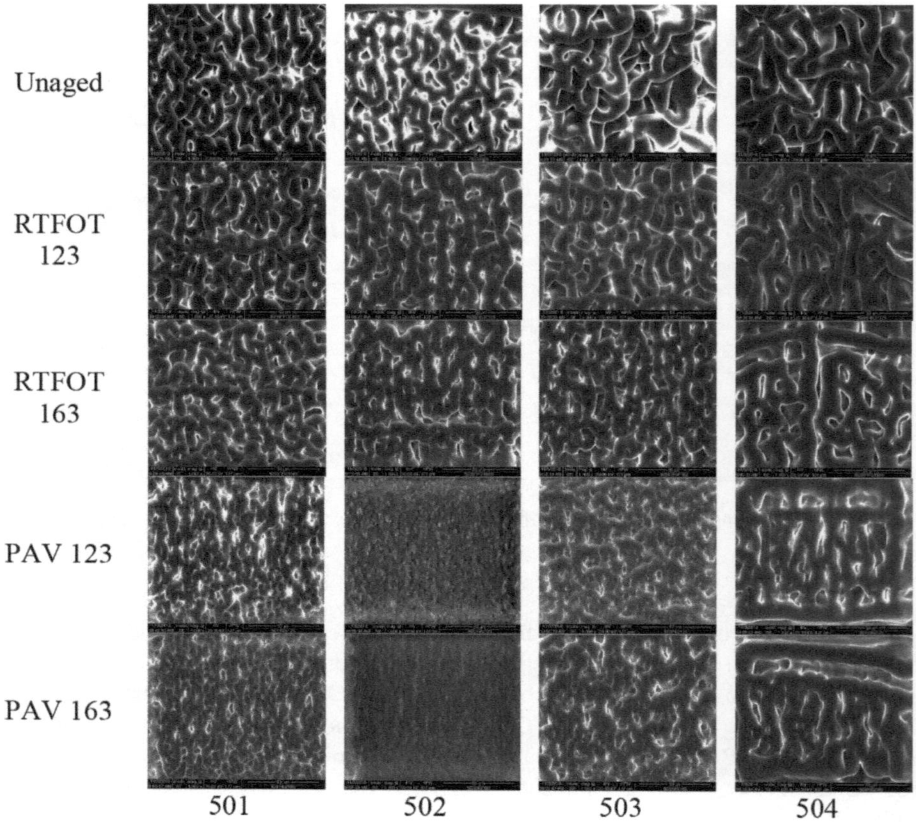

Fig. 3. ESEM images of asphalt binders under different aging conditions (scale bar 50 μm)

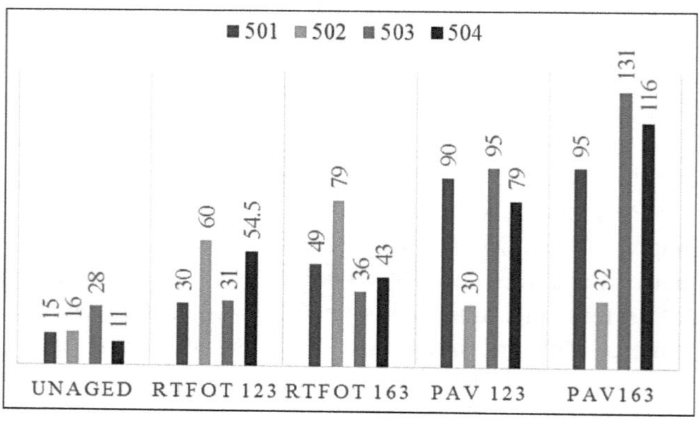

Fig. 4. ESEM fibril microstructure forming (irradiation) time (based on one video for each binder)

The formation time of the fibril structure is the shortest with the virgin binder (10–30 s), and increases a moderate amount with RTFOT aging (30–60 s) and then more significantly with the PAV (80–130 s). It also increased with the increase of RTFOT aging temperature. Binder 502 was an exception to this, as the evolution time increased after the RTFOT aging but was very short after PAV aging, indicating that this parameter cannot be applied in the same way to some binders.

4 Conclusions

The principal conclusions for the 'comparison of ESEM and physical properties in virgin and laboratory aged asphalt binders' study are as follows:

- The evolution of asphalt binder physical properties under aging correspond to an evolution in the ESEM microstructure of the binders, as stiffer binders correspond to a denser microstructure. PAV aging also results in smaller fibril diameters.
- RTFOT aging temperature does not have a significant effect on the microstructure evolution. The main microstructural evolution appears after PAV aging.
- Asphalt binders with similar physical properties can have different ESEM 'fingerprints', which will also evolve with aging in different ways.
- The parameters of 'fibril area' and 'formation time' in ESEM generally have a good correlation with the evolution of physical properties in the asphalt binder, although the 'formation time' is not applicable for some binders.

References

Rozeveld, S.J., Shin, E.E., Bhurke, A., et al.: Network morphology of straight and polymer modified asphalt cements. Microsc. Res. Tech. **38**, 529–543 (1997). https://doi.org/10.1002/(SICI)1097-0029(19970901)38:5%3c529:AID-JEMT11%3e3.0.CO;2-O

Stangl, K., Jager, A., Lackner, R.: Microstructure-based identification of bitumen performance. Road Mater. Pavement Des. J. **7**, 111–142 (2006)

Gaskin, J.: On bitumen microstructure and the effects of crack healing. Ph.D. Thesis, University of Nottingham (2013)

Mikhailenko, P., Kadhim, H., Baaj, H., Tighe, S.: Observation of asphalt binder microstructure with ESEM. J. Microsc. **267**, 347–355 (2017a). https://doi.org/10.1111/jmi.12574

Mikhailenko, P., Kou, C., Baaj, H., Tighe, S.: Observation of polymer modified asphalt microstructure by ESEM. In: 6th International Conference on Engineering Mechanics and Materials CSCE 2017, Vancouver, Canada (2017)

Mikhailenko, P., Kadhim, H., Baaj, H.: Observation of bitumen microstructure oxidation and blending with ESEM. Road Mater. Pavement Des. **18**, 216–225 (2017b). https://doi.org/10.1080/14680629.2017.1304251

Hofko, B., Falchetto, A.C., Grenfell, J., et al.: Effect of short-term ageing temperature on bitumen properties. Road Mater. Pavement Des. **18**, 108–117 (2017). https://doi.org/10.1080/14680629.2017.1304268

Investigation of the Asphalt Binder Sample Preparation Methods Based on AFM

Zhijun Wang[1(✉)], Rong Chang[1], Zhenyu Zhou[1], Yongchun Qin[1], and Gaochao Wang[2]

[1] Research Institute of Highway Ministry of Transport, Beijing, China
wzjzjy@126.com, {r.chang, zy.zhou, yc.qin}@rioh.cn
[2] ChangSha University of Science and Technology, Changsha, China
wang190808@126.com

Abstract. The preparation methods of asphalt binder samples have significant influences on the experimental results when the microscale analysis of asphalt binder was conducted by the Atomic Force Microscope (AFM) devices. In this paper, an unmodified asphalt binder was selected and then, the materials were prepared with three different preparation methods: the solvent method, the alcohol lamp method, and the oven method, respectively. Then, the phase analysis was chosen as the key parameter for comparison purpose; with following the concluded conclusion. Results indicate that the solvent method may partial damage the structure of asphalt binder and, potentially leads to the misestimation. Besides, the alcohol lamp method burned sample with a flame for the preparation; therefore, the fluctuated heating temperature will age the material is easily and the repeatable is not good. Consequently, the oven method was recommended due to it has the least effect on the materials' structure and chemical property, and further investigation will be conducted.

Keywords: Asphalt binder · Atomic force microscopy (AFM)
Sample preparation methods · Microscopic analysis · Evaluation parameter

1 Introduction

Asphalt binder is one of the most important components of asphalt pavement which is widely used for the pavement construction; hence, it will be important to understand the properties of asphalt binders. Recently, an emerged technology Atomic Force Microscope (AFM) was introduced to study the microstructure of asphalt binder. AFM technology can quantitatively analyses the microstructure and mechanical properties of asphalt binders. Hence, it became one of the dominate tools in the area of microstructure of asphalt binder (Wang 2015); However, there is not a general standard of the sample preparation method. Currently, several preparation methods are used to prepare the testing samples: solvents method, alcohol lamp method, and the oven method (Loeber et al. 1996; Pauli et al. 2013; Arifuzzaman 2010; Fan 2016). In this paper, three methods were used to prepare the asphalt binder samples, respectively; then the AFM results were used to quantitatively evaluate these methods and a reliable preparation method was recommended.

L. D. Poulikakos et al. (Eds.): RILEM 252-CMB 2018, RILEM Bookseries 20, pp. 156–161, 2019.
https://doi.org/10.1007/978-3-030-00476-7_25

2 Materials and Testing

The base asphalt binders 70#, which is according to the Chinese national standard 'Technical Specifications for Construction of Highway Asphalt Pavements (JTG F40 2004) was selected to use in this study. Then the samples were prepared with three different preparation methods, respectively; at least two replicate tests were performed for each sample preparation method.

(1) Solvent method: firstly, the asphalt binders were put into the oven at 135 °C for melting. Next, use a glass stick to mix the binder with organic solvents evenly; 1.5 g binder and different volume of organic solvents were selected, respectively. For example, 3 ml, 10 ml and 15 ml trichloroethylene. Then, condition the mixed samples at room temperature for around 20 min until the dissolution finished completely. Finally, drip three drops of mixed samples onto a cover glass and place them into an oven at 105 °C for 20 min until the solvent is evaporates. Then the samples can be taken out for further study (Allen et al. 2012).
(2) Alcohol lamp method: firstly, the asphalt binders were put into the oven at 135 °C for melting. Next, drip one drop of the melted binder onto a cover glass. Finally, heat the sample by an alcohol lamp with a vertical distance of 10 cm, and shake it to reduce the thickness. Then, it was prepared for further study (Loeber et al. 1996).
(3) Oven method: firstly, the asphalt binders were put into the oven with 135 °C until the materials were melt. Next, tinfoil was taped around the cover glass to form a rectangular sample area, then a few of uniformly mixed asphalt binder sample were dropped on one side, and the samples can be settled on the stand. Then, the samples were then put into the oven, with a settlement scope 25%, for condition 30 min at a constant temperature 130 °C. Finally, the samples can be taken out for further study.

The tapping mode was used for the AFM test, and the 3-dimensional topographic map and phase diagrams can be obtained. In the phase diagram, different graphs represent different substances states, it can be used for further comparison. The available device is Agilent 5500 AFM in this study, the probe elasticity coefficient is 46.0 N/m, while resonance frequency is 182 kHz, and the test temperature is 27.0 °C. Four to six random points were measured for each sample.

3 Results and Analysis

3.1 Solvent Method

As shown in Fig. 1, is the AFM testing results which based on the solvent method, the trichloroethylene was used as the solvent.

According to Fig. 1(a), (b) and (c), the bee-structure will be vanished when the

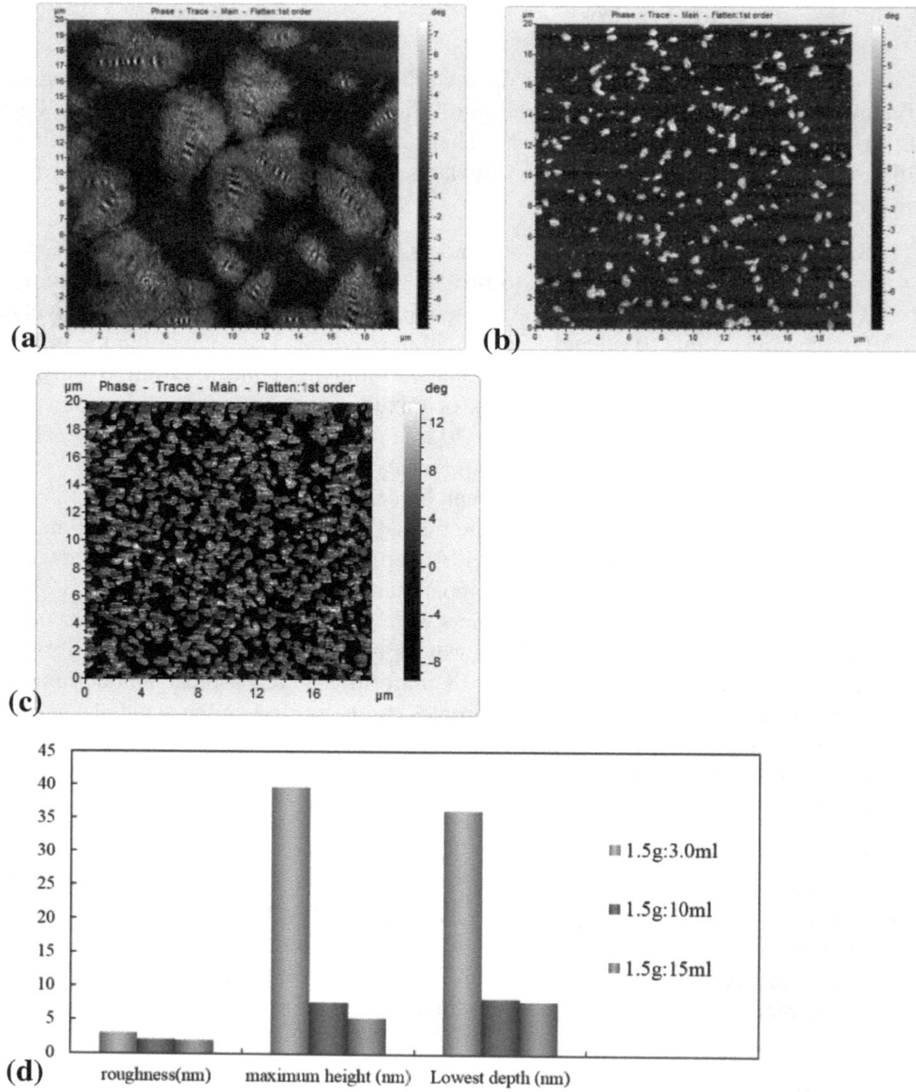

Fig. 1. AFM testing results based on solvent method: (a) 1.5 g: 3 ml phase diagram, (b) 1.5 g: 10 ml phase diagram, (c) 1.5 g: 15 ml phase diagram, (d) surface parameters

solvent content increases; and from the surface parameters, it can be seen that the convex and concave values is gradually reduced, it means that the solvent can damage the surface of asphalt binders. The same sample was then tested one week later; however, same result was observed which means that the damage is an irreversible procedure. Therefore, the solvent method is not recommended as the preparation method for AFM testing.

3.2 Alcohol Lamp Method

In Fig. 2, is the asphalt binder surface parameter which based on the alcohol lamp method.

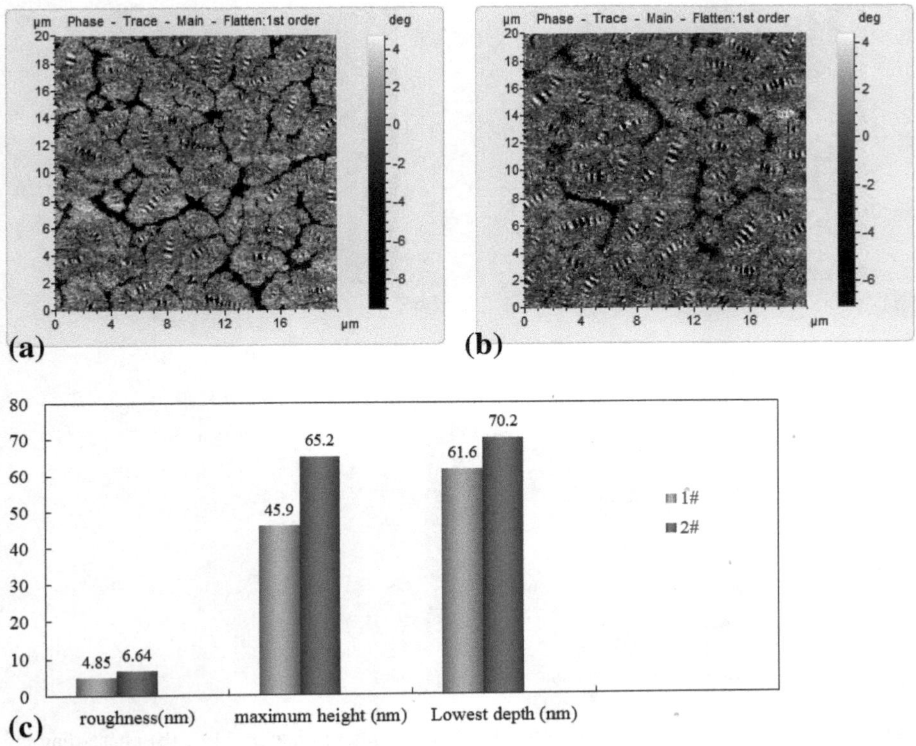

Fig. 2. AFM testing results based on alcohol lamp method: (a) phase diagram (1#), (b) phase diagram (2#), (c) surface parameters

As shown in Fig. 2, the roughness, maximum height, and lowest depth from sample 1# are very different with the one from sample 2#. Comparative analysis to its diversity, it results by the uncontrollable heating process. The asphalt binder samples can be aged to different aging conditions when heated by the alcohol lamp (Fischer et al. 2013). The paving asphalt quality, heating area, asphalt film thickness are not uniform in the process of sample preparation, and asphalt film thickness of asphalt microscopic molecular distribution has a certain influence (Pauli et al. 2011).Therefore, this method is also not recommended to use to prepare the AFM samples.

3.3 Oven Method

As shown in Fig. 3, is the AFM testing results which based on the oven method.

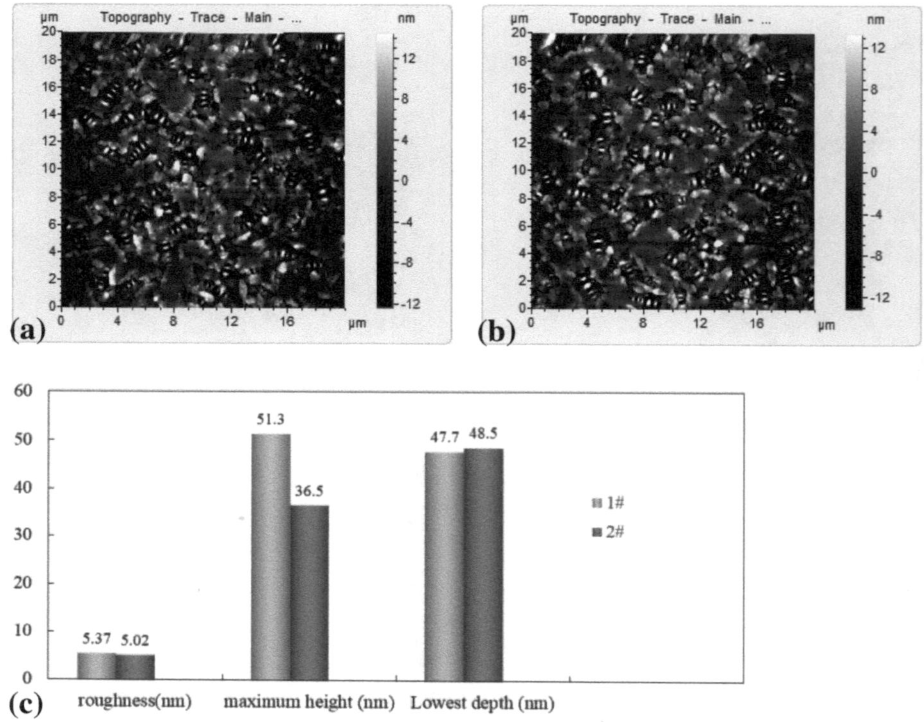

Fig. 3. AFM testing results based on oven method: (a) phase diagram (1#), (b) phase diagram (2#), (c) surface parameters

As shown in Fig. 3, very similar bee-structure can be observed in the phase diagrams in different binder samples. Further observation shown that the bee-structure is in small dimension and dense (Yang et al. 2015), which means less aging was conducted on the samples when the oven method was performed. Then, ten random points were selected to measure the surface parameter, a quite small variance value equals to 0.075 was obtained, it means that the overall roughness difference is small. Hence, it can be concluded that there is very good replication of oven method prepared sample, a constant heating temperature can be achieved, it leads more accurate results. Therefore, this preparation method is recommended for the AFM test.

4 Conclusion

(1) The solvent preparation method has a destructive effect on the molecular structure of asphalt binder. The higher the amount of solvent, the greater influence will be obtained. Moreover, the solvent contains toxic chemical components; hence, it is not appropriate to use.
(2) The temperature of alcohol lamp method is not easy to control, it will lead aged the asphalt binder. Besides, the sample thickness is not always uniform. Therefore, it is not recommended to use.
(3) The oven method has a constant temperature and good parallelism. The asphalt sample thickness is thin and uniform. When the samples were tested in the tapping mode, the qualitative analysis of the composition of the microscopic state of the asphalt and the quantitative analysis of the surface texture conditions shown a good result. Different heating temperature and heating time have a certain influence on the thickness and uniformity of the asphalt samples, and it is recommended to use a 130 °C oven for heating for 30 min.

References

Allen, R.G., Little, D.N., Bhasin, A.: Structural characterization of micromechanical properties in asphalt using atomic force microscopy. J. Mater. Civ. Eng. **24**(10), 1317–1327 (2012)
Arifuzzaman, M.: Nano-scale evaluation of moisture damage in asphalt. Doctoral dissertation, The University of New Mexico (2010)
Fan, Z.F.: Study on basic theory of asphalt micro-characteristics based on asphalt aging. Master dissertation, South China University of Technology (2016). (In Chinese)
Fischer, H., Poulikakos, L.D., Planche, J.P., Das, P., Grenfell, J.: Challenges while performing AFM on bitumen. Multi-scale Modeling and Characterization of Infrastructure Materials, pp. 89–98. Springer, Dordrecht (2013)
JTG F40: Technical Specifications for Construction of Highway Asphalt Pavements. Ministry of Communications of the People's Republic of China, Beijing (2004). (In Chinese)
Loeber, L., Sutton, O., Morel, J.V., Valleton, J.M., Muller, G.: New direct observations of asphalts and asphalt binders by scanning electron microscopy and atomic force microscopy. J. Microsc. **182**(1), 32–39 (1996)
Pauli, A.T., Grimes, R.W., Beemer, A.G., Turner, T.F., Branthaver, J.F.: Morphology of asphalts, asphalt fractions and model wax-doped asphalts studied by atomic force microscopy. Int. J. Pavement Eng. **12**(4), 291–309 (2011)
Pauli, T., Grimes, W., Cookman, A., Huang, S.C.: Adherence energy of asphalt thin films measured by force-displacement atomic force microscopy. J. Mater. Civ. Eng. **26**(12), 04014089 (2013)
Wang, Z.P.: Surface physical Chemistry, Tongji University Press (2015). (In Chinese)
Yang, J., Wang, X.T., Gong, M.H.: Study on microscopic image of asphalt atomic force microscopy. Acta Pet. Sin. **31**(5), 1110–1115 (2015)

Precision of Iatroscan Method for Assessment of SARA Compounds in Bitumen

Diana Simnofske(✉) and Konrad Mollenhauer

University of Kassel, Kassel, Germany
{d.simnofske,k.mollenhauer}@uni-kassel.de

Abstract. Bitumen consists of thousands of different chemical molecules which are difficult to distinguish. Iatroscan method using thin layer chromatography with flame ionization detection (TLC/FID) is a comparably simple test method to characterize bituminous binder with regard to the colloidal bitumen structure in saturates, resins and asphaltenes (SARA). In a round robin test, the general precision of the test procedure was confirmed. However, especially the reproducibility of the test procedure is comparably especially in terms of the proportion of aromatic compounds. Within a sensitivity study, important test parameters were identified which will be further controlled in new RRT studies in order to improve the test precision.

Keywords: Iatroscan · SARA-analysis · Round robin test · Bitumen

1 Introduction

The mechanical performance and durability properties of asphalt roads highly depend on the quality of the bituminous binder. The relevant binder properties (rheology, adhesion, cohesion and ageing characteristics) are defined by the chemical composition of the binder. Due to the natural origin of bitumen of source oils, the chemical structure is complex based on the fact that bitumen consists of thousands of different molecules. For better understanding the effect of chemical composition, it becomes preferable to divide bitumen in four chemical fractions which are saturates, aromatics, resins and asphaltenes (SARA). Generally asphaltenes can be defined as high aromatic ring structures with high content of heteroatoms. Resins are similar to asphaltene but with less polar aromatic structures and heteroatoms. Aromatics are naphtene aromatics molecules and saturates non-polar bitumen phase (Lesueur 2009). According the colloidal bitumen model, asphaltenes are considered as solid particles which agglomerate with resins to micelles dispersed in the aromatics and saturates. The four colloidal fractions are usually defined according their solubility in various solvents (Redelius 2006).

2 Motivation and Objectives

Based on the common colloidal bitumen structure, bitumen can be easily described by Iatroscan method (Horváthné and Lvey 2000) compared to alternative methods (e.g. column chromatography). There are lots of publications using Iatroscan method to

L. D. Poulikakos et al. (Eds.): RILEM 252-CMB 2018, RILEM Bookseries 20, pp. 162–167, 2019.
https://doi.org/10.1007/978-3-030-00476-7_26

investigate bitumen aging (Isacsson and Zeng 1997; Horváthné and Lvey 2000; Le Guern et al. 2010) or chemical differences between bitumen of different proveniences (León et al. 2000), quality control by Gaestel- Index (Horváthné and Lvey 2000; Hubner et al. 2009; Esnouf and Khoo 2012; Paliukaitė et al. 2014) or the compatibility between bitumen and polymers (Airey 2003).

However, publications report about a poor repeatability and reproducibility of this test procedure (Oliver and Liddle 2006; Holleran and Holleran 2010). Therefore, a round robin test was initiated in order to check the test precision of Iatroscan method. Furthermore, a sensitivity study was conducted in order to identify relevant test parameters which affect the test result and which have to be controlled carefully.

Following objectives should be reached:

- Analyse the effect of poor test precision in a parameter study
- Preparing a work construction based on the parameter study
- Initiating a round robin test (rrt) based on the work construction
- Evaluation the results of rrt with regard to repeatability and reproducibility and comparing with results of IP469

3 Materials and Methods

Iatroscan method is a comparably fast test procedure to identify the four fractions saturates, aromatic CHs, resins and asphaltenes in a bitumen sample. By Thin-Layer Chromatography (TLC), the bitumen fractions are separated by applying a bitumen solution onto a silica quartz rod (so-called chromarods) which is successively exposed to three solvent-saturated atmospheres. Within each so-called development, the compounds in the bitumen, which are soluble in the specific solvent will capillary spread on the chromarods. The separated bitumen fractions can afterwards be quantified by using Flame Ionization Detection (FID) device called "Iatroscan".

According to IP 469, the bitumen is solved in dichloromethane (DCM) or Toluene (Paliukaitė et al. 2014) for applicating the sample droplet onto the chromarods. For separation of the colloid fractions, the saturates are extracted from the sample spot by nHeptan. Within the second development step, aromatic binder compounds are extracted by a mix of nHeptan:Toluene (20:80). Resins are lastly separated by a mix of methanol: dichloromethane (5:95). The asphaltenes compounds will stay at the application spot on the chromarod.

Afterwards the chromarods are subjected to a moving hydrogen flame within the Iatroscan device. The burned residues from the individual spots from the chromarod are assessed by flame ionization detection (FID). The resulting ion count is plotted versus the length of the chromarod which results in a chromatogram, on which the four SARA fractions are identified by four peaks. By integration, the peak areas are calculated and used as a mean for the volumetric composition of the bitumen sample. A test set-up as well as chromatograms for the assessment of the SARA composition is shown in Fig. 1.

From 2015 to 2017 a round robin test (RRT) was organized by University of Kassel with participation of variable international laboratories (bitumen producers, research

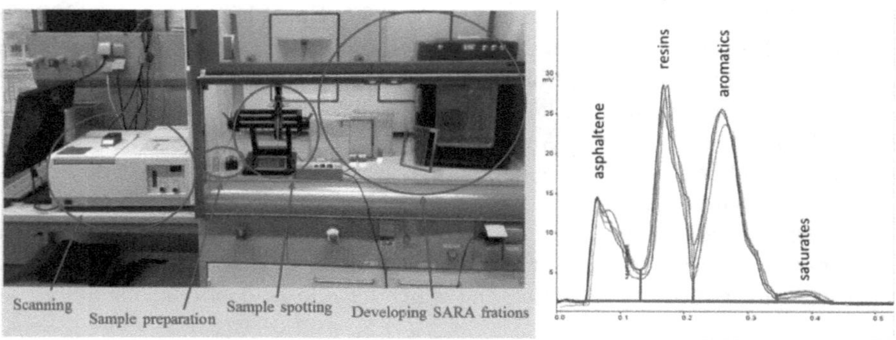

Fig. 1. Iatroscan test setup and example for chromatograms resulting after FID of chromarods

centers, universities). In total three rounds of the RRT were conducted, in which the test procedure after IP 469 were further specified and narrowed. Two bitumen samples were tested. Each laboratory conducted two complete measurements including the test of 5 single chromarods per sample and measurement. The statistical evaluation of the SARA fractions is based on FGSV paper no. 926 part 1–6 for determining averages, repeatability r and reproducibility R.

In parallel a sensitivity study was conducted with the device of University of Kassel in order to quantify the effect of selected test parameters (development solutions, sample application, FID parameters) to the test results.

4 Results

4.1 Results of Round Robin Test

The resulting test precision values reached within each of the three RRT rounds for the proportions of the four SARA fractions are given in Table 1 and plotted in Fig. 2. Further, the precision values as given in IP 469 are added both in the table and figure.

It can be observed that the repeatability r reached in the third RRT round is similar to the precision statement in IP 469 for asphaltene and aromatic content. Acceptable values could reach only for saturates and resins. With increasing number of test round the reproducibility R becomes better. However, despite very precise test specification, still the reproducibility especially for the content of aromatic fractions are very sensitive to test parameters.

Table 1. Precision values for repeatability r and reproducibility R for the proportions of SARA fractions (in percent by mass)

Repeatability/Reproducibility (r/R)[%]	Saturates	Aromatics	Resins	Asphaltenes	Laboratories
RRT round 1	1.3/2.9	6.5/26.8	3.7/11.4	4.7/15.0	2
RRT round 2	0.6/2.3	3.6/20.1	4.8/37.1	4.8/35.1	7
RRT round 3	0.3/1.1	6.8/11.3	3.7/7.5	4.5/9.1	5
IP 469	1.1/3.3	3.8/9.4	8.0/18.8	4.0/11.6	

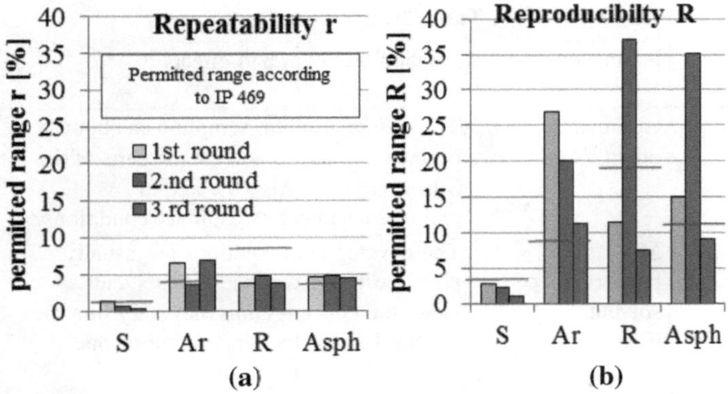

Fig. 2. Results of RRT for repeatability r (a) and reproducibility R (b) comparing to permitted range according to IP 469

4.2 Parameters Influencing Iatroscan Results

As especially during the 1st and 2nd round of the RRT low reproducibility results were obtained, a sensitivity study was conducted in order to identify relevant test parameters with a significant effect on the test results. Therefore, some test parameters were varied by purpose in order to quantify their effect onto the resulting SARA-composition. The results are summarized in Table 2.

Table 2. Precision values for repeatability r and reproducibility R according to IP 469

Test stage	Parameter	Applied variation and effects	Effect on test result
Sample preparation	Type of solvent	DCM/Toluene	High
	Concentration	0.1 g of bitumen on 5 ml or 10 ml solvent	High
	Age of solution before test	0 to 9 days	High
	Asphaltene separation	By the procedure to apply the bitumen sample by means of a solution onto the chromarods, non-soluble asphaltenes won't be applied to the chromarods and therefore are missing in the SARA compounds.	High
Sample application on chromarod	Slow/Fast injection	The injection speed affects the sample spot size on the chromarods. With slow application, a higher binder concentration is reached	High
	Spot volume	1 µl/2 µl	

(continued)

Table 2. (*continued*)

Test stage	Parameter	Applied variation and effects	Effect on test result
Development phase	Chromarod quality/Age	The syringe used for sampling can bounce off the chromarods and abrade parts of the Silica particles. Also the age of chromarods affect its capillary conditions.	High
	Mix proportions of solvent mixtures	The development solutions are usually prepared by mixing the two solvents. Here, the concentration may vary due to mixing fails or by evaporation of one solvent during the day.	High
	Humidity and air temperatur	The actual air humidity and temperature during measurement (room conditions) can usually not be controlled within specific ranges.	Low
	Air pressure	The air pressure affects the solvent steam concentration within the development tanks	Unknown
FID	Hydrogene rate	The hydrogene rate (140, 160 and 180 ml/min) affects the flame heat during burning of the chromarods.	Low
	Scan speed	The speed (30, 40, 50 s/chromarods) at which the hydrogene flame moves below the chromarods. A slower flame move will improve the ion detection.	High
Post processing	Integration procedure	Assessment of the peak areas by "valley to valley" or "baseline" integration	Low

5 Conclusions and Next Steps

Iatroscan procedure as described in IP 469 is a feasible test procedure for a comparably fast assessment of SARA fractions of a bitumen sample. The precision data given in IP 469 can generally be confirmed by the conducted RRT. However, several test parameters not fully defined within IP standard have a significant effect onto the test results. Especially in order to improve the reproducibility of the test, further narrowing of tests conditions are required for new RRT. As asphaltene fractions might not be identified correctly because of their reduced solubility during the sample preparation phase, the test of maltene phase should be considered. As the chromarods quality also affects the test results strongly, an assessment procedure to identify aged chromarods is required.

References

Paliukaitė, M., Vaitkus, A., Zofka, A. (eds.): Evaluation of bitumen fractional composition depending on the crude oil type and production technology. Paper presented at the 9th International Conference Environmental Engineering, Vilnius, Lithuania, 22–23 May 2014 (2014)

Esnouf, J., Khoo, K.Y.: Fourth national survey of Australian bitumens. In: Austroads Technical Report (2012). https://www.onlinepublications.austroads.com.au/items/AP-T209-12

Holleran, G., Holleran, I.: Bitumen chemistry using cheaper sources – an improved method of measurement by TLC-FID and the characterisation of bitumen by rheological and compositional means. Paper presented at the 24th ARRB Confernce – Building on 50 years of road and transport research, Melbourne, Australia (2010)

Le Guern, M., Chailleux, E., Farcas, F., Dreessen, S., Mabille, I.: Physico-chemical analysis of five hard bitumens: identification of chemical species and molecular organization before and after artificial aging. Fuel (2010). https://doi.org/10.1016/j.fuel.2010.04.035

Hubner, D., Oliver, J., Chin, C.: First National Survey of Australian Bitumens. In: Austroads Technical Report (2009). https://www.onlinepublications.austroads.com.au/items/AP-T124-09

Lesueur, D.: The colloidal structure of bitumen: consequences on the rheology and on the mechanisms of bitumen modification. Adv. Coll. Interface Sci. **145**, 42–82 (2009). https://doi.org/10.1016/j.cis.2008.08.011

Oliver, J., Liddle, G: Development of a procedure to chemically characterise bitumen using the iatroscan. In: Austroads Technical Report (2006). https://www.onlinepublications.austroads.com.au/items/AP-T52-06

Redelius, G.: The structure of asphaltenes in bitumen. Road Mater. Pavement Des. **7**, 143–162 (2006)

Airey, G.: Rheological properties of styrene butadiene styrene polymer modified road bitumens⋆. Fuel **82**, 1709–1719 (2003). https://doi.org/10.1016/S0016-2361(03)00146-7

Horváthné, E., Lvey, J.: Structure analysis of road-building bitumens: the effect of ageing for the structure. Paper presented at 2nd Eurasphalt & Eurobitume Congress, Barcelona, Spain (2000). Accessed 13 Dec 2015

León, O., Rogel, E., Espidel, J., Torres, G.: Asphaltenes: structural characterization, self-association, and stability behavior. Energy Fuels **14**, 6–10 (2000). https://doi.org/10.1021/ef9901037

Isacsson, U., Zeng, H.: Relationships between bitumen chemistry and low temperature behaviour of asphalt. Constr. Build. Mater. **11**, 83–91 (1997)

Visualization and Chemical Analysis of Bitumen Microstructures

Xiaohu Lu[1(✉)], Peter Sjövall[3], Hilde Soenen[2], Johan Blom[4], and Martin Andersson[5]

[1] Nynas AB, SE 149 82 Nynäshamn, Sweden
Xiaohu.lu@nynas.com
[2] RISE Research Institutes of Sweden, SE 501 15 Borås, Sweden
Hilde.soenen@nynas.com
[3] Nynas NV, 2000 Antwerp, Belgium
Peter.sjovall@ri.se
[4] Antwerp University, 2000 Antwerp, Belgium
Johan.blom@uantwerpen.be
[5] RISE, SE 114 86 Stockholm, Sweden
Martin.andersson@ri.se

Abstract. Microstructures of bitumen were investigated using atomic force microscopy (AFM) and environmental scanning electron microscopy (ESEM), and a chemical characterization was successfully carried out using time-of-flight secondary ion mass spectrometry (TOF-SIMS). The bee structures were observed by AFM, for which a chemical explanation by wax was confirmed by the TOF-SIMS analysis. A tube pattern or worm structures were generated and visualized by ESEM on bitumen surfaces. Chemical differences between the structured and unstructured areas, as well as between different areas of the structure, were observed. A mechanism for the structure formation on bitumen surface during ESEM analysis is suggested.

Keywords: Bitumen structure · AFM · ESEM · TOF-SIMS

1 Introduction

Various microstructural models for bitumen have been proposed over the years. Long time ago, a colloidal system was proposed, consisting of high molecular weight asphaltene micelles dispersed in a lower molecular weight maltene matrix. Later, during the Strategic Highway Research Program (SHRP), a microstructural model was proposed, according to which bitumen consists of a solvent moiety. The moiety is composed of relatively aliphatic and non-polar molecules that are low in heteroatoms and disperses microstructures (structural units formed from molecular associations) consisting of more polar, aromatic and asphaltene-like molecules. More recently, in the Yen-Mullins model, basic chemical and structural properties of asphaltenes are codified, and three distinct structures, namely asphaltene molecules, asphaltene nanoaggregates, and clusters of nanoaggregates, are proposed for petroleum oils. In addition to the models containing certain structures, a thermodynamic model was proposed for

L. D. Poulikakos et al. (Eds.): RILEM 252-CMB 2018, RILEM Bookseries 20, pp. 168–173, 2019.
https://doi.org/10.1007/978-3-030-00476-7_27

bitumen based on an assumption that different molecules of different size and polarity are kept in solution due to their mutual solubility. The structural models mentioned are summarized in Fig. 1 (references are omitted).

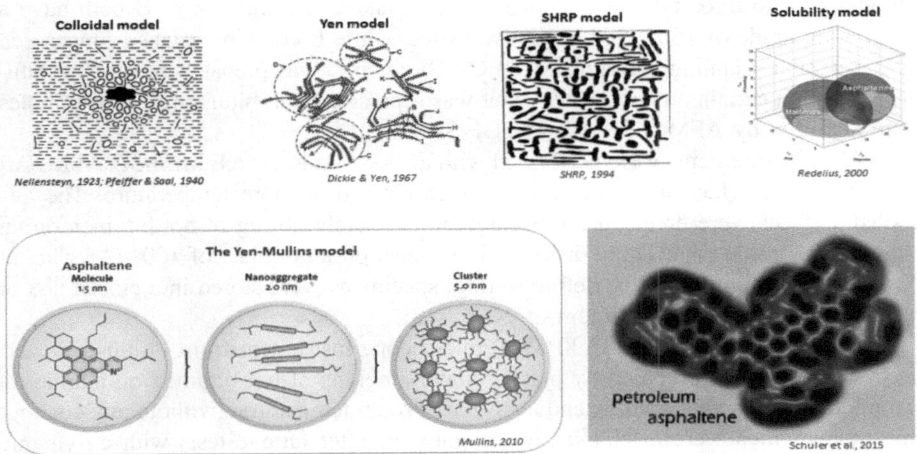

Fig. 1. Structure models proposed for petroleum bitumen

Natural wax, which is present in most bitumen types, is defined as the fraction that can crystallize at lower temperatures. On a molecular level, it is associated with regular, saturated molecules or molecular fragments. This crystallization, which is in fact a solid-liquid type of phase separation, is contributing to the bitumen microstructure. Moreover, one should realize that these waxy type of compounds, when large enough, may become part of the asphaltenes, as they become insoluble in n-heptane.

Experimentally, bitumen structures have been visualized using different microscopic techniques, such as environmental scanning electron microscopy (ESEM) and atomic force microscopy (AFM). By AFM and for certain bitumen, a kind of "bee" structure was observed, which in most cases are claimed to be attributed to wax crystallization (Hung and Fini 2015), and sometimes also explained by asphaltenes. Even less well understood are the tube patterns or worm structures visualized by ESEM. In spite of the great number of research activities, from a chemistry point of view, direct evidence has not been provided to what was claimed.

This paper investigates the microstructures of bitumen using AFM and ESEM. For the chemical characterization, time-of-flight secondary ion mass spectrometry (TOF-SIMS) was used. TOF-SIMS is a powerful surface analysis technique which provides chemical information about the outermost molecular layers of a sample surface.

2 Samples and Analysis Methods

A series of bituminous binders were selected, and air-cooled and fractured surfaces were analysed. In this paper, results are limited to the observations made on air-cooled surfaces, and to three bitumens, coded as A, B, and C. Bitumen A & B both have a penetration grade of 160/220, but A is wax-free, while B contains waxes as measured by differential scanning calorimetry (DSC). Bitumen C was prepared by mixing bitumen A with a certain amount of wax that was separated from bitumen B. The samples were analyzed by AFM, ESEM, and TOF-SIMS.

AFM measurements were performed with an Asylum Research MFP-3D AFM. All scans were performed using tapping mode in air and at room temperature. The air-cooled surfaces were prepared by placing approximately 20 mg of hot bitumen on an aluminium sample disk. These disks were reheated on a hot plate of 100 °C for about 30 s to create a thin flat film. Before testing, specimens were stored in a closed box at room temperature for a maximum of 24 h.

For the ESEM tests, an XL30 TMP microscope instrument from Philips/FEI was used. Samples were prepared by placing small amounts of hot bitumen on aluminum sample holders using a spatula, and analyzed at room temperature without any coating. Specimens which were mechanically elongated or after fatigue tests with a dynamic shear rheometer (DSR) were also analyzed. For the subsequent analysis by TOF-SIMS, samples were prepared by generating ESEM-induced structures and then immediately placed into separate plastic tubes in a styrofoam box with dry ice, to keep the sample temperature at about −80 °C.

TOF-SIMS analysis was carried out at about −80 °C in a TOF-SIMS IV instrument (ION-TOF GmbH, Germany) using 25 keV Bi_3^+ primary ions and low energy electron flooding charging compensation. Positive and negative data were acquired separately with the instrument optimized for high mass resolution or for high lateral resolution. Mass spectra were acquired from selected analysis areas up to 500 × 500 μm^2, and ion images of selected characteristic ions were created on different areas of the samples. For the mass spectra data, principal components analysis (PCA) was carried out. More details on the sample preparations, test procedures or analysis details can be found in (Lu et al. 2017) for TOF-SIMS, in (Lu et al. 2018) for ESEM and in (Blom et al. 2018) for AFM.

3 Structures Observed by AFM and TOF-SIMS

Figure 2 represents phase images from AFM scans, as well as TOF-SIMS scans taken from the air-cooled surfaces of the three bitumen samples. In addition, positive ion mass spectra, obtained from TOF-SIMS are also shown. The AFM scans clearly show so-called bee structures for bitumen B and C, but not for bitumen A.

TOF-SIMS was applied to air-cooled bitumen surfaces and also to the undiluted wax, separated from bitumen B. All major peaks of the positive ion spectra correspond to $C_xH_y^+$ fragment ions. The distribution of these ions, the ratio carbon to hydrogen and the total molecular mass, provide details on the molecular structures they are derived from, in particular as regards the aromatic nature. Clear differences were observed in

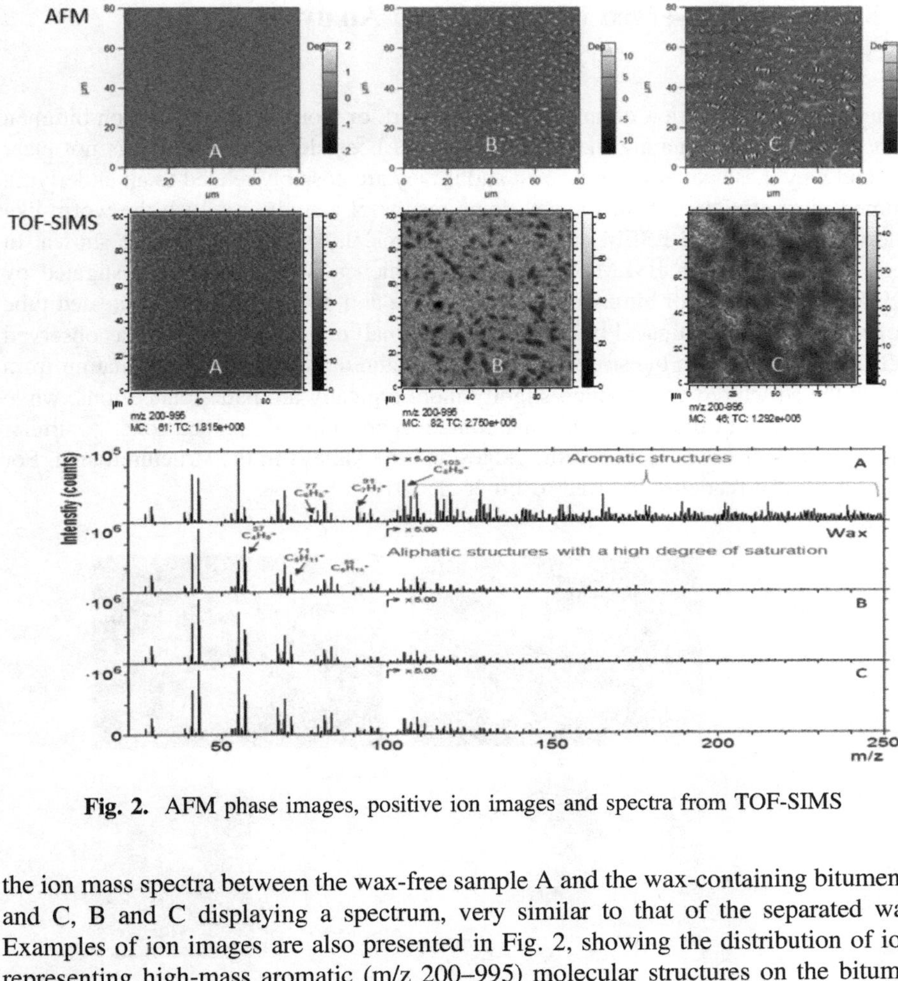

Fig. 2. AFM phase images, positive ion images and spectra from TOF-SIMS

the ion mass spectra between the wax-free sample A and the wax-containing bitumen B and C, B and C displaying a spectrum, very similar to that of the separated wax. Examples of ion images are also presented in Fig. 2, showing the distribution of ions representing high-mass aromatic (m/z 200–995) molecular structures on the bitumen surface. Bitumen A displays a fully homogeneous distribution of all ions on its surface. For bitumen B phase separations on the surface are observed, resulting in elongated structures with a size range 1–8 μm, these are consistent with the bee structures revealed by AFM. This separation was most clear in the image based on the high-mass aromatic ions, showing a reduced signal inside the bee structures, indicating a lower concentration of these ions from these places, whereas the saturated aliphatic ions displayed weakly increased signals inside the bee structures. Sample C also produced inhomogeneous and complementary distributions of high-mass aromatic and saturated aliphatic ions, but the distributions were more diffuse and without the bee-like pattern.

4 Structures Observed by ESEM and Analyzed by TOF-SIMS

In literature, the formation of an entangled network or worm-like structure on bitumen surfaces, when investigating these by ESEM has been described. But, it is not clear how and why these patterns are formed and if they are possibly related to an underlying bitumen microstructure. In this paper, some results of a study in which the worm-like patterns induced in the ESEM are evaluated chemically by TOF-SIMS are shown. In Fig. 3, examples of an ESEM-tested sample, which is subsequently investigated by TOF-SIMS is shown for bitumen A. Places inside and outside the ESEM created tube like pattern were investigated by TOF-SIMS. A small molecular contrast was observed by TOF-SIMS between the structured and unstructured areas. Signals originating from outside the structured area contain slightly more typically aromatic-related ions, while signals from within the structured area contain more aliphatic-related ions. A similar difference was observed between the ridges and the valleys in the structured area. For more details the reader is referred to Lu et al. 2018.

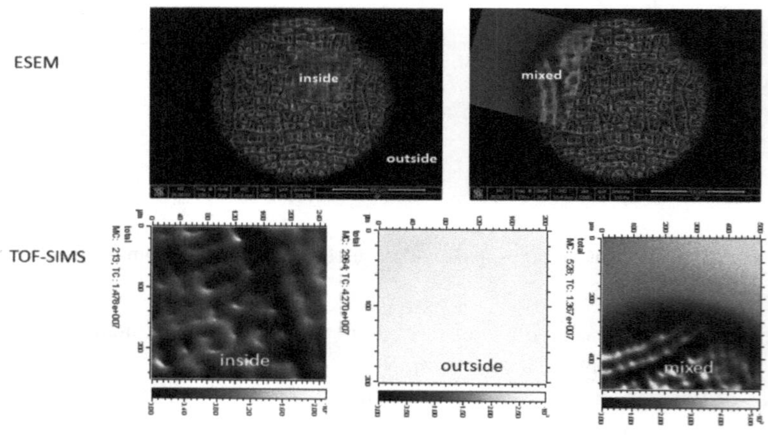

Fig. 3. Total ion TOF-SIMS images overlayed on ESEM image and total positive ion images acquired in the different areas of Bitumen A

Phenomena that may help to explain the tube-pattern formation in ESEM are further discussed. The temperature on the bitumen surface during ESEM analysis was estimated to be significantly higher than the initial boiling point of the bitumen, causing evaporation of volatiles, mass loss, surface hardening and local expansion. These changes can generate mechanical instabilities which result in a micro topography, which is further enhanced by the edge effect under ESEM (Stokes 2008). The edge effect is an enhanced emission of electrons from edges and peaks within the specimen, so that these areas will appear brighter. Furthermore, it is well known that an electron-beam irradiation can induce certain chemical reactions, including breaking or rearrangement of chemical bonds, chain scission and crosslinking. The chemical contrast observed by TOF-SIMS between structured and unstructured areas is reflecting these effects, the consequences of evaporation of volatile components, and also possible chemical changes induced by the electron beam. This study suggests that the structures observed by ESEM are unlikely present in the bitumen before irradiation.

5 Conclusions

Structures of bitumen were visualized using AFM and ESEM, and chemically analyzed using TOF-SIMS. The bee structures observed by AFM and the explanation by wax are confirmed by the TOF-SIMS analysis. For a wax-free bitumen sample, the surface is characterized by a homogeneous distribution without chemical variations or phase structures. Under ESEM, the tube or worm structures observed are speculated to result from edge effects in ESEM, and these effects probably start with a local heating of the bitumen surface by the electron-beam irradiation leading to the evaporation of volatiles, mass loss, surface hardening and local expansion. Under the electron beam exposure, certain chemical reactions may also take place. The chemical contrasts observed by TOF-SIMS possibly reflect differences between the structured and unstructured areas, as well as between the different areas of the structure. As the structures observed by ESEM are unlikely present in the bitumen, one should be cautious with any further speculation on their relation to bitumen bulk properties and performance.

References

Blom, J., Soenen, H., Katsiki, A., Van den Brande, N., Rahier, H., van den Bergh, W.: Investigation of the bulk and surface microstructure of bitumen by atomic force microscopy. Constr. Build. Mater. **177**, 158–169 (2018)

Hung, A.M., Fini, E.H.: AFM study of asphalt binder "bee" structures: origin, mechanical fracture, topological evolution, and experimental artifacts. RSC Adv. **5**, 96972–96982 (2015)

Lu, X., Sjövall, P., Soenen, H.: Structural and chemical analysis of bitumen using time-of-flight secondary ion mass spectrometry (TOF-SIMS). Fuel **199**, 206–218 (2017)

Lu, X., Sjövall, P., Soenen, H., Andersson, M.: Microstructures of bitumen observed by environmental scanning electron microscopy (ESEM) and chemical analysis using time-of-flight secondary ion mass spectrometry (TOF-SIMS). Fuel **229**, 198–208 (2018)

Stokes, D.J.: Principles and Practice of Variable Pressure/Environmental Scanning Electron Microscopy (VP-ESEM). Wiley, Chichester (2008)

Chemo-Mechanical Characterization of Bituminous Materials: Recycling and Rejuvenation

A New Green Rejuvenator: Evaluation of Structural Changes of Aged and Recycled Bitumens by Means of Rheology and NMR

Cesare Oliviero Rossi[1(✉)], Paolino Caputo[1], Valeria Loise[1], Saltanat Ashimova[1,2], Bagdat Teltayev[2], and Cesare Sangiorgi[3]

[1] Department of Chemistry and Chemical Technologies, University of Calabria, 87036 Arcavacata di Rende, CS, Italy
{cesare.cliviero,polino.caputo, valeria.loise}@unical.it, salta_32@mail.ru

[2] Kazakhstan Highway Research Institute, Nurpeisova Str., 2A, Almaty 050061, Kazakhstan

[3] DICAM-Roads, Department of Civil, Chemical, Environmental and Materials Engineering, University of Bologna, V.le Risorgimento 2, 40136 Bologna, Italy
cesare.sangiorgi4@unibo.it

Abstract. The functionality of a green additive, acting as bitumen rejuvenator was considered in the presented experimental work. The additive's effects on aged bitumen have been investigated through advanced rheological and NMR-relaxometry measurements. Bitumen ageing encompasses volatilization and oxidation which enable changes in the material molecular structure. Volatilization occurs mainly at high temperatures during production, transport and laying of the asphalt concrete. The oxidation, also caused by atmospheric oxygen and UV radiation, leads to an increased fragility and development of cracks in the asphalt layer. Fresh, aged, and doped recycled bitumens were tested. Rheology and NMR have been used to assess the structural differences between the bitumens and to understand the role of the proposed additive. A real rejuvenator helps to rearrange the colloidal structure of the oxidized bitumen, thus recreating one similar to the fresh bitumen. As a novel approach to bitumen characterisation, an inverse Laplace transform of the NMR spin-echo decay (T2) was here applied.

Keywords: Bitumen · Rejuvenator · Nuclear Magnetic Resonance Rheology

1 Introduction

Bitumen's organic complex are easily oxidized during paving and pavement service life, especially under thermal and/or ultraviolet radiation (UV) conditions (Hu et al. 2018). In general, an aged bitumen has higher reprocessing temperature because some of the aromatic components and resins, which are responsible for a certain grade of mobility, are oxidized to asphaltenes and reduced to saturates. Hence asphaltene micelles become larger so that the fluidity of the system is reduced. Compared with virgin bitumen, the aged bitumen is more brittle and has worse relaxation characteristics that make it

L. D. Poulikakos et al. (Eds.): RILEM 252-CMB 2018, RILEM Bookseries 20, pp. 177–182, 2019.
https://doi.org/10.1007/978-3-030-00476-7_28

vulnerable to cracking (Baldino et al. 2012). Once removed and processed, bituminous layers become Reclaimed Asphalt Pavement (RAP), which contains valuable asphalt binder and aggregates (Baldino et al. 2017). Over the past decades, researchers have conducted many investigations on the use of RAP materials in the production of recycled asphalt. As a result, rejuvenators are a solution that can be utilized to restore RAP binder properties towards its original state (Dinis-Almeida et al. 2016; Zaumanis et al. 2014). Today, rejuvenators play a crucial role in bitumen recycling methods aiming to an optimized performance of the reclaimed bitumen. Nevertheless, the rejuvenator affects are still not well understood and especially its impact on bitumen's supramolecular structure arrangement has not been fully investigated.

This research describes the physical-chemical characteristics of a new green rejuvenator, using the potentiality of Nuclear Magnetic Resonance (NMR) techniques to identify the main effect of chemicals on the regeneration process of the aged bitumen. The vegetable oils are a common flux for bitumen and many times, the flux of oil action has been mistakenly considered as regenerating operation. This confusion arises from the fact that oils simply soften the hard bitumen to match the macroscopic mechanical parameters according to specific requirements.

2 Experimental Work

2.1 Chemicals, Materials and Sample Preparation

A 100/130 pen virgin bitumen was sourced from Kazakhstan and supplied by Highway Research Institute (Almaty, Kazakhstan). The Vegetable Flux Oil (VO) and the green rejuvenator (HR) were provided by KimiCal s.r.l. (Rende, Italy).

The transformed bitumen was prepared with a high shear mixing homogenizer (IKA model, USA). Firstly, bitumen was heated up to 150 ± 5 °C until it fully flowed, then a given part of HR or VO (2% of the weight) was added to the melted bitumen under a high-speed shear of 400 to 600 rpm/min. Subsequently, the mixture was kept under mechanical stirring at 150 °C for 10 min in a closed beaker to avoid any oxidation process. After mixing, the resulting bitumen was poured into a sealed container and stored in a dark chamber at 25 °C to retain the obtained morphology. The in-service aging of the base bitumen was simulated with the Pressure Aging Vessel (PAV) according to the AASHTO/ASTM T179 standard.

All the prepared mixtures are listed and labelled in the paper as follows: Virgin bitumen: Sample A, PAV bitumen: Sample B, PAV bitumen + 2 wt% VO: Sample C and PAV bitumen + 2 wt% HR: Sample D.

2.2 Rheology, NMR Tests and Inverse Laplace Transform (ILT)

The rheological behavior at different temperatures was investigated by a Dynamic Shear Rheometer (DSR) time cure test at 1 Hz with a ramp rate of 1 °C/min (from 25 °C to 120 °C) within the linear viscoelastic range of the binders.

Relaxation experiments were performed by means of a purposely-built NMR equipment that operates at a proton frequency of 15 MHz. Those experiments were done at temperatures lower than 15 °C which correspond to the respective temperature transition from viscoelastic to liquid (the temperature is chosen in order to standardize the structure of all samples). Inhomogeneity of field and surface effects usually causes the T_2 relaxation times to vary in the sample (Oliviero Rossi et al. 2015). The T_2 parameter is the spin-spin relaxation time as the relaxation relates to the exchange of energy only among spins and not with the surrounding environment. Hence, if inside the sample a continuous distribution of relaxation time exists, the amplitude A_n of the nth echo in the echo train is given by:

$$A_n = A_0 \int_0^{\infty} P(T_2) e^{-2n\tau/T_2} dT_2 \tag{1}$$

where A_0 is a constant, τ is the half echo time and $P(T_2)$ is the ILT of the unknown function that fits the echo amplitude curve. Furthermore, $P(T_2)$ can be agreed upon as a distribution of rate (inverse of time) constant. $P(T_2)$ can be related to probability density function (PDF) accounting different macro-structures that compose the bitumen binder (Oliviero Rossi et al. 2015). In this work, ILT computation was performed by means of UpenWin (Bortolotti et al. 2009).

3 Results and Discussion

Rheology temperature-sweep tests were performed to collect information on the structural changes induced by temperature, trying to define a transition temperature range better than usual empirical tests (i.e. Ring and Ball) (Baldino et al. 2013). The elastic modulus (G′) is continuously monitored during a temperature ramp at a constant heating rate (1 °C/min) and at a frequency of 1 Hz (Fig. 1). The transition temperature is evidenced when G′ is plotted as a function of temperature. The initial trend is almost linear with temperature, when the material mainly behaves like a viscoelastic system. The subsequent decrease occurs in correspondence of the transition towards a liquid-like behavior (G′ plot disappears).

The aged bitumen (sample B) shows much higher transition temperatures than the unaged fresh material (Romera et al. 2006). This effect is due to an increased fraction of asphaltenes resulted from oxidation processes of the soft unsaturated organic part. The higher asphaltene fraction causes a hardening of the bitumen and higher inner connectivity, if compared to the less dense and weaker network of the virgin bitumen where the asphaltene domains are less connected. Both additives shift at lower temperatures the transitions from the viscoelastic to the liquid material. HR which is a surfactant, shows a stronger effect. Authors believe that its presence might reduce the associative interactions amid the asphaltene particles by interposition between them and the maltenes. As a result, the colloidal network can be weakened, which in turn may correspond to a reduction of the transition temperature.

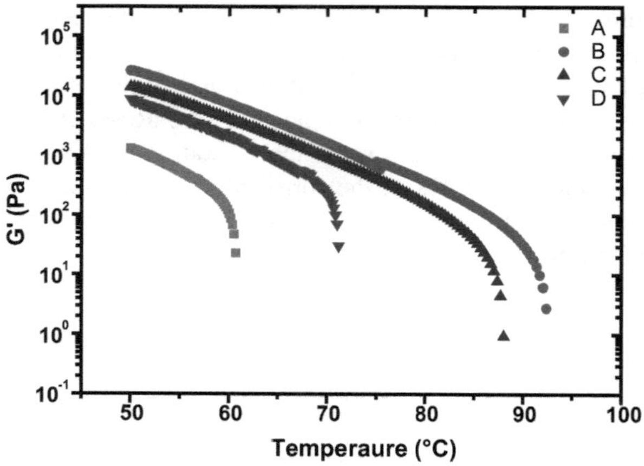

Fig. 1. Semi-log plot of high temperature ramp test for the A, B, C and D samples

3.1 NMR Study

The ILT analysis of the NMR echo signal decay was used to obtain the T2 relaxation time distributions. This technique allows finding the PDF distribution which associates to relaxation times that correspond to unrelated molecular aggregates inside the samples (Gentile et al. 2012). Results are presented in Fig. 2, where the time relaxation distributions PDFs are plotted as a function of the relaxation time. The T2 relaxation time distribution shows two peaks. The shorter T2 times correspond to more rigid supra-molecular aggregates, ascribed to asphaltenes, while longer T2 times are attributed to maltene fractions. For the virgin bitumen, one peak falls around 10 ms and it is due to asphaltene fraction; while the one centered at around 100 ms refers to maltenes.

The ILT of the aged bitumen again exhibits two peaks shifted towards shorter times and present very characteristic shapes. This more likely indicates a gradually increase of the material rigidity with the oxidation process. In particular, the asphaltene peaks are now closer to 1 ms for the aged bitumen. The hard consistency of the samples is strongly affected by the aging processes. During the oxidative aging, the concentration of polar functional groups becomes sufficiently high to immobilize an excessive number of molecules through intermolecular association. What is more, the molecules or molecular agglomerates lose sufficient mobility to flow past one another under thermal or mechanical stresses. The resulting embrittlement of the asphalt makes it susceptible to fracturing or cracking and resistant to healing. This presence of two peaks also supports the colloidal model of the bitumen. All experiments are performed at temperatures lower than 15 °C which is the respective temperature transition from solid to liquid (the temperature is chosen in order to standardize the structure of all samples). On the other hand, it is evident that the addition of VO and HR to the aged bitumen results in the asphaltene peaks shift to longer T2 times. The HR sample shows time distributions similar to the virgin bitumen, although VO simply shifts the distribution to longer times evidencing only its softening action.

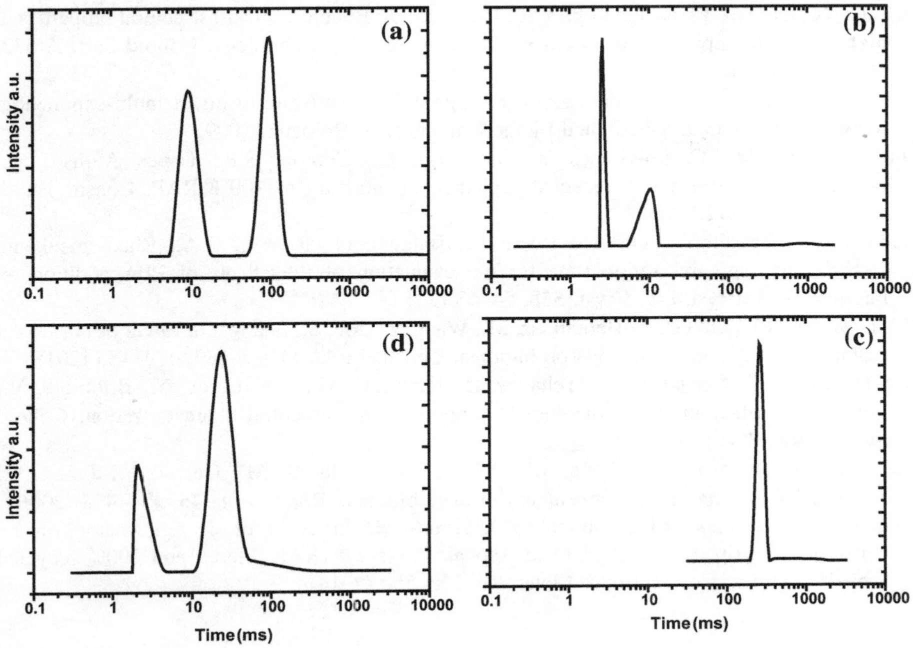

Fig. 2. ILT relaxation time distributions of bitumen samples at 15 °C lower than transition temperature (solid-liquid) determined by dynamic temperature ramp test experiment for each sample

4 Conclusions

This work shows the effectiveness of the HR additive in restoring both bitumen rheological (DSR) and physical (NMR) properties. This green additive tends to restore the mechanical properties of the oxidized bitumen. Moreover, the article aims to demonstrate the importance of testing the regenerated bitumen using structural techniques in order to distinguish between fluxed bitumen and real regenerated compound. Thus, bituminous systems can have alike macroscopic (ring and ball) or rheological properties, but unique supra-molecular structure. The bitumen with flux can be mistakenly considered as a real regenerated one according to the ring and ball test or to other simple rheological investigations.

References

Baldino, N., Gabriele, D., Lupi, F.R., Oliviero Rossi, C., Caputo, P., Falvo, T.: Rheological effects on bitumen of polyphosphoric acid (PPA) addition. Constr. Build. Mater. **40**, 397–404 (2013)

Baldino, N., Gabriele, D., Oliviero Rossi, C., Seta, L., Lupi, F.R., Caputo, P.: Low temperature rheology of polyphosphoric acid (PPA) added bitumen. Constr. Build. Mater. **36**, 592–596 (2012)

Baldino, N., Oliviero Rossi, C., Lupi, F.R., Gabriele, D.: Rheological and structural properties at high and low temperature of bitumen for warm recycling technology. Colloid Surf. A **532**, 592–600 (2017)

Bortolotti, V., Brown, R.J.S., Fantazzini, P.: UpenWin: a software to invert multi-exponential relaxation decay data. Distributed by the University of Bologna (2009)

Dinis-Almeida, M., Castro-Gomes, J., Sangiorgi, C., Zoorob, S.E., Lopes Afonso, M.: Performance of warm mix recycled asphalt containing up to 100% RAP. Constr. Build. Mater. **112**, 1–6 (2016)

Gentile, L., Filippelli, L., Oliviero Rossi, C., Baldino, N., Ranieri, G.A.: Rheological and H-NMR spin-spin relaxation time for the evaluation of the effects of PPA addition on bitumen. Mol. Cryst. Liq. Cryst. **558**, 54–63 (2012)

Hu, J., Wu, S., Liu, Q., García Hernández, M., Wang, Z., Nie, S., Zhang, G.: Effect of ultraviolet radiation in different wavebands on bitumen. Constr. Build. Mater. **159**, 479–485 (2018)

Oliviero Rossi, C., Spadafora, A., Teltayev, B., Izmailova, G., Amerbayev, Y., Bortolotti, V.: Polymer modified bitumen: rheological properties and structural characterization. Colloid Surf. A **480**, 390–397 (2015)

Romera, R., Santamarìa, A., Peña, J.J., Muñoz, M.E., Barral, M., Garcia, E., Jañez, V.: Rheological aspects of the rejuvenation of aged bitumen. Rheol. Acta **45**, 474–478 (2006)

Zaumanis, M., Mallick, R.B., Poulikakos, L., Frank, R.: Influence of six rejuvenators on the performance properties of Reclaimed Asphalt Pavement (RAP) binder and 100% recycled asphalt mixtures. Constr. Build. Mater. **71**, 538–550 (2014)

A Rheological Study on Rejuvenated Binder Containing Very High Content of Aged Bitumen

Marco Pasetto(✉), Andrea Baliello, Giovanni Giacomello, and Emiliano Pasquini

Department of Civil, Environmental and Architectural Engineering (ICEA), University of Padua, Via Marzolo 9, 35131 Padua, Italy
{marco.pasetto, andrea.baliello, giovanni.giacomello, emiliano.pasquini}@unipd.it

Abstract. Hot recycling of reclaimed asphalt pavement (RAP) coming from flexible pavement rehabilitation is a technique able to ensure real economic and environmental benefits related to the reduction of virgin bitumen and aggregate supply and to the reuse of a recycled aggregate. Otherwise, adequate performance of recycled materials with high amount of RAP must be guaranteed since the recycling involves the presence of old and stiffened aged binders within the mixture. In this perspective, the present experimental study was aimed at verifying in the laboratory the performance at mid and high-service temperatures of bituminous blends composed by 40% of virgin binder and 60% of old rejuvenated bitumen (simulating aged bitumen coming from RAP). Rheological properties of materials were studied through the dynamic shear rheometer, testing unaged, short-term aged and long-term aged samples. Viscosity, stiffness and permanent deformation resistance of recycled blends seemed to guarantee comparable behavior with those of original bitumens, regardless the aging condition of the materials. These findings seemed to demonstrate the effectiveness of the hot recycling procedure using rejuvenators to obtain suitable bituminous binder containing high amount of aged bitumen.

Keywords: Hot recycling · Reclaimed asphalt · Rejuvenator · DSR

1 Introduction and Problem Statement

Hot recycling of reclaimed asphalt pavement (RAP) in renewed bituminous mixtures is a technique dated back 30 years so that, nowadays, is a standard procedure adopted in many countries (Hugener and Kawakami 2017). The most significant reasons are connected to the well-documented economic and environmental savings due to the reduction of virgin bitumen and aggregate supply and the prevention of waste transportation and disposal (Newcomb et al. 2007). Actually, huge research efforts are spent worldwide for the increasing of RAP amount within mixes, as well as paving industry and agencies are strongly applying political pressure towards this direction (McDaniel et al. 2012). Regardless recycling rate, appropriate bitumen grades for adequate performance of pavements must be guaranteed to meet technical prescriptions (Izaks et al. 2015). This fact

L. D. Poulikakos et al. (Eds.): RILEM 252-CMB 2018, RILEM Bookseries 20, pp. 183–188, 2019.
https://doi.org/10.1007/978-3-030-00476-7_29

must be considered in view of the re-use of RAP, which involves the inclusion in mixes of relevant quantity of aged bitumen at the end of its life characterized by significantly different mechanical and chemical properties (Chen et al. 2007). For these reasons, rejuvenators are recommended to face lacks in chemical composition and rheology of oxidized bitumen coming from RAP and restore their initial properties (Pradyumna and Jain 2016). In this sense, even if some literature exists about the analysis of the rheological proprieties of blends composed by virgin bitumen and aged binders coming from RAP (Shen and Ohne 2002; Chen et al. 2007; Riccardi et al. 2017), further studies are still needed, in particular regarding the aging of rejuvenated bitumens (Grilli et al. 2017).

Given this introduction, the present paper illustrates an experimental study aimed at verifying the feasibility of recycling high amount of RAP (up to 60%), focusing on rheological properties of binders. To accomplish this objective, a comparative analysis of virgin bitumens with respect to blends constituted by virgin bitumen, aged bitumen (simulating bitumen coming from RAP) and rejuvenator was carried out.

2 Materials and Methods

2.1 Materials

In order to replicate all processes involved in hot recycling, a virgin bitumen (V) was short-term aged (V_S) and long-term aged (V_L) in the laboratory simulating the RAP bitumen. A commercial rejuvenator (R) was blended with the long-term aged binder V_L to restore the original properties of virgin bitumen. The selected rejuvenator is a chemical additive at liquid state formed by a package of specific compounds. Based on the producer's specifications, three dosages (3, 6 and 9% by bitumen weight) were used in this study to identify the optimum R content. Indeed, such dosages are included in the range suggested by the producer (0.15÷0.45% by the RAP weight). Then, a blend (B) composed by the virgin bitumen V, the long term-aged bitumen V_L and the rejuvenator R (optimally dosed) was prepared to replicate the final binder obtained during hot recycling of RAP. Short and long-term aging were finally executed also on blend B to assess the behavior of the recycled binder in different service life conditions.

Tables 1 reports the basic properties of the tested binders: preliminary results indicated an optimum R content close to 6% (with respect to the bitumen weight), since consistencies of rejuvenated binder V_LR_6 and original (unaged) one (V) were very close. In this regard, B blend was manufactured introducing the rejuvenator at 6% by weight with respect to the V_L binder.

2.2 Methods

Blending of the different components were performed though a mechanical stirring instrument, heating the bitumens at 160 °C and adding the cool liquid rejuvenator. Simulating 60% RAP in the recycled mixture, a blend of 40% V and 60% V_L was prepared (the exact recycling rate will depend on the real amount of bitumen within the RAP). Short-term aging simulation was achieved in the laboratory through the well-known Rolling Thin Film Oven (RTFO) procedure (EN 12607-1). Then, long-term

Table 1. Summary of material codifications and basic properties

Code	Material description	Penetration at 25 °C	Softening point
V	virgin bitumen	53 mm · 0.1	49.0 °C
V_S	short-term aged virgin bitumen	32 mm · 0.1	53.7 °C
V_L	long-term aged virgin bitumen	15 mm · 0.1	68.3 °C
R	rejuvenator	–	–
V_LR_3	V_L + 3% rejuvenator by bitumen weight	33 mm · 0.1	53.7 °C
V_LR_6	V_L + 6% rejuvenator by bitumen weight	56 mm · 0.1	47.8 °C
V_LR_9	V_L + 9% rejuvenator by bitumen weight	84 mm · 0.1	40.2 °C
B	blend V + V_L + R (at 6% to V_L weight)	49 mm · 0.1	46.8 °C
B_S	short-term aged B blend	30 mm · 0.1	54.6 °C
B_L	long-term aged B blend	18 mm · 0.1	65.3 °C

aging was obtained lengthening the RTFO procedure to 325 min according to Muller and Jenkins (2011) who indicated similar effects for the abovementioned extended RTFO (ExRTFO) and the standardized Pressure Ageing Vessel procedure.

According to Fig. 1, the assessment of the rheological properties of materials was achieved using a Dynamic Shear Rheometer (DSR).

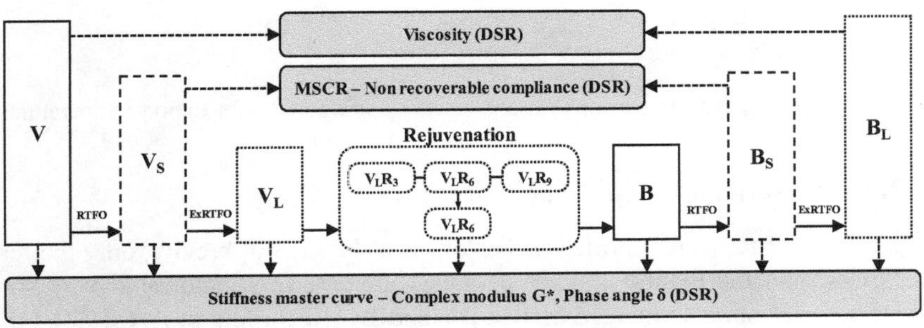

Fig. 1. Summary of experimental program

Viscosity tests were performed on unaged binders (V, B) according to EN 13702 with a 50 mm cone-plate geometry. Tests were carried out at 60 °C (shear rate: 0.05 s^{-1}), 100 °C (shear rate: 50 s^{-1}) and 150 °C (shear rate: 500 s^{-1}). Short-term aged binders (V_s, B_s) were subjected to Multiple Stress Creep Recovery (MSCR) tests (EN 16659) at high temperatures (50 to 80 °C) determining the non-recoverable creep compliances J_{nr} at shear stress of 0.1 and 3.2 kPa. Percent $J_{nr\text{-}diff}$ (i.e. the difference of J_{nr} at 3.2 and 0.1 kPa, normalized with respect J_{nr} at 0.1 kPa) was also calculated. Frequency sweep tests (EN 14770) were finally performed on V, V_S, V_L, V_LR_6, B, B_S and B_L to construct the complex modulus (G*) master curves at 20 °C based on the time-temperature superposition principle using the well-known Williams-Landel-Ferry (WLF) theory (Williams et al. 1955). Tests were executed in a wide range of temperatures (from 10 to 80 °C) and frequencies (from 0.1 to 10 Hz) in strain-controlled mode (0.05% shear deformation).

3 Results and Analysis

3.1 Shear Viscosity

Figure 2a shows that unaged V and B bitumens did not exhibit significant variations of shear viscosities thanks to the rejuvenation process suggesting almost the same mixing and compactions temperatures according to AASHTO T 312 (Fig. 2b). Thus, R seems able to restore viscosity of original binder despite the significant presence (60%) of aged hard bitumen coming from RAP.

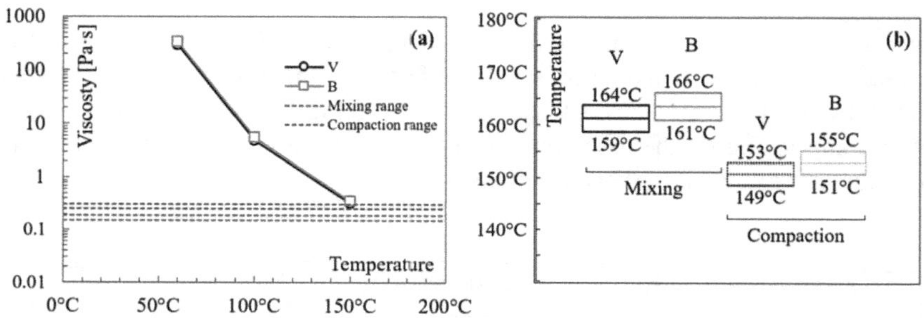

Fig. 2. V and B viscosity test results (a) and related mixing and compaction temperatures (b)

3.2 Non-recoverable Creep Compliance

Results of MSCR tests are reported in Table 2 (for the sake of brevity, only the results at the lowest and the highest test temperatures are listed). J_{nr} indicated very similar behavior between short-term aged binders V_S and B_S. According to AASHTO MP 19, J_{nr} and $J_{nr\text{-}diff}$ parameters revealed that blend B_S is characterized by the same upper limit of the Performance Grade (equal to 50 or 60 °C depending on the traffic class considered) of that demonstrated by V_S. Thus, similar permanent deformation resistance can be hypothesized between recycled blend B and the reference virgin binder V.

Table 2. MSCR test results for RTFO-aged original bitumen and blend

Material	V_S	V_S	V_S	V_S	B_S	B_S	B_S	B_S
T [°C]	50	50	80	80	50	50	80	80
τ [kPa]	0.1	3.2	0.1	3.2	0.1	3.2	0.1	3.2
J_{nr} [kPa^{-1}]	0.34	0.36	20.99	23.13	0.31	0.32	21.27	22.62
$J_{nr\text{-}diff}$ [%]	4.3		10.2		3.1		6.4	

3.3 Complex Modulus

Representative G* master curves were obtained through the recognition of specific frequency shift factors according to the WLF formulation (Fig. 3).

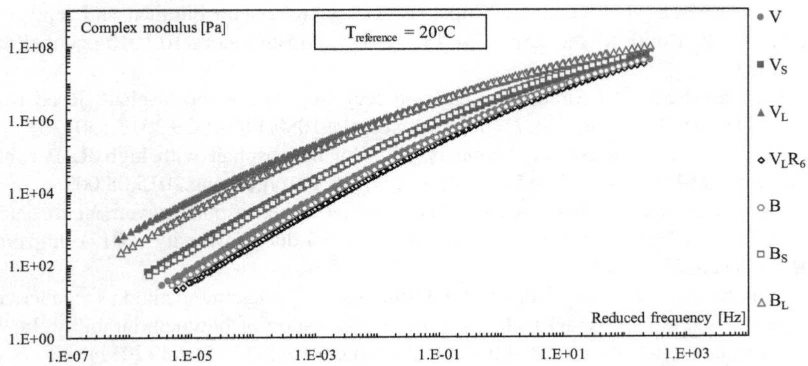

Fig. 3. Stiffness master curves: comparison between original bitumens and blends

As known, aging processes were responsible of G* increases because of the material hardening after RTFO and ExRTFO procedures. Rejuvenator dosed at 6% by VL bitumen weight allowed a downward translation of master curve towards that of virgin binder V (V and VL data were approximately overlapped) confirming an efficient recovery of mechanical performance. Moreover, results of aged blends (BS, BL) and corresponding original bitumens (VS, VL) were very similar proving once again the effectiveness of the hot recycling. Indeed, it is worth remembering that severe stiffening caused by long-term aging might lead to brittleness and fatigue failure of material. Otherwise, considering the PG classification on the basis on frequency sweep responses, BL displayed the same fatigue temperature limits that those of traditional long-term aged binder VL (25 °C regardless the traffic grade).

4 Conclusions

Rheological properties of a recycled bituminous blend containing long-term aged bitumen were studied to assess the feasibility of hot recycling high amount (about 60%) of reclaimed asphalt pavement. Rejuvenation of aged binders resulted able to restore main rheological properties of tested bitumens. In fact, the comparison between virgin binders and corresponding recycled blends indicated very similar performance in terms of viscosity, stiffness and permanent deformation resistance. Thus, the feasibility of hot recycling high amounts of reclaimed asphalt pavement was demonstrated in terms of main performance of the recycled binder at mid and high-service temperatures.

References

Chen, J.S., Huang, C.C., Chu, P.Y., Lin, K.Y.: Engineering characterization of recycled asphalt concrete and aged bitumen mixed recycling agent. J. Mater. Sci. **42**(23), 9867–9876 (2007). https://doi.org/10.1007/s10853-007-1713-8

Grilli, A., Iorio Gnisci, M., Bocci, M.: Effect of ageing process on bitumen and rejuvenated bitumen. Constr. Build. Mater. **136**, 474–481 (2017). https://doi.org/10.1016/j.conbuildmat.2017.01.027

Hugener, M., Kawakami, A.: Simulating repeated recycling of hot mix asphalt. Road Mater. Pavement Des. **18**(2), 76–90 (2017). https://doi.org/10.1080/14680629.2017.1304263

Izaks, R., Haritonovs, V., Klasa, I., Zaumanis, M.: Hot mix asphalt with high RAP content. Procedia Eng. **114**, 676–684 (2015). https://doi.org/10.1016/j.proeng.2015.08.009

McDaniel, R., Kowalski, K., Shah, A.: Evaluation of reclaimed asphalt pavement for surface mixtures. Indiana Department of Transportation and Purdue University (2012). https://doi.org/10.5703/1288284314665

Muller, J., Jenkins, K.J.: The use of an extended rolling thin film ageing method as an alternative to the pressurised ageing vessel method in the determination of bitumen durability. In: 10th Conference on Asphalt Pavements for Southern Africa, KwaZulu Natal (2011)

Newcomb, D.E., Brown, E.R., Epps, J.A.: Designing HMA mixtures with high RAP content. A Practical Guide. Quality Improvement Series 124. National Asphalt Pavement Association, US Department of Transportation, Federal Highway Administration (2007)

Pradyumna, T.A., Jain, P.H.: Use of RAP stabilized by hot mix recycling agents in bituminous road construction. Transp. Res. Procedia **17**, 460–467 (2016). https://doi.org/10.1016/j.trpro.2016.11.090

Riccardi, C., del Barco Carrión, A.J., Lo Presti, D., Cannone Falchetto, A., Losa, M., Wistuba, M.: A new procedure to determine the rheological properties of RAP binder and corresponding bituminous blends. Constr. Build. Mater. **154**, 361–372 (2017). https://doi.org/10.1016/j.conbuildmat.2017.07.204

Shen, J., Ohne, Y.: Determining rejuvenator content for recycling reclaimed asphalt pavement by SHRP binder specifications. Int. J. Pavement Eng. **3**(4), 261–268 (2002). https://doi.org/10.1080/10298430210000083685

Williams, M.L., Landel, R.F., Ferry, J.D.: The temperature dependence of relaxation mechanisms in amorphous polymers and other glass-forming liquid. J. Am. Chem. Soc. **77**, 3701–3707 (1955). https://doi.org/10.1021/ja01619a008

An Examination of Property Changes of Repeatedly Recycled Asphalt Bitumen Using Rejuvenator with High Aromatic Content

Atsushi Kawakami(✉), Yoko Kawashima, Hiroyuki Nitta, and Masayuki Yabu

Public Works Research Institute, Tsukuba, Japan
kawakami@pwri.go.jp

Abstract. In Japan, recycling of asphalt mixtures has been practiced for more than 40 years. Since the mixing ratio of recycled aggregate is increasing every year, use of asphalt mixtures containing repeatedly recycled aggregate is expected to increase. However, sufficient knowledge regarding the repeated recycling has not been accumulated, and no clear understanding of the changes has been established. In this study, asphalt bitumen was aged in a laboratory and then applied a rejuvenator to restore its penetration in order to understand the property changes occurring in repeatedly aged and recycled asphalt bitumen and asphalt mixtures. This process was repeated multiple times and thereby an understanding of the properties of recycled asphalt materials was gained. In the results, although recycled asphalt bitumen recovered its penetration, its softening point shifted to a higher-temperature range in third recycling. Ductility only reached a level less than half of the initial ductility even after the first recycling process. Moreover, it was revealed that ductility exhibited almost no improvement in and after third recycling. In the black diagram curves from the DSR test, the phase angle transitioned to a smaller value range when asphalt bitumen progressed to recycle three times and five times.

Keywords: Asphalt bitumen · Repeated recycling · Rejuvenator
Asphalt mixture

1 Introduction

During the high economic growth period from 1954 to 1973 in Japan, the paved road ratio of the national roads of Japan dramatically increased from 15.7% to 92.6% (MLIT 2016). In this period, road pavement replacement work increased in addition to work for newly constructed roads, and an enormous amount of asphalt concrete waste must have been generated. Then, the Waste Management and Public Cleansing Act was enacted in 1970. Around the same time, the first oil crisis in 1973 further boosted the necessity for effective use of residual asphalt paving materials. Against this backdrop, studies on the recycling of asphalt concrete waste started in the 1970s, and various recycling technologies were developed and have since improved. Thus, in Japan the recycling of asphalt concrete has a history of more than 40 years. Nowadays, the shipment of recycled asphalt mixtures accounts for 75% of the total shipment of hot

L. D. Poulikakos et al. (Eds.): RILEM 252-CMB 2018, RILEM Bookseries 20, pp. 189–194, 2019.
https://doi.org/10.1007/978-3-030-00476-7_30

asphalt mixtures (hereafter, mixtures), and the national average of the mixing ratio of recycled aggregates has reached 50% (JAMA 2016).

The above fact suggests that the use of mixtures containing repeatedly recycled aggregates will increase. However, sufficient knowledge for clarifying the phenomena of repeated recycling asphalt bitumen has not yet been accumulated.

In this study, aging of asphalt bitumen was conducted in a laboratory and then some kind of rejuvenators were added to restore its penetration in order to understand the property changes occurring in repeatedly aged and recycled asphalt bitumen (Kawashima et al. 2017) and asphalt mixture. This process was repeated multiple times to analyze the effects of recycling. This paper reports the property changes of repeatedly aged and recycled asphalt bitumen using a rejuvenator with high aromatic content from the study.

2 Examination Methods

2.1 Material Properties of Asphalt Bitumen

Virgin asphalt bitumen (hereafter, ORG) was repeatedly aged and recycled in a laboratory, and material property tests were performed to examine its property changes. Petroleum bitumen for road construction 60/80 was used as the ORG. This ORG exhibits the properties listed in Table 1.

Table 1. Properties of asphalt bitumen (ORG) and rejuvenator (RejA)

Asphalt bitumen (ORG)	Density [g/cm^3]	Penetration [1/10 mm]	Softening point [°C]		Ductility [cm]
	1.037	70	46.5		Over 100
Rejuvenator (RejA)	Density [g/cm^3]	Components [%]			
		Asphaltenes	Resins	Aromas	Saturates
	0.975	0.1	4.0	70.5	25.4

ORG was aged by TFOT with 165 °C for five hours and by PAV for 54 h to drop the penetration value (PEN) to 20. Then, this aged bitumen was recycled using rejuvenator (hereafter, RejA) that was a commercially available to PEN 70. This RejA is made of the components listed in Table 1 and contains a relatively large amount of aromatics. Since the tests focused on ascertaining the effects of the rejuvenator, they were performed on what is called 100% recycled materials, to which no new asphalt bitumen was added. This process was repeated five times (hereafter, recycled asphalt bitumen that has undergone repeated aging and recycling N times is expressed as Age-N and Recycle-N).

2.2 Evaluation Methods for Asphalt Bitumen

The material property tests on the asphalt bitumen included a penetration test (JRA-A041), a softening point ring and ball test (JRA-A042), and a ductility test (JRA-A043) to examine physical properties, and a dynamic shear rheometer (DSR) test (JRA-A062) was performed to examine dynamic viscoelastic properties. In the DSR test, the same conditions as those listed in Table 2 were applied to ORG, Recycle-3, and Recycle-5. Each test was performed according to the Handbook on Pavement Survey and Test Methods (JRA 2007). In addition, a composition analysis test (TLC/FID method) (Yamaguchi et al. 2003) were performed to examine the chemical properties.

Table 2. Test conditions for DSR

Diameter [mm]	Gap width [mm]	Strain [%]	Temperature [°C]	Angular frequency [rad/s]
8	2	1.0	−10 to 30	0.1 to 100
25	1	10.0	40 to 100	0.1 to 100

As a results of repeated recycling, the proportion of asphalt bitumen in bitumen recycled from recycle-0 (ORG) to recycle-5 were 100%, 81.7%, 67.4%, 61.2% 53.3%, and 46.5% respectively. The effects of these matters were examined in the following tests for material properties.

3 Examination Results

3.1 Basic Physical Properties

Figure 1 show the changes in softening point and the ductility. Although recycled asphalt bitumen recovered penetration to about 70, the softening point of Recycle-3 shifted to a higher value and that trend continued thereafter.

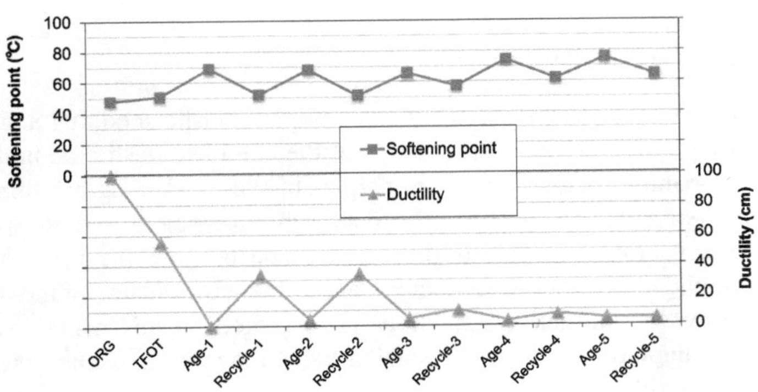

Fig. 1. Results of Softening point test (upper) and ductility test (bottom)

Regarding the results of ductility measurement, even Recycle-1 did not recover a ductility of 100 cm, and from Recycle-3 on there was almost no recovery. This suggests that repeated aging and recycling may prevent recycled asphalt bitumen from sufficiently recovering the softening point and ductility in contrast with its penetration recovery.

3.2 Chemical Properties

Figure 2 shows the results of the composition analysis test performed by the TLC/FID method. Saturates exhibited almost no changes, and asphaltenes exhibited only slight changes. Resins and aromatics repeatedly increased and decreased every time asphalt bitumen was aged and recycled; resins generally increased, whereas asphaltenes first decreased and later fluctuated around a certain level. Asphalt bitumen under aging generally decreases in aromatics and increases in resins and asphaltenes. An increase in resins was also observed in this test. The increases in aromatics in the recycling processes are presumed to have been brought about by the rejuvenator that was relatively rich in aromatics.

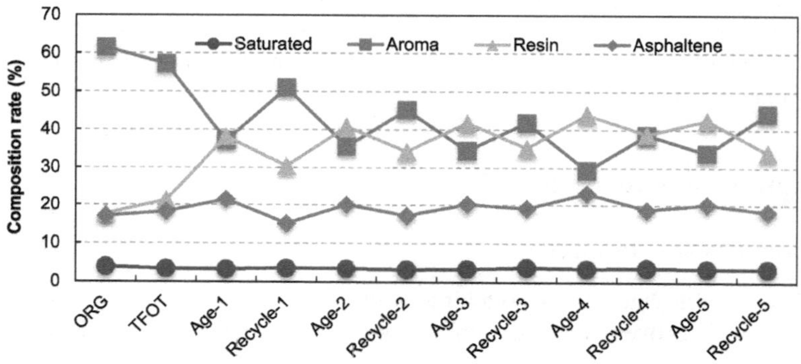

Fig. 2. Results of composition analysis

3.3 Dynamic Viscoelastic Properties

Figure 3 shows the relationship between the complex elastic modulus and the phase angle (Black diagram) and the master curves of the complex elastic modulus obtained by the time-temperature superposition principle. In the black diagram, the curve for ORG formed a smooth line, whereas the curves for Recycle-3 and Recycle-5 were made up of many branches. This is presumed to be because the proportion of the rejuvenator got higher in the high-frequency range where the change in the phase angle got smaller. However, further examinations are required on this matter. As a whole result, the phase angle transitioned to a smaller value range when asphalt progressed to Recycle-3 and Recycle-5.

Regarding the master curve, the position of the curve for ORG and that of the curves for Recycle-3 and Recycle-5 change between the low-frequency range (high-temperature range) and the high-frequency range (low-temperature range).

In the high-temperature range, the value of the complex elastic modulus increased with the increase in the number of recycling operations. In the low-temperature range,

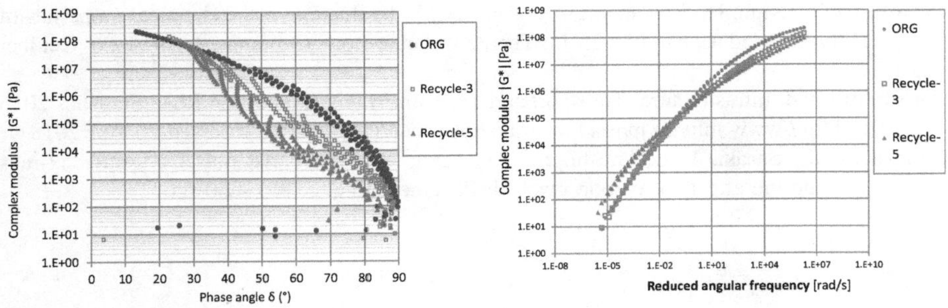

Fig. 3. Results of DSR test, black diagram (left) and master curve (right)

in contrast, the value of the complex elastic modulus decreased with the increase in the number of recycling operations. The slope of master curves decreased with this increase in the number of recycling operations; this is presumed to show a decrease in temperature susceptibility.

4 Summary

In this study, aging of asphalt bitumen was conducted in a laboratory and this asphalt bitumen was recycled by applying only a rejuvenator without adding any new material (what is known as 100% recycling). After repeating this process multiple times, the asphalt material properties were examined. The following summarizes our findings.

(1) Although recycled asphalt bitumen recovered its penetration, its softening point shifted to a higher-temperature range in third recycling and later and stayed in same range. Ductility only reached a level less than half of the initial ductility even after the first recycling process. Moreover, it was revealed that ductility exhibited almost no improvement in and after third recycling.

(2) The results of the composition analysis of asphalt bitumen showed that resins and aromatics repeatedly increased and decreased with every iteration of the aging and recycling process. As a whole, resins showed an increasing tendency, and aromatics first decreased and then leveled off. This change in aromatics is presumed to have been brought about by the components of the rejuvenator.

(3) In the analysis of the dynamic viscoelastic properties, the relationship between the complex elastic modulus and the phase angle formed a smooth curve for ORG but formed branched curves for Recycle-3 and Recycle-5. This is presumed to be because the proportion of the rejuvenator became high.

References

Japan Asphalt Mixture Association: Annual Report on Asphalt Mixtures (2016)
Japan Road Association: Handbook on Pavement Survey and Test Methods, June 2007

Kawashima, Y., Nitta, H., Kawakami, A., Nishizaki, I.: Evaluation of the chemical properties of repeatedly recycled asphalt binder. In: The 32nd Japan Road Conference, vol. 3056, October 2017

Ministry of Land, Infrastructure, Transport and Tourism: Annual Report on Road Statistics 2016 (2016). http://www.mlit.go.jp/road/ir/ir-data/tokei-nen/index.html. Accessed 26 Apr 2018

Yamaguchi, K., Sasaki, I., Meiarashi, S.: The aging of asphalt materials by ultraviolet-rays radiation and the effects of carbon black. J. Pavement Eng. **8**, 251–260 (2003)

Effects of Rejuvenator on Reclaimed Asphalt Binder: An Exploratory Study of the RILEM TC 264-RAP Task Group 3

Augusto Cannone Falchetto[1(✉)], Laurent Porot[2], Chiara Riccardi[1], Martin Hugener[3], Gabriele Tebaldi[4], and Eshan Dave[5]

[1] Technische Universität Braunschweig, Brunswick, Germany
{a.cannone-falchetto, chiara.riccardi}@tu-bs.de
[2] Kraton Chemical B.V., Almere, The Netherlands
Laurent.porot@kraton.com
[3] Empa, Dübendorf, Switzerland
martin.hugener@empa.ch
[4] Università di Parma, Parma, Italy
gtebaldi@unipr.it
[5] University of New Hampshire, Durham, USA
eshan.dave@unh.edu

Abstract. This paper presents the preliminary experimental activity conducted to develop the round robin test (RRT) plan on the use of rejuvenators for the Task Group (TG) 3 on Asphalt Binder for Recycled Asphalt Mixtures as part of the RILEM TC 264-RAP. For this purpose, a reference Reclaimed Asphalt (RA) binder was extracted from a single RA source. This material was then fully characterised with conventional experimental methods and with Fourier Transform Infrared Spectroscopy (FTIR) for evaluating the chemical structure. The RA binder was found to be very hard compared to target binder of PEN50/70 grade and the use of virgin binder cannot restore its properties. In this condition, a rejuvenator dosage of 9% per weight of RA binder is foreseen as optimum value to restore the binder properties close to target PEN50/70 grade for the final binder-rejuvenator blend over a RA content ranging from 60% and 100%.

Keywords: Asphalt binder · Reclaimed asphalt pavement · Rejuvenator RILEM

1 Introduction

In the last decades, the use of Reclaimed Asphalt (RA) from pavement has seen a continues increase due to the economic and environmental benefit. This is especially true when considering the new approach to pavement construction practices that need to be oriented toward the concept of circular economy while balancing reduced construction costs and enhanced performance (Büchler et al. 2018).

However, the presence of RA binder may be source of different effects on the recycled asphalt mixture. This includes an increase in stiffness and a decrease in penetration grade (EN 12591 2009) for higher RA contents (Mangiafico et al. 2012;

L. D. Poulikakos et al. (Eds.): RILEM 252-CMB 2018, RILEM Bookseries 20, pp. 195–200, 2019.
https://doi.org/10.1007/978-3-030-00476-7_31

Moon et al. 2014). In addition, the aged RA binder may ultimately limit the allowable amount of recycled material in the mix design due to a potential embrittlement of the mixture. This represents a considerable constrain for pavement engineers and road authorities as more and more RA will have to be recycled in the upcoming years, possibly affecting the pavement performance beyond acceptable service levels. In order to address such a potential limitation, binder additives were developed over the years to restore the characteristics of the aged binder contained in the reclaimed material. These products, known as rejuvenators, aim to restore the flexibility without compromising the rutting resistance (Shen et al. 2007).

In view of the fundamental importance on the use of RA in asphalt pavement, a specific Technical Committee (TC) was established: TC 264-RAP. Within the framework of this TC, the Task Group 3 on Asphalt Binder for Recycled Asphalt Mixtures is devoting its activities on the investigation of binders and additives that can rejuvenate the aged RA binder. The initial goal of TG3 consists in the implementation of an experimental plan to organise a round robin test (RRT) among different laboratories to address and to characterise the effect of rejuvenators. This paper reports the preliminary work conducted by the TG3 leaders and focuses on extraction/recovery of aged binder from a pre-selected RA material, its characterisation and the determination of the optimum dosage of rejuvenator to meet the targeted rejuvenation.

2 Experimentation

Granulated RA material was collected from a stockpile consisting of an asphalt concrete (AC) having a maximum aggregate size of 22 mm (AC22). RA binder was then extracted and recovered according to EN 12697-3 (2013) in a sufficient quantity to implement a comprehensive experimental plan and to supply twenty RRT participants. Then, RA binder was characterised based on penetration value (EN 1426 2015) and softening point temperature (EN 1427 2015) methods.

The rejuvenating effect of the additive was evaluated with two initial dosages, 5% and 10% by weight of RA binder. Based on penetration value, a dosage chart was generated to determine the optimum content of rejuvenator to target a 50/70 pen grade asphalt binder, which is commonly used in Europe for asphalt mixtures. A final blend with the optimum dosage level was made to reproduce a combination of 60% RAP mix with virgin binder, up to a 100% RAP without virgin binder. The rejuvenator used is a bio-based agent, which is a liquid additive that, with its specific amphipathic chemical structure, disperses the highly polar fractions limiting the agglomeration of asphaltenes (Porot and Grady 2016). Table 1 presents the main properties.

Table 1. Properties of the rejuvenating agent

Flash point	Viscosity at 60 °C	Density	Cloud point
>260 °C	22 Cst	0.93	< –25 °C

3 Results

3.1 Characterisation of the Reclaimed Asphalt Binder

Two samples of RA binder were extracted from the reference AC22 mixture. Both extractions provided the same results with a binder content of 4.4%, a penetration value of 15 0.1 × mm and a softening point temperature close to 70 °C. These values are far away of the targeted 50/70 pen grade binder. Under these conditions, a conventional binder is alone unable to restore the original material properties, especially flexibility at intermediate temperature.

The RA binder was then subject to FTIR to identify if any residue of solvent was present after extraction and next compared with a reference virgin Polymer modified Binder (PmB). Figure 1 displays the FTIR profile.

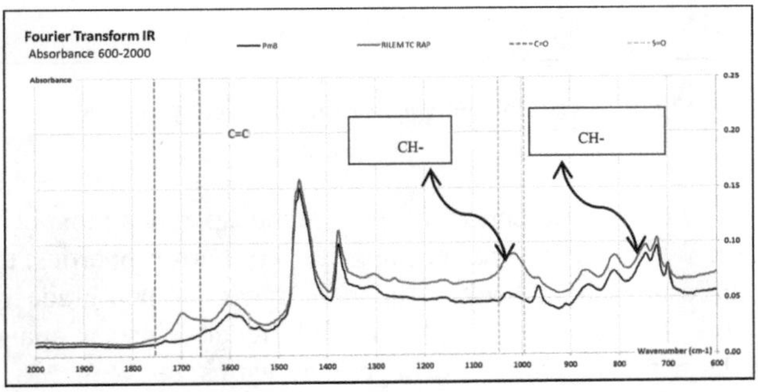

Fig. 1. FTIR profile of the RAP binder compared to PmB

Compared to the PmB binder, the RA binder did not show significant presence of polymer. On the other hand, it distinguishes with a clear peak in the carbonyl zone as a result of oxidation.

3.2 Evaluation of the Effect of the Rejuvenator

A dosage study was conducted to evaluate how the additive affects the aged binder properties at two dosage levels, 5% and 10%. Aging and oxidation cause hardening of asphalt binders. In this first part of the work the evaluation was based on the basic properties, such penetration value at 25 °C, and softening point temperature, as a quick and simple approach to quantify the rejuvenation of the binder.

Aged recovered binder and additive were blended based on the method described in ASTM D4887 (2016) standard; neither high shear mixing nor maturation time were required.

Figure 2 displays the impact of the additive at different dosages on penetration value reported in the vertical axis and on the softening point temperature on the horizontal axis.

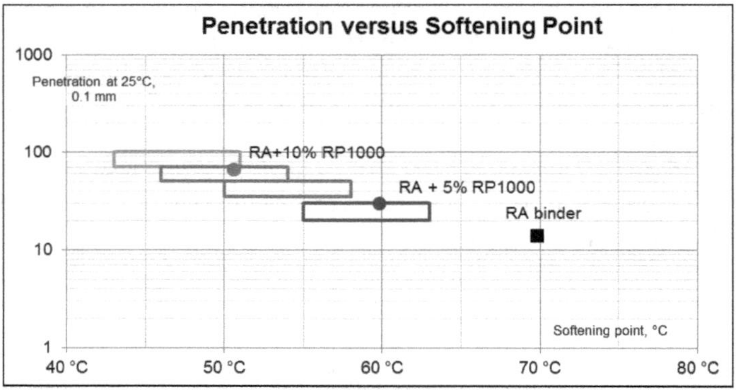

Fig. 2. Effect of rejuvenator on empirical properties

The boxes provide an indication of pen grade specifications in accordance to EN 12591 (2009). While the RA binder presents very hard properties, the effect of the additive helps to restore them back to a wider range of pen-grade ranges even at dosages below 10%, based on the weight of aged binder. This can span from 20/30 to 70/100 pen grade with an additive dosage that varies between 3% and 10%. On average, a dosage of 5% restores the properties by 2 pen grades, which is in agreement with previous studies (Porot and Grady 2016).

Due to aging, asphalt binders become harder. This has a positive effect for the high temperature behaviour against rutting; however, at the same time, the binder becomes more brittle especially at low and intermediate temperatures. Therefore, to determine the optimum dosage, the main focus was to restore the consistency at intermediate temperature based on the penetration value while still verifying that the softening point temperature is not adversely reduced. Figure 3 displays a dosage chart to predict the optimum rejuvenator content. Targeting the mid-point of a 35/50 pen grade binder, the optimum dosage would be about 8%. For the mid-point of a 50/70, it would be 10%. Since in the planned RILEM RRT a conventional virgin 50/70 binder was selected to be blended with different amount of RA ranging between 60% and 100% to target a 50/70 pen grade binder blend, an optimum rejuvenator dosage of 9% was determinate to encompass the two RA percentages limits (Fig. 3).

Based on the defined rejuvenator dosage of 9%, a detailed RRT plan was implemented to include three different RA percentages (60%, 80% and 100%). The experimental RRT procedure consists of both traditional penetration and softening point tests as well as on Dynamic Shear Rheometer and Bending Beam Rheometer measurements on asphalt binder and rejuvenator blends under different conditions.

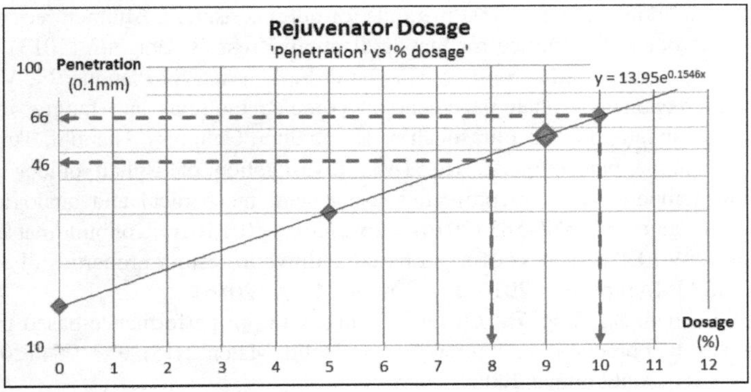

Fig. 3. Dosage chart for rejuvenator

4 Summary and Conclusions

In this paper the simple experimental procedure used to determine the optimum rejuvenator content to be used in the planned round robin test as part of the activities of the Task Group 3 on Asphalt Binder for Recycled Asphalt Mixtures within the RILEM TC 264-RAP is presented. RA binder was first extracted and recovered from a preselected source of reclaimed material. Then, penetration and softening point methods were used to characterise the recovered binder and to determine the rejuvenator dosage for a range of blends having a RA binder content between 60% and 100%. In addition, Fourier Transform Infrared Spectroscopy was used to characterise chemically the RA binder.

Based on the experimentation presented in this paper, it was observed that the RA binder is considerably hard but with no significant presence of polymer detected. Blending RA with a conventional pen grade virgin binder cannot restore the original material properties and therefore an optimum rejuvenator content of 9% is required to achieve the 50/70 blend targeted for the RRT.

References

ASTM D4887/D4887 M – 11: Standard Practice for Preparation of Viscosity Blends for Hot Recycled Asphalt Materials, USA (2016)

Büchler, S., Falchetto, A.C., Walther, A., Riccardi, C., Wang, D., Wistuba, M.: Wearing course mixtures prepared with high reclaimed asphalt pavement content modified by rejuvenators. Transp. Res. Record. (2018). https://doi.org/10.1177/0361198118773193

EN 1426: Bitumen and bituminous binders - Determination of needle penetration. European Committee for Standardization, Brussels, Belgiu (2015)

EN 1427: Bitumen and Bituminous Binders – Determination of the Softening Point - Ring and Ball Method. European Committee for Standardization, Brussels, Belgium (2015)

EN 12591: Bitumen and Bituminous Binders – Specifications for Paving Grade Bitumens. European Committee for Standardization, Brussels, Belgium (2009)

EN 12697-3: Bituminous mixtures. Test methods for hot mix asphalt. Bitumen recovery: Rotary evaporator. European Committee for Standardization, Brussels, Belgium (2013)

Mangiafico, S., Di Benedetto, H., Sauzeat, C., Olard, F., Dupriet, S., Planque, L., Van Rooijen, R.: Effect of reclaimed asphalt pavement on complex modulus and fatigue resistance of bitumens and asphalts. In: 5th Eurasphalt & Eurobitume Congress, Istanbul, Turkey (2012)

Moon, K.H., Cannone Falchetto, A., Hu, J.W.: Investigation of asphalt binder and asphalt mixture low temperature creep properties using semi mechanical and analogical models. Constr. Build. Mater. **53**, 568–583 (2014). https://doi.org/10.1016/j.conbuildmat.2013.12.022

Porot, L., Grady, W.: Effectiveness of a bio-based additive to restore properties of aged asphalt binder. In: ISAP Symposium 2016, Jacksonhole, USA (2016)

Shen, J., Amirkhanian, S., Tang, B.: Effects of rejuvenator on performance-based properties of rejuvenated asphalt binder and mixtures. Constr. Build. Mater. **21**(5), 958–964 (2007). https://doi.org/10.1016/j.conbuildmat.2006.03.006

New Binders Using Natural Bitumen Selenizza

Edith Tartari(✉)

Selenice Bitumi Sha, Tirana, Albania
e.tartari@selenicebitumi.com

Abstract. The purpose of this paper is to highlight the benefits of using natural bitumen as part of the emerging environmental sustainability innovations and trends in asphalt mixture design. An overview of a recent research work focusing for the first time on the use of vegetable oils and natural bitumen to produce a new type of binder for asphalt mixes is followed by a brief description of the experimental investigation of a new Job Mix Formula, aiming to evaluate the potential use of 100% RAP and a rejuvenated asphalt binder, based on waste vegetable oil and natural bitumen.

Keywords: Natural bitumen · Aging · Vegetable oil · Rejuvenators

1 Hardening Effect and Anti-aging Properties of Natural Bitumen Selenizza

Various petroleum refining techniques are available for the production of hard bitumen. With the ever increasing use of high modulus asphalt mix technology, there is a strong demand for hard bitumen. An alternative is the modification of soft bitumen with natural bitumen additive. For many years now, in several construction projects such as highways, ports, airports, the natural bitumen Selenizza, has proved to be a worthwhile additive for obtaining hard bitumen.

The impact of adding different percentages of Selenizza to the mechanical properties of modified bitumen is reflected by the decrease of penetrability and increase of softening point temperature, (Santarelli and Scarsella 2005). Usually, by adding 8–10% of Selenizza to the base bitumen, it decreases by one penetration grade (e.g., 50/70 base bitumen modified with 8–10% Selenizza, will be equivalent to 35/50 paving grade bitumen).

Considering that the aging of the binder is of key importance for pavement durability, a recent research study (Themeli et al. 2015) examined the evolution of bitumen modified with natural bitumen during artificial aging process. In this study, a specimen of 50/70 bitumen, modified with different percentages of Selenizza, as well as the respective equivalent normal paving grade bitumen, were submitted to accelerated artificial ageing RTFOT followed by PAV.

It was observed that after the ageing, the changes that occurred in the modified bitumen were lower compared to the changes in the initial bitumen 50/70. Also, for the modified specimen, the changes are attenuated with the increase of the modification rate. The comparison between the modified bitumen and the equivalent normal bitumen, show that the modified bitumen are characterized by minor changes, which means that Selenizza, acts as an aging inhibitor (Table 1).

L. D. Poulikakos et al. (Eds.): RILEM 252-CMB 2018, RILEM Bookseries 20, pp. 201–207, 2019.
https://doi.org/10.1007/978-3-030-00476-7_32

Table 1. Penetration and softening temperature before and after aging

Description	Penetration (0.1 mm)			R&B (°C)		
	New binder	After RFTOT	After PAV	New binder	After RFTOT	After PAV
50/70 bitumen	54	37	19	49	53.4	61.4
50/70 with 5% Selen	38	27	15	52.6	57.2	66.0
50/70 with 10% Selen	28	21	13	56.2	60.8	68.8
50/70 with 15% Selen	20	14	11	61.6	65.4	72.2
Normal 35/50	40	27	12	52.6	56.8	66.2
Normal 20/30	23	12	7	60.0	67.0	78.8
Normal 10/20	18	9	5	65	72.6	86.0

2 Use of Waste Vegetable Oil and Natural Bitumen for Developing a New Type of Binder

Recently published papers indicate that waste oils have the capability to restore the properties of aged asphalt binder providing more flexibility to the final binder.

On the other hand, the high asphaltene content in the natural bitumen Selenizza along with its anti-aging property, make it a suitable choice to combine with the rejuvenating properties of waste vegetable oils for the development of innovative binders.

In a research study conducted jointly by the French Centre for Studies and Expertise CEREMA and Institute for Science and Technology IFSTTAR, (Somé et al. 2016), was developed a new type of binder using 71.4% natural bitumen Selenizza, blended with 17.9% of rapeseed or sunflower waste vegetable oil and 10.7% hard bitumen 15/25.

The analysis of engineering properties of the blended binders showed that in terms of penetration, the new binders are close to P35/50 petroleum bitumen, but with higher values of softening temperature compared to conventional bitumen (Fig. 1).

The differential scanning calorimeter analysis highlighted the fact that the new produced binders were characterized by the increase of low temperature performance due to the waste vegetable oil's glass transition temperatures that are lower than those of bitumen.

Using Metravib analyser, frequency sweep tests have been conducted at 10 different frequencies in the range from 1 to 80 Hz. From –20 °C to 20 °C, the tensile-compression complex modulus E* was determined, and from 20 °C to 60 °C, the shear complex modulus G* was measured. Master curves were built at a reference temperature of 15 °C using a software developed by IFSTTAR. Binder's complex modulus and phase angle master curves (Fig. 2) indicate that the reference bitumen is slightly stiffer than the new binders in temperatures that range between –20 °C and 60 °C. The phase angles of the new binders, are lower than those of reference bitumen for the reduced

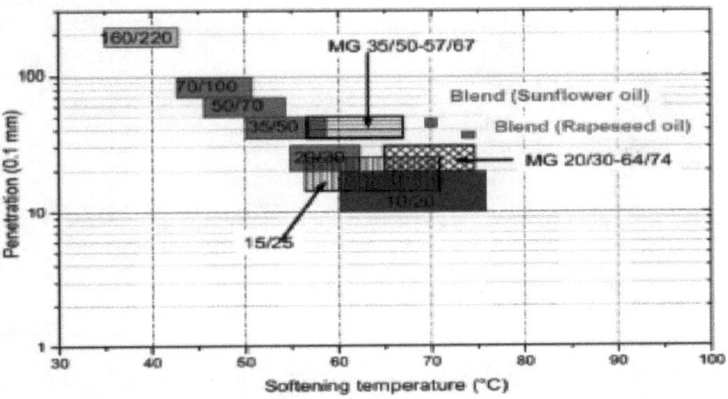

Fig. 1. Penetration versus softening temperature

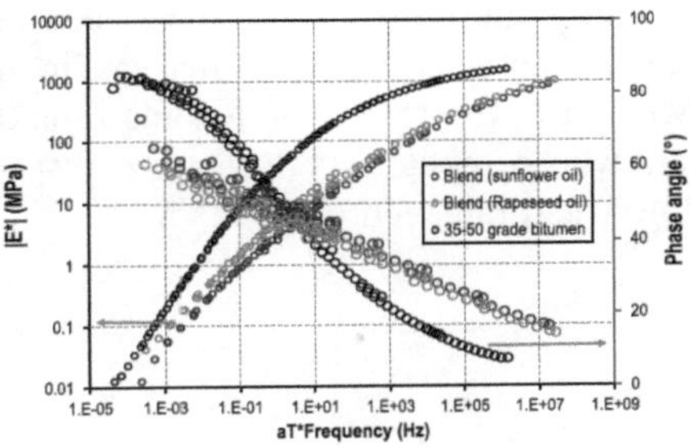

Fig. 2. Binders complex modulus and phase angle master curves at 15 °C

frequency $a_T \times f \leq 2.5$ Hz (e.g. T $\geq$ 20 °C) and higher for the reduced frequency $a_T \times f \geq 2.5$ Hz (e.g. T $\leq$ 20 °C).

Also, it was noticed that at very low temperatures, the phase angles of the new produced binders are not equal to zero, which means that the viscous effects are not negligible compared to reference bitumen. Consequently, at low temperatures, the new produced binders' behaviour cannot be assumed to be that of a purely elastic material and this property may be advantageous for low temperature stress relaxation.

To determine the behaviour of hot mix asphalt designed with the new produced binders, a Semi Coarse Asphalt Concrete (BBSG 3, 0/10) with 6.3% binder content was manufactured.

The resistance to the permanent deformation was determined using the wheel tracker large device. For each asphalt concrete, test has been conducted on two plates, compacted by a rolled compactor. The rut depth has been measured using a depth gauge as a function of rolled passes. The rut depths (at 60 °C) of the mixes manufactured with the new produced binders are lower than the control mix (with 35/50 bitumen) rut depth (see Fig. 3 below). The better resistances to the permanent deformation obtained with the new produced binders are probably due to the high asphaltene content in natural bitumen, but the real mechanism that occurs, is not known yet.

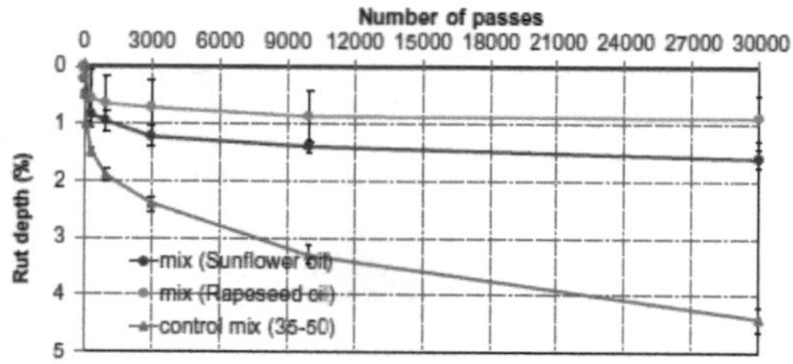

Fig. 3. Evolution of the rut depth

The mixtures stiffness modulus was determined using DT-CY (direct tensile test). The experiment was conducted at –10 °C, 0 °C, 10 °C, 15 °C, 25 °C, 40 °C, and at 1 s, 3 s, ., 300 s loading times. Beforehand, was determined the strain amplitude to be applied during the tests in order to preserve the linear elastic behaviour of the samples. The results of each temperature and time sweeps have been used to build the master curves at the reference temperature 15 °C. As shown in Fig. 4, the reference asphalt mix was stiffer 13171 MPa (at T_{ref} = 15 °C and loading time 0.02 s) compared to the two other asphalt mixes, whose modulus values were respectively 8233 MPa and 5678 MPa for the sunflower and rapeseed oil.

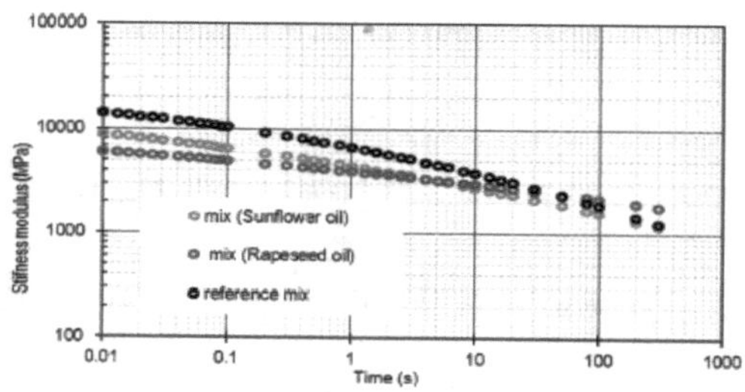

Fig. 4. Master curves of the stiffness modulus of asphalt mixes at 15 °C

This study has to be completed and further developed, especially focusing on the fatigue resistance, aging and low temperature cracking of the asphalt mixes with the new binder.

3 Example of Innovative Asphalt Mix Design for Surface Layer Using 100% RAP Aggregates and a Binder Composed Only of Vegetable Oil and Natural Bitumen

A recent study carried out by Prof. Dr. Ing. Steffen Riedl and M. Eng. Ronny Sorge from Erfurt University, (Riedl and Sorge 2017), as a part of a national innovation program, proposed and evaluated an innovative asphalt mix using 100% RAP aggregates with the addition of a rejuvenator. The new developed rejuvenator, which aims to restore the original characteristics of the fresh bitumen and its effectiveness, is composed of waste vegetable oil and natural bitumen Selenizza. For this project, 12 variants of an asphalt concrete AC 11 DN and the associated binders, without a rejuvenator and the same aged mixtures with 3, 4 and 8% rejuvenator content by mass of the bitumen in the asphalt, were investigated. In the Table 2, JA refers to reference variants of asphalt mixtures, JB to the aged variants and JC, to the aged asphalt mixes added with a rejuvenator.

Table 2. Different variants of AC 11 DN

Variant	Asphalt mix	Binder	Binder content [M-%]	Additive content [M-%]
JA1	AC 11 DN	Shell B 50/70	6.2	–
JA2	AC 11 DN	BP3 B 50/70	6.2	–
JA3	AC 11 DN	Olexobit PmB 25/55-55	6.2	–
JB1	AC 11 DN	Shell B 50/70 - BSA	6.2	–
JB2	AC 11 DN	BP3 B 50/70 - AASHTO R 30	6.2	–
JB3	AC 11 DN	Olexobit PmB 25/55-55 - AASHTO R 30	6.2	–
JB4	AC 11 DN	RC - Elxleben	6.2	–
JC 1	AC 11 DN	Shell B 50/70 - BSA	6.2	4.0
JC 2	AC 11 DN	BP3 B 50/70 - AASHTO R 30	6.2	8.0
JC 3	AC 11 DN	Olexobit PmB 25/55-55 - AASHTO R 30	6.2	8.0
JC 4.1	AC 11 DN	RC - Elxleben	6.2	3.0
JC 4.2	AC 11 DN	RC - Elxleben - BSA	6.2	3.0

During the binder characterization tests, it was observed that due to ageing, the softening temperature of aged binders (JB1, JB1.2 and JB2) increased in comparison with (JA1, JA2) reference variants and the penetration decreased. The addition of the additive, leads to a significant reduction of softening point (JC1, JC2) as well as a significant increase of the penetration.

The results of Dynamic Shear Rheometer analysis at a load frequency of 1.59 Hz and temperature range of 20 °C to 65 °C for binder investigation, showed that aged variants (JB) have a greater rigidity compared to reference variant (JA) over the entire temperature range (Fig. 5). The stiffness modulus of rejuvenated variants (JC), are again in the range of the initial values. The phase angle values (Fig. 6) of the aged variants (JB), in particular compared to the reference variant (JA), decreased over the entire temperature range and came back to the initial values for the rejuvenated variants (JC).

Iatroscan SARA analysis showed that the addition of the additive, leads to a difference in the percentage distribution of the main SARA groups of the rejuvenated binder compared to the aged binder with an increase of the polarizable fractions (resins and asphaltenes) and the reduction of the aromatics and saturates.

Comparison tests were carried out to evaluate the changes of aged and rejuvenated mixture performance (JB and JC variants) with respect to the reference asphalt mixture

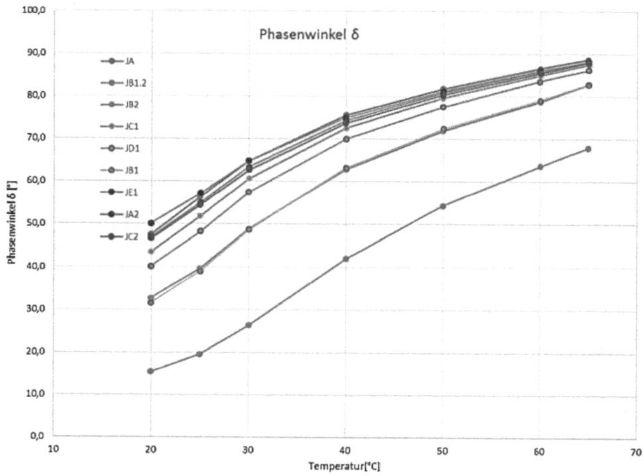

Fig. 5. Temperature-sweep G* test

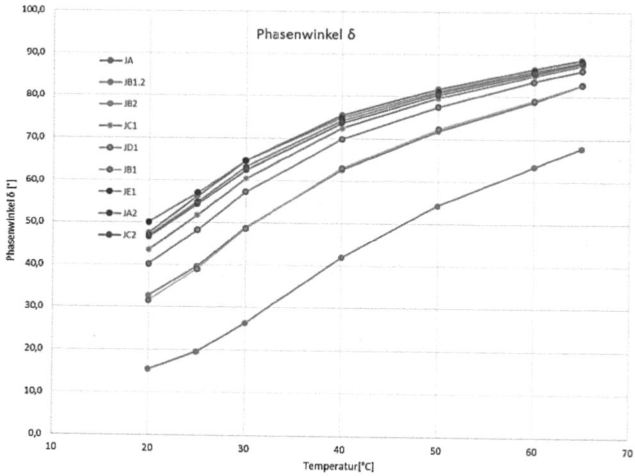

Fig. 6. Temperature-sweep phase angle test

(JA). Fatigue functions of dynamic indirect tensile test indicate that the rejuvenated variants (JC variants) compared to the aged (JB) and reference variant (JA), for the same elastic initial strain, endure more load changes up to the macro cracking.

Stiffness-temperature curves, at 10 Hz in the temperature range between –20 °C to 30 °C, showed an increase of the stiffness modulus values of the aged variants JB. After rejuvenation, these values return again to the initial values of reference variants or below.

4 Conclusion

The addition of the natural bitumen Selenizza, strongly affects the mechanical behavior of modified bitumen. Also, it has been demonstrated that Selenizza decreases the susceptibility to ageing of modified bitumen as the percentage of natural bitumen content increases.

The hardening and anti-aging properties of natural bitumen, can be advantageously used to develop new binders combining the high performance mechanical and durability properties of Selenizza, with the rejuvenating capability of waste vegetable oils, opening new perspectives for innovative concepts in the Hot Mix Asphalt technology. More particularly, in a context where great emphasis is placed on the need to use ever increasing content of reclaimed asphalt pavement in asphalt mixtures, these new types of binders, could play an essential role to restore performance properties of oxidized RAP binders and to improve the mixture resistance to cracking without adversely affecting its resistance to rutting.

References

Santarelli, M.L., Scarsella, M.: Natural asphalts as modifiers of distillation bitumen: thermo-rheological characterization, Rassegna del Bitume, pp. 21–29 (2005)

Somé, S.C., et al.: Int. J. Pavement Res. Technol. **9**, 368–375 (2016)

Riedl, S., Sorge, R.: Research Institute FH-Erfurt Project AiF-Juvenate: The Innovative Contract on Asphalt Production (2017)

Themeli, A.: Université de Strasbourg, Ecole doctorale MSII, Thèse «Etude du potentiel d'emploi des bitumes naturels dans la production des liants bitumineux durs et des enrobés à module élevé» (2015)

Rejuvenated Binders, Reclaimed Binders and Paving Bitumens, Are They Any Different?

Tomas Koudelka(✉), Pavel Coufalik, Michal Varaus, and Iva Coufalikova

Brno University of Technology, Veveri 331/95, 602 00 Brno, Czech Republic
Koudelka.tl@gmail.com, paja.coufalik@gmail.com, varaus.m@fce.vutbr.cz, coufalikova.iva@gmail.com

Abstract. Asphalt binder hardens over time; therefore, its chemical and rheological properties are altered considerably in the course of the time. The change of properties starts during the production process and continues throughout pavement's lifetime. Further, the process of rejuvenation is supposed to restore lost properties of aged binders. Additionally, a hypothesis that rejuvenated binders behave similarly to paving bitumens have been suggested by researchers and rejuvenator producers many times. However, there is still a limited number of studies comparing behaviour of paving bitumens, rejuvenated and reclaimed binders altogether. This study investigates the influence of long term laboratory ageing on empirical properties of paving bitumens and rejuvenated binders. In total, behaviour of 10 paving bitumens available on the Middle European market was compared with the behaviour of the laboratory aged paving bitumen combined with 10 different rejuvenators. Binders were aged by RTFOT (Rolling Thin Film Oven Test) and further by up to two cycles of PAV (Pressurized Ageing Vessel). The properties of the aged binders were thereupon compared with 10 characteristics of reclaimed binders. The results indicate that RTFOT+PAV procedure ages the binder less in comparison to what is observed in the field in some cases. Aged rejuvenated binders exhibit higher penetrations and penetration indexes compared to aged paving bitumens and to reclaimed binders. Additionally, it was found out that the properties of rejuvenated binders upon ageing are similar to air rectified bitumens rather than to reclaimed binders. Aging of rejuvenated binders thus follows a different mechanisms compared to paving bitumens.

Keywords: Rejuvenated binders · Reclaimed binders · Paving bitumens Ageing

1 Introduction

Binders used for road applications are acquired through refining processes of crude oils. As such, properties of asphalt binders depend strongly on the source of and on the exact processing techniques of crude oils. These are the main reasons why asphalt binders available on the market can display different long-term behaviour caused by ageing susceptibility (Lesuer 2008). At the point when the material is unable to withstand the induced stresses due to loss of engineering properties, the material is ready to be milled off and reused as reclaimed asphalt.

L. D. Poulikakos et al. (Eds.): RILEM 252-CMB 2018, RILEM Bookseries 20, pp. 208–214, 2019.
https://doi.org/10.1007/978-3-030-00476-7_33

To be able to incorporate aged binder in the form of reclaimed asphalt during the production of new asphalt mixtures, it is necessary to restore aged binder properties so that the resulting new mixture demonstrates at least the same performance as the virgin ones. The aged binder properties are altered in order to improve low temperature behaviour along with fatigue characteristics at intermediate temperatures while not negatively affecting high temperature performance. Only then reclaimed asphalt can be reused in high amounts without causing the danger of occurrence of premature failures (Zaumanis and Mallick 2014).

The process of selecting a rejuvenator and evaluating its dosage play a crucial role in the reclaimed asphalt mixture design. It was demonstrated that in some cases the rejuvenators can improve the aged binder characteristics (Nayak and Sahoo 2016; Ongel and Hugener 2015) while other researchers proved that both bio and crude oil based rejuvenators can pose some detrimental effects. (Hesp and Shurvell 2010; Golalipour 2013). On top of that, it was noted that behavior of rejuvenated binders (RB) may differ from that of paving bitumens (PB) (Ali and Mohammadafzali 2015). Hence, the asphalt binder can behave upon rejuvenation ambivalently.

2 Motivation and Objectives

The aim of this study was to compare long term behaviour of PB and RB. Rejuvenators which differ in chemical composition were used to rejuvenate laboratory long term aged (3xRTFOT) 50/70 PB. The individual rejuvenators' dosages were calculated according to penetration. Thereafter, fresh PB and RB were aged by means of Rolling Thin Film Oven Test (RTFOT) and of Pressurized Ageing Vessel (PAV). In the end, the properties of laboratory aged binders were compared to those of reclaimed binders. The main objective of the paper was then to confirm or reject the assumption that the behaviour of rejuvenated binders is the same in the case of PB and/or reclaimed binders. All binders were characterized in a standard way using Penetration and Ring and Ball tests as these are normally required for testing of PB (EN 12591) and reclaimed binders (EN 13108-8). Furthermore, penetration indexes (PI) were used to classify binder rheological behaviour. Originally, PI was derived in a way to have a value of about zero. Negative values exhibit binders which are more temperature susceptible while higher PI display rather blown (oxidize) bitumens. The lower the PI, the quicker the binder changes its behaviour. (Van der Poel 1954). The actual range of PI for paving bitumens is −1.5 to 0.7 and is set in EN 12591.

3 Materials and Methodology

3.1 Rejuvenators

This study includes testing of 10 various rejuvenators which were labeled A to J. Most rejuvenators (A to H) were commercial products which are sold on the US and European market, while I and J were alternative products. The rejuvenators were characterized by dynamic viscosities at 25 °C. The products are described in Table 1.

Table 1. Characteristics of rejuvenators and their optimal dosages calculated according to penetration

	Composition	Dynamic viscosity, 25 °C, mPa.S	Optimal dosage, % of aged binder mass
A	Polyol ester, tall and turpentine oils	87	5.6
B	Mixture of resins, waxes, polymers	1129	6.5
C	Residues (petroleum), vacuum	191	8.5
D	Fatty acid methyl ester	58	5.9
E	Mixture of vegetable oils	62	5.2
F	Fatty acid derivates	118	5.5
G	Rosins esters, fatty acid and vegetable oil	169	5.9
H	Mixture of modified alkylamidopolyamine and vegetal oils	183	6.5
I	Refined rapeseed oil	68	6.2
J	Aromatic extract	4013	10.9

3.1.1 Determination of the Rejuvenator's Dosage

The optimal rejuvenator dosages added to the laboratory aged bitumen 50/70 were calculated with respect to their individual effectiveness. It was established that the dosages ranged from 5.2% to 10.9% with a mean value of 6.7% depending on the type of rejuvenator. An example of the optimal dosage evaluation can be seen in (Koudelka et al. 2018). The exact calculated dosages for each rejuvenator are given in Table 1.

3.2 Binders

In the first step, one paving bitumen categorized as 50/70 was chosen for the preparation of laboratory aged binder (No. 1 in Table 2). Afterwards, this binder was aged by means of a prolonged RTFOT procedure - 3xRTFOT. The test was conducted at temperature 163 °C for 225 min. The resulting properties after ageing were: penetration 23 (0.1 mm), softening point 61.6 °C and PI −0.34. As the following step, additional 9 PB categorized as 50/70 were gathered to obtain a representative sample. These varied in binder source. Binders were produced in the following refineries: Litvinov (CZ), Paramo (CZ), Schwechat (AT), Leuna (DE), Godorf (DE), Plock (PL) and Gdansk (PL). Finally, 10 reclaimed binders labeled 1# to 10 #were extracted from reclaimed asphalts collected at numerous asphalt plants in the Czech Republic.

3.3 Test Methods and Procedures

Short term ageing was simulated by RTFOT procedure according to EN 12607-1. Long term in service ageing was conducted using PAV according to EN 14769. RTFOT tests were carried out at 163 °C for 75 min while PAV tests were conducted at 100 °C for 20 h and further for 40 h in case of RB. Recovery processes were carried out according to EN 12697-1 and were followed by binder distillations according to EN 12697-3. The

Table 2. Binders' characteristics, penetration and penetration indexes

PAVING BITUMENS											
	1	2	3	4	5	6	7	8	9	10	**Mean**
Fresh [0.1 mm]	56	52	57	64	56	57	54	47	55	63	**56**
PI $_{Fresh}$ [-]	−1.5	−0.9	−1.6	−1.5	−1.3	−1.6	−1.5	−1.1	−1.1	−1.3	**−1.3**
PAV [0.1 mm]	21	19	20	19	18	21	14	18	19	24	**19**
PI $_{PAV}$ [-]	−0.1	0.8	−0.5	−1.1	−0.6	−0.1	−0.6	0.3	0.3	−0.6	**−0.2**
REJUVENATED BINDERS											
	A	B	C	D	E	F	G	H	I	J	**Mean**
Reju. [0.1 mm]	59	58	58	61	61	61	59	61	61	58	**60**
PI $_{RB}$ [-]	−0.4	−0.7	−0.2	−0.6	−0.6	−0.6	−0.7	−0.8	−0.3	−0.7	**−0.6**
PAV [0.1 mm]	29	20	32	30	27	26	27	26	31	27	**28**
PI $_{PAV}$ [-]	0.2	0.6	0.8	0.2	0.1	0.0	0.3	0.3	0.4	0.2	**0.3**
2xPAV [0.1 mm]	22	13	24	22	20	19	19	20	23	19	**20**
PI $_{2xPAV}$ [-]	0.9	2.2	1.5	1.0	1.0	1.0	1.1	0.8	1.4	0.6	**1.1**
RECLAIMED BINDERS											
	1#	2#	3#	4#	5#	6#	7#	8#	9#	10#	**Mean**
Reclaim [0.1 mm]	12	25	15	12	14	14	19	21	12	21	**17**
PI [-]	0.1	−0.9	−0.1	−0.6	−0.6	−0.1	−0.4	−0.1	−0.3	−0.4	**−0.4**

solvent used was Perchlorethylene. Each of the binders was tested in order to determine the needle penetration (as specified by EN 1426) and the softening point (using the EN 1427 Ring and Ball method). Having the results of penetration and softening point temperature allows for the calculation of the PI. PI determination is standardized and its calculation is described in an annex of EN 12591.

4 Results and Discussion

Table 2 summarizes the results of the needle penetration tests and calculated PI whereas Fig. 1 displays the difference in behaviour between all binders graphically (mean values with standard deviations).

Penetrations of fresh PB fluctuated from 47 (0.1 mm) to 64 (0.1 mm) with average value of 56 (0.1 mm). Mean PI for fresh PB was −1.3 (temeparture susceptible behaviour), the actual values varied from −1.6 to −0.9. It can be noticed that binders 3, 6 and 8 would not meet the criteria on fresh PB, although PI is not always required to meet. Remaining penetration upon RTFOT+PAV ranged from 26% to 38%. PI then varied from −1.1 to 0.8 with a mean value settled at −0.2.

In case of RB the penetration ranged from 58 (0.1 mm) to 61 (0.1 mm) while the target was 56 (0.1 mm). RB, thus, displayed slightly higher penetration compared to the original binder (No. 1 in Table 2). RB displayed in general higher PI, in average −0.56 and the value fluctuated from −0.3 to −0.8. The values are somewhat higher than in case of PB because the initial softening point temperatures were not reached upon

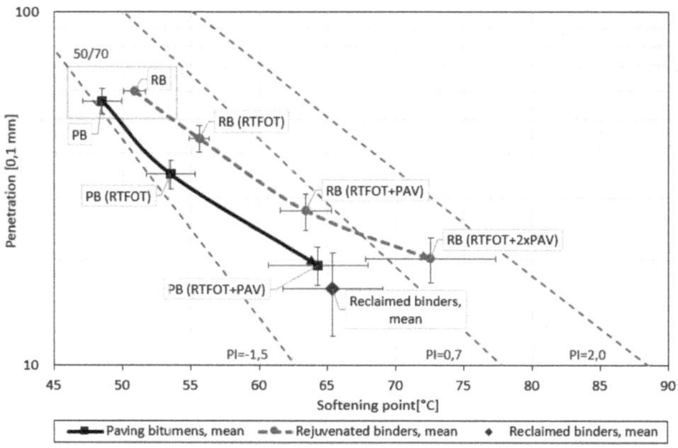

Fig. 1. The comparison of behaviour of PB along with RB and reclaimed binders

the rejuvenation process. Moreover, the results suggest that it is impossible to maintain same penetrations and corresponding softening points after rejuvenation independently of the rejuvenator used. Either penetration or softening point values can be targeted. Remained penetrations after RTFOT+PAV ranged from 34% to 55%. RB are, thus, able to sustain higher penetration after conventional long term ageing protocol compared to PB. Average PI after RTFOT+2xPAV ranged from 0.6 to 2.2 with an average PI 1.1 Such values are typically attributed to blown or polymer modified bitumens. Calculated PI of RB were certainly higher in comparison to PB regardless on the state of ageing.

Measured penetrations of reclaimed binders demonstrated the fact that in field ageing is slightly harsher compared to that of RTFOT+PAV although some overlap exists. Penetration values ranged from 12 (0.1 mm) to 25 (0.1 mm) with mean value of 17 (0.1 mm) while mean PI was −0.4.

As can be perceived from Fig. 1 RB stand a bit on the right side compared to PB and reclaimed binders. Accordingly, it can be concluded that behaviour of RB is not the same as in the case of PB and reclaimed binders.

5 Summary and Conclusion

This study assessed the ageing behavior of paving bitumens, rejuvenated binders and compared them with that of reclaimed binders. The tests were performed on original and rejuvenated binders as well as on long term aged binders. The following conclusions can be drawn from the study:

- It was not possible to restore penetration and softening point of the aged binder at the same time, no matter what rejuvenator was used. Only one characteristic can be target at a time.

- RB sustained higher penetrations compared to PB after long term ageing. This in turn means that RB display higher PI than PB. Therefore, it seems that RB have better temperature susceptibility in comparison to PB.
- After prolonged ageing (2xPAV) all RB except one had higher PI than 0.7 Thus, after deep ageing RB behave more as air rectified (blown) bitumens or polymer modified bitumens rather than PB.
- RB behaviour differed from what was observed in case of PB and reclaimed binders.

Commentary

In order to fully fathom and more in depth describe the effect of rejuvenators on aged binders it would be vital to carry out functional testing using DSR or BBR equipment. However, this paper aims to describe the behaviour in a standard way using procedures stated in EN 12591. In many cases, penetration and softening points are the only parameters tested at asphalt plants and rejuvenator dosages are derived accordingly. Moreover, control testing of bituminous binders is more often than not limited only to empirical tests. Therefore, the authors believe the information pointing out the differences in the long-term behaviour between abovementioned binders is of high importance no matter what method to assess it is being used.

Acknowledgements. This paper was prepared with the support of the Czech Technology Agency, project TJ01000248 "Rejuvenating agents for the extension of pavement lifetime with the use of a higher amount of Reclaimed Asphalt Pavement materials".

References

Ali, H., Mohammadafzali, M.: Long-Term Aging of Recycled Binders, Final report: BDV29 Two 977-01. Florida Department of Transportation (2015). DIALOG. http://www.fdot.gov/research/Completed_Proj/Summary_SMO/FDOT-BDV29-977-01-rpt.pdf. Accessed 22 Mar 2018

Golalipour, A.: Investigation of the effect of oil modification on critical characteristics of asphalt binders. Dissertation, University of Wisconsin Madison, June 2013

Lesuer, D.: The colloidal structure of bitumen: consequences on the rheology and on the mechanisms of bitumen modification. Adv. Colloid Interface Sci. **145**(1–2), 42–82 (2008). https://doi.org/10.1016/j.cis.2008.08.011

Hesp, A.M.S., Shurvell, H.F.: X-ray fluorescence detection of waste engine oil residue in asphalt and its effect on cracking in service. Int. J. Pavement Eng. **11**(6), 541–553 (2010). https://doi.org/10.1080/10298436.2010.488729

Koudelka, T., Porot, L., Coufalik, P., Varaus, M.: The use of rejuvenators as an effective way to restore aged binder properties. In: TRA 2018, Article 10783 (2018)

Nayak, P., Sahoo, U.C.: Rheological, chemical and thermal investigations on an aged binder rejuvenated with two non-edible oils. Road Mater. Pavement Des. **18**(3), 612–629 (2016). https://doi.org/10.1080/14680629.2016.1182058

Ongel, A., Hugener, M.: Impact of rejuvenators on aging properties of bitumen. Constr. Build. Mater. **94**(2015), 467–474 (2015). https://doi.org/10.1016/j.conbuildmat.2015.07.030
Van der Poel, C.: A general system describing the visco-elastic properties of bitumens and its relation to routine test data. J. Appl. Chem. **4**(1954), 221–236 (1954)
Zaumanis, M., Mallick, R.: Review of very high-content reclaimed asphalt use in plant-produced pavements: state of the art. Int. J. Pavement Eng. **16**(1) (2014). https://doi.org/10.1080/10298436.2014.893331

Study on the Mechanical Properties of Waste Cooking Oil Modified Asphalt Binder

Xin Qu[1(✉)], Dawei Wang[1], Quan Liu[1], Markus Oeser[1], and Chao Wang[2]

[1] Institute of Highway Engineering, RWTH Aachen University, D52074 Aachen, Germany
{qu, wang, q.liu, oeser}@isac.rwth-aachen.de
[2] Department of Road and Railway Engineering, Beijing University of Technology, Beijing 100124, People's Republic of China
wangchao@bjut.edu.cn

Abstract. Conventional asphalt binder derived from the petroleum refining process is widely used in pavement engineering. However, asphalt binder is a non-renewable material, while waste cooking oil (WCO) is a renewable original material, so it can be used as an additive for asphalt binder. Several studies found that the WCO changes mechanical properties of asphalt binder. However, the micro investigation of the modification of WCO is not studied. In this paper, A reasonable molecular structure for WCO is created based on previous research findings firstly, which are fundamental and crucial for establishing the molecular model of bio-asphalt binder with various WCO contents. Then the molecular dynamics technology is employed to investigate the mechanical properties of asphalt binder with different WCO contents. The results show that the WCO softens the asphalt binder increasing with its content, which is consistent with the previous research, indicating that the simulated results characterize the mechanical properties of asphalt binder effectively and scientifically.

Keywords: Waste cooking oil · Asphalt binder · Mechanical properties
Molecular dynamics simulation

1 Introduction

Different types of oil and fat are used in restaurants, hospitals, food packagers and the many households all over the world every day. Common vegetable oils such as olive oil, palm oil, rapeseed oil, soybean oil and many more are liquid at room temperature. They find application in baking, frying and other ways of cooking. After application they remain as so called waste cooking oil (WCOs). WCOs are composed of various fractions of fatty acids. In China, for example, more than five million tons of WCOs are generated every year. WCOs have very little water solubility and change into solid state as temperature drops. Therefore oil can congeal in pipes provoking blockages. Because of this, WCOs should be collected and recycled. There are different possible ways of reuse such as applications in fuel and cosmetic industry. Recently, WCO has caught attention as a possible modifier of asphalt binder. Replacing the conventional

L. D. Poulikakos et al. (Eds.): RILEM 252-CMB 2018, RILEM Bookseries 20, pp. 215–219, 2019.
https://doi.org/10.1007/978-3-030-00476-7_34

petroleum based asphalt binder as a renewable biomass component, the reused WCO could be a promising modifier. However, the addition of WCO to asphalt binder will, in consequence, change its material properties; above all, the important high and low temperature behavior.

Several studies have been carried out on the asphalt binder properties yet using WCO based bio-oil. Wen et al. studied the influence of bio-oil obtained from WCO polymerization on conventional asphalt binder and found that the bio-oil improved thermal cracking resistance but reduced rutting resistance (Wen et al. 2012). Sun et al. obtained similar results, stating that adding WCO based bio-oil to asphalt binder reduced the deformation resistance but improved the low temperature behavior (Sun et al. 2017). Recently Guarin et al. proposed that the acid value of the WCO was a critical parameter for the performance of bio-asphalt. The permanent deformation resistance of bio-asphalt can be improved reducing the acid values of WCO by chemical pre-treatments (Guarin et al. 2016). Since the addition of WCO based bio-oil can soften the petroleum asphalt, this kind of bio-oil for long-term aged asphalt binders as a rejuvenating agent were also utilized (Kowalski et al. 2017; Peralta et al. 2012; Yang et al. 2014).

Even though there are many studies focusing on the performance of WCO based bio-asphalt, the fundamental modification mechanism of bio-oil is still not completely understood on micro scale. To further sustainable future road building, alternative binding materials are in demand. More detailed and proceeding studies in this field can make significant contribution to future innovation development. Purpose of this study is to investigate the effects of WCO on the mechanical properties of asphalt binder by means of micro methodology. Therefore a molecular dynamic model of asphalt binder and bio-oil is created to simulate the material behavior in a realistic way. To observe the differences for various ratios of WCO, models with three different contents of WCO were taken into consideration. These models help characterizing the material properties such as Young's modulus, Poisson's ratio, bulk and shear modulus can specify and substantiate the existing results at the micro level.

2 Molecular Dynamics Simulation

2.1 Creation of Molecular Structure Model

Based on the research of Li et al. and Qu et al. (Li and Greenfield 2014; Qu et al. 2018), the molecular structures for asphalt binder and WCO are created. And then three molecular models of asphalt binders with 0, 2, 4 WCO molecules are constructed, in order to create the models of asphalt binder and two bio-binders respectively with 5%, 10% WCO according to the molar mass of the WCO, which is 993.6 g/mol.

2.2 Mechanical Properties Simulation

Every material has its corresponding specific temperature susceptibility, so 25 °C is selected as the typical value for the initial investigation due to our limited computing power.

2.2.1 Young's Modulus and Poisson's Ratio

The Young's modulus (E) is a kind of mechanical parameter, which means the resistance against deformation for solid materials. While the Poisson's ratio (v) is the ratio between the strain value in horizontal direction and the strain value in vertical direction, when the material is pulled or compressed in a single direction, which is an elastic constant that reflects the lateral deformation of the material, it reflects the lateral deformation of the material, because the Poisson's ratio is always an elastic constant when the material is pulled or compressed.

The Young's modulus and Poisson's ratio are calculated by the MD simulation. As starting the calculation, strain in 6 different directions are imposed on the model, and then corresponding stresses are obtained.

To improve the accurateness of the simulated mechanical properties of the asphalt binder model, 3 frames are selected for each model to calculate the average value for the Young's modulus and Poisson's ratio.

The Young's modulus and the Poisson's ratio are obtained through calculations and shown in Fig. 1. The Young's modulus and Poisson's ratio decrease with the WCO content, the Young's modulus falls by 8.9% as 5% WCO is added to the binder, and it decreases by 20.2% as 10% WCO is added, which indicates that the asphalt binder with WCO would be softer. While the Poisson's ratio falls by 16.7% and 41.8%, as 5% and 10% WCO are added into the asphalt binder, respectively, indicating that the horizontal deformation of the asphalt binder with WCO would be smaller under a certain load.

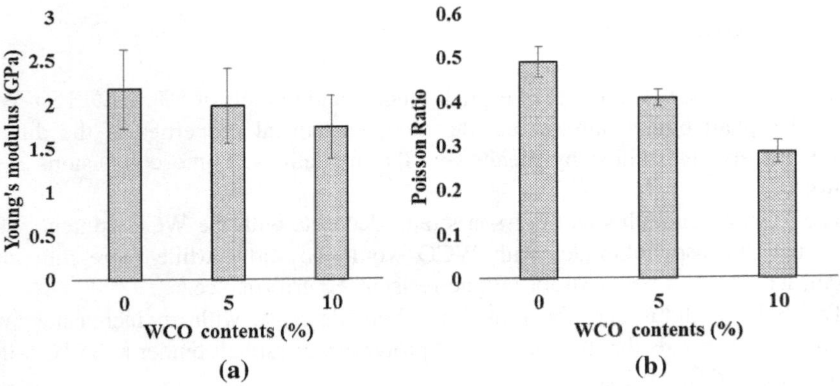

Fig. 1. (a) Young's modulus results; (b) Poisson's ratio results of these three models

2.2.2 Bulk Modulus and Shear Modulus

The MD simulation is also employed to calculate the corresponding elastic compliance matrix of these models. The bulk modulus is the physical quantity between the body strain and the average stress, which is a kind of elastic modulus, reflecting the macroscopic characteristics of the material.

The shear modulus (G) characterizes the ability of materials to resist shear strain, which is the ratio of shear stress to shear strain when the material is under shear stress.

The bulk modulus K and shear modulus G of these models are shown in Fig. 2.

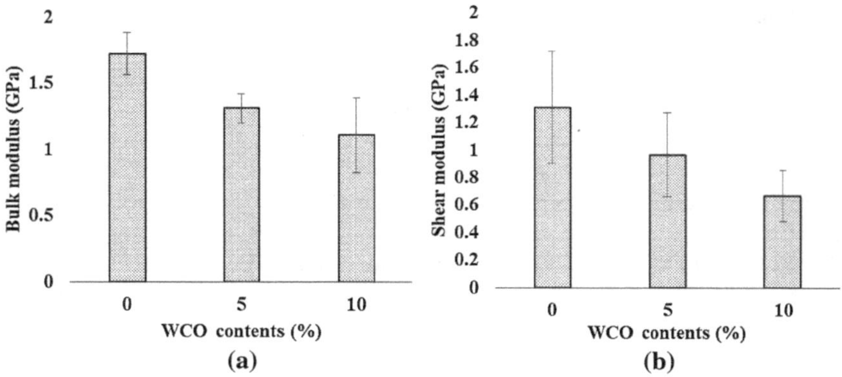

Fig. 2. (a) Bulk modulus results; (b) Shear modulus results of these three models

It is seen that the two modulus of WCO modified asphalt binder decrease with an increasing WCO content. The bulk modulus falls by 23.8% as 5% WCO is added to the binder, and falls by 35.6% as 10% WCO is added. While the shear modulus falls by 26.3% and 49.2%, when 5% and 10% WCO are added into the asphalt binder, respectively, indicating that the both modulus decrease significantly with the WCO content.

3 Conclusion

Based on previous work, model of asphalt binders and models of 5%, 10%, 15% WCO modified asphalt binder are created, then the mechanical properties of the different asphalt ·binders are studied by means of MD simulations. Some conclusions are as follows:

The Young's modulus and Poisson's ratio decrease with the WCO content, which means that the asphalt binder with WCO would be softer while more difficult to deform, leading to a better strain fatigue resistance performance.

The bulk modulus and shear modulus both decrease with an increasing WCO content, which reveals that the mechanical properties of asphalt binder at 25 °C will be worse with modifier WCO.

The effect of WCO on the mechanical properties of asphalt binder has been investigated. In the near future the micro modifying mechanism of the WCO will be further studied.

References

Guarin, A., Khan, A., Butt, A.A., Birgisson, B., Kringos, N.: An extensive laboratory investigation of the use of bio-oil modified bitumen in road construction. Constr. Build. Mater. **106**, 133–139 (2016)

Kowalski, K.J., Król, J.B., Bańkowski, W., Radziszewski, P., Sarnowski, M.: Thermal and fatigue evaluation of asphalt mixtures containing RAP treated with a bio-agent. Appl. Sci. **7**, 216 (2017)

Li, D.D., Greenfield, M.L.: Chemical compositions of improved model asphalt systems for molecular simulations. Fuel **115**, 347–356 (2014). https://doi.org/10.1016/j.fuel.2013.07.012

Peralta, J., Raouf, M.A., Tang, S., Williams, R.C.: Bio-renewable asphalt modifiers and asphalt substitutes. In: Sustainable Bioenergy and Bioproducts, pp. 89–115. Springer (2012)

Qu, X., Liu, Q., Wang, C., Wang, D., Oeser, M.: Effect of co-production of renewable biomaterials on the performance of asphalt binder in macro and micro perspectives. Materials **11**, 244 (2018). (Basel)

Sun, D., Sun, G., Du, Y., Zhu, X., Lu, T., Pang, Q., Shi, S., Dai, Z.: Evaluation of optimized bio-asphalt containing high content waste cooking oil residues. Fuel **202**, 529–540 (2017)

Wen, H., Bhusal, S., Wen, B.: Laboratory evaluation of waste cooking oil-based bioasphalt as an alternative binder for hot mix asphalt. J. Mater. Civ. Eng. **25**, 1432–1437 (2012)

Yang, X., You, Z., Dai, Q., Mills-Beale, J.: Mechanical performance of asphalt mixtures modified by bio-oils derived from waste wood resources. Constr. Build. Mater. **51**, 424–431 (2014)

The Effect of the Nature of Rejuvenators on the Rheological Properties of Aged Asphalt Binders

Raúl Tauste(✉), Fernando Moreno-Navarro, Miguel Sol-Sánchez, and Ma Carmen Rubio-Gámez

Department of Construction Engineering and Engineering Projects, Construction Engineering Laboratory, University of Granada, Granada, Spain
{rtauste, fmoreno, msol, mcrubio}@ugr.es

Abstract. Pavement rehabilitation will become the main activity in the road engineering sector in the majority of developed countries. Thus, large quantities of reclaimed asphalt pavements (RAP) will be generated and their reuse in asphalt mixtures would provide considerable economic and environmental benefits. To overcome the problems caused by using an excessive amount of RAP in asphalt mixtures, the use of rejuvenators will play a key role. Recent years have seen the emergence of many types of asphalt rejuvenators. The main objective of this study is to characterize the effect of different rejuvenators (varying in nature) on the rheological properties of aged asphalts binders. For this purpose, an aged binder extracted from RAP was blended with four rejuvenators at different dosages. Their rheological properties were then evaluated using frequency sweep tests and several rheological parameters that assess the brittleness of aged binders ("G*·sin δ"; R-value and crossover frequency; G-R parameter). The results indicate that the rejuvenator based on oleyldiamine ethoxylate showed the best performance. However, similar results can also be achieved using bio-rejuvenators from plant-based ester resins.

Keywords: RAP · Rejuvenator · Rheology · Aging

1 Introduction

Asphalt mixtures are the most common paving materials used worldwide. At the end of the service life of the pavement, these materials are removed to produce reclaimed asphalt pavement (RAP) that can be added to new mixtures, giving both economic and environmental benefits. Nevertheless, the use of a high percentage of these materials has certain limitations, the most important of which is related to the aged asphalt binder contained in the RAP (Zaumanis et al. 2014a). In particular, oxidation of the aged binder produces a hardened mixture that loses some of the viscoelastic properties that would otherwise give cohesion to the mixture (Durand et al. 2016). In addition, the aged binder renders the mixture more prone to cracking failure at low temperatures (Ongel and Hugener 2015; Al-Qadi et al. 2014; Miró et al. 2015).

The successful use of high percentages of RAP in asphalt mixtures should reverse this, and therefore restoration of the original properties of the aged binder must be

L. D. Poulikakos et al. (Eds.): RILEM 252-CMB 2018, RILEM Bookseries 20, pp. 220–225, 2019.
https://doi.org/10.1007/978-3-030-00476-7_35

achieved (Zaumanis et al. 2013). When the RAP content is low, softer virgin binder can compensate for the aging effect. However, with higher quantities of RAP, the use of rejuvenators is required (Moghaddam and Baaj 2016; Zaumanis et al. 2014a). These additives compensate for the effects of oxidative ageing by adding aromatic and resinous compounds so that the binder reaches similar chemical and rheological characteristics to those of the original material (Dony et al. 2013). In addition, the resulting binder should provide sufficient adhesion and cohesion between the aggregates to prevent moisture damage and raveling (Zaumanis et al. 2014b).

Traditionally, the majority of asphalt rejuvenators are composed of the lowest-weight molecular fractions of asphalt binders (saturates and aromatics). Nonetheless, recent years have seen the emergence of new types of rejuvenators from other sources (most notably from plants). Depending on their nature, these additives interact differently with the aged asphalt contained in RAP, leading to variations in the performance of the mixture (Huang et al. 2014). Therefore, whilst all of these additives could potentially reduce the penetration or viscosity of aged binder (since they are oil-based), not all rejuvenators would exert the same effect on the rheological properties of the aged binder, which will govern the final mechanical performance of the asphalt mixture. The main objective of this paper was to assess the effect of rejuvenators of varying nature on the rheological properties of aged bitumen extracted from RAP. In this regard, it is expected that the chemical characteristics and dosages of these additives will determine their effect on the mechanical behavior of this material.

2 Materials and Methods

The present study examines the effect of four different rejuvenators on the rheological properties of aged asphalt binders: R1, composed mainly of oleyldiamine ethoxylate; R2, mainly composed of non-recycled bio-oils; and R3 and R4, mainly composed of plant-based ester resins. These additives were blended with aged asphalt extracted from RAP milled in road after 20 years of service life (the original binder in the asphalt layer milled was a B 50/70).

The aged binder (AB) was extracted from the RAP was obtained using a centrifugal extractor and rotary evaporator equipment. This is performed at temperatures below the mixing temperature of the binder and for a limited period making its contribution to the ageing negligible. Following this, the rheological properties of the binder were compared with those found in different mixtures and dosages of AB (by weight of binder) with the rejuvenators studied (3%, 6% and 9% of R1, R2 and R3; 3%, 5% and 7% of R4), and with those found in unaged B 50/70 (without any additive). The dosages were selected according to the recommendations of the additive providers, and they were blended with the AB at a temperature of 90 °C, allowing its application to warm-asphalt mixtures if desired.

For the evaluation of the rheological properties of these materials, frequency sweep tests were conducted over a range of frequencies (0.1–20 Hz) and different temperatures (10–80 °C) so to cover its performance conditions. Based on the results obtained, the evolution of the properties due to the effect of the rejuvenators was analyzed through the representation of several parameters that assess the brittleness of aged binders: "G*·sin δ" from SHRP at different temperatures and a frequency of 5 Hz; R-value and Crossover frequency (Mogawer et al. 2015) at a temperature of 20 °C; and G-R (Glove-Rowe) parameter "G*·[cos δ]^2/sin δ" at 15 °C and an angular frequency of 0.005 rad/s (Rowe and Sharrock 2016).

3 Results and Discussion

Figure 1 shows the average values obtained in "G*·sin δ" parameter at the different temperatures tested. In general terms, as the dosage of rejuvenator is increased in the AB extracted from RAP, the curve obtained tends to approach that of the original binder (non-aged) B 50/70. In addition, it is shown that as the test temperature increases, the effect of the rejuvenators on the rheological response of the aged asphalt becomes weaker. This indicates that the action of the rejuvenator would be less effective in RAP materials that operate at higher service temperatures. Finally, Fig. 1 also demonstrates that the nature of the rejuvenator influences the degree of modification of the rheological response of the aged asphalt.

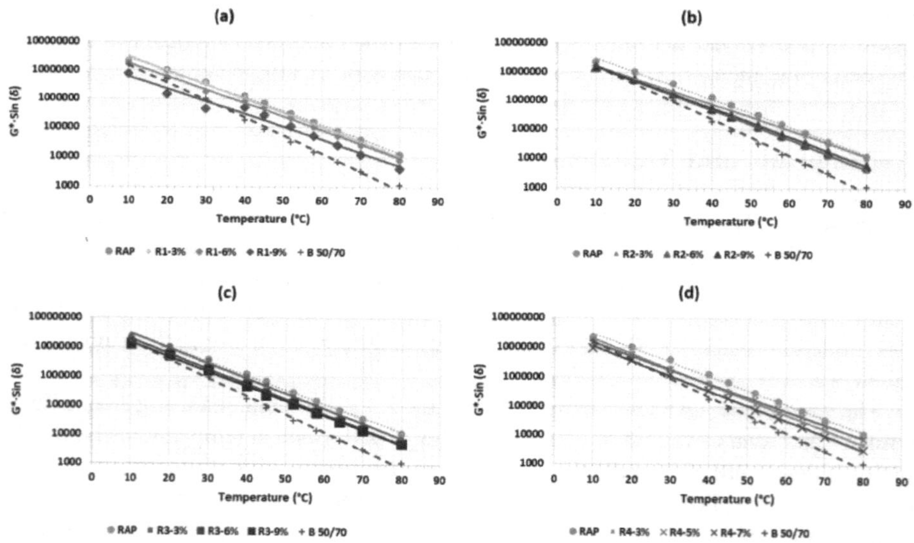

Fig. 1. Average values of the "G*·sin δ" parameter for the different temperatures tested.

In this respect, Fig. 2 (which shows the average values of R and crossover frequency, obtained using a reference temperature of 20 °C) clearly represents the effect of each type and dosage of rejuvenator on the mechanical properties of the aged binder. Based on these results, it appears that the rejuvenator R3 (composed of plant-based ester resins) modifies the rheological response of the AB to restore its properties to those of the original B 50/70. The curve described by the increment in dosage appears to be similar to that shown for the aging process (but iin the opposite direction) and the use of 9% of this rejuvenator offers similar rheological parameters to the unaged B 50/70. When rejuvenator R1 is used, the curve shown for the increment in dosage is different to that of R3 (which clearly indicates that different effects are exerted on the binder properties depending on the nature of the rejuvenator). The R-values and crossover frequencies achieved at the highest dosage of R1 are higher than those shown by the original B 50/70 (which implies that the material remains stiffer for a given degree of visco-elastic response, but that the rejuvenators can produce this visco-elastic response at higher frequencies). Finally, R2 and R4 appear to have a similar effect on the rheological properties of the AB, which are intermediate to the effects produced by R1 and R3, and appear to be less effective (since there was a large discrepancy between the final rheological parameters and those obtained in the unaged B 50/70).

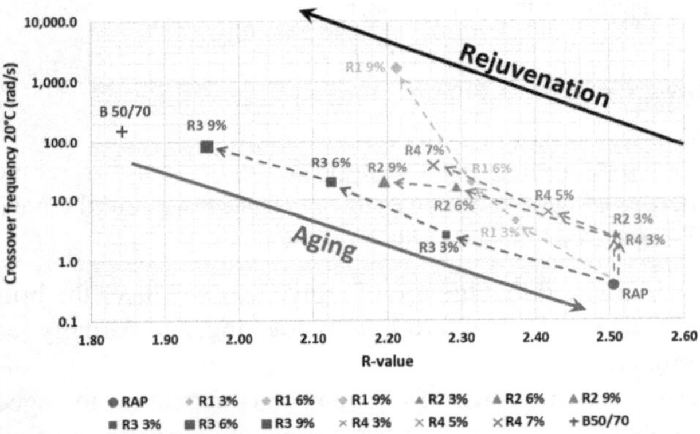

Fig. 2. Average values of R and crossover frequencies of the binders studied.

Finally, Fig. 3 shows the average complex modulus and phase angle for an angular frequency of 0.005 rad/s tested at a temperature of 15 °C (Fig. 3a), along with the average values of the G-R parameter as a function of the dosage of additive used (Fig. 3b). These results suggest that the use of rejuvenators clearly reduces the brittleness of aged asphalts and reduces their risk of suffering from cracking. Nonetheless, the results again reveal that the degree of rejuvenation varies as a function of the nature of the additive. In this respect, it is again demonstrated that the effect produced by R2 and R4 is very similar, and less effective in restating the original rheological properties of the binder in comparison with R1 and R3 (which clearly offer the most effective rejuvenation of the AB). In any case, it is important to note that the dosage used plays an important role in the final G-R parameter obtained, and that similar values can be achieved for any additive by varying their dosage (which indicates that not all the rejuvenators have the same optimal dosage).

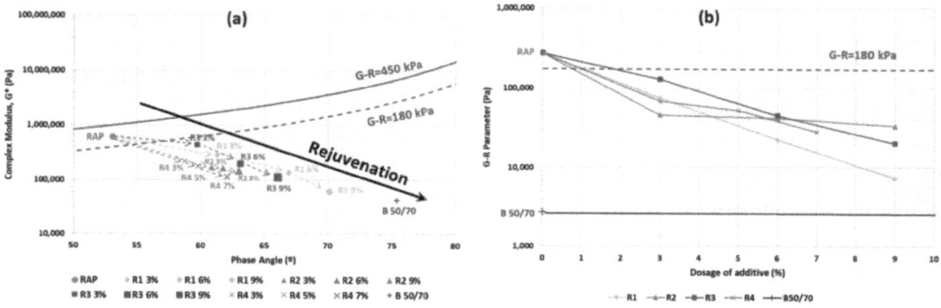

Fig. 3. Average complex modulus and phase angle for an angular frequency of 0.005 rad/s tested at a temperature of 15 °C (a) and average values of the G-R parameter under these circumstances as a function of the additive dosage (b).

4 Conclusions

The present paper has studied the effect produced by various rejuvenators on the rheological properties of aged bitumen extracted from RAP. On the basis of the results obtained, the following conclusions can be drawn:

- It has been demonstrated that the use of rejuvenators reduces the brittleness of aged asphalts, lowers their risk of suffering from cracking, and partially restores their lost rheological properties.
- In general terms, as the dosage of rejuvenator is increased, the aged binder of the RAP tends to obtain rheological properties that are increasingly similar to those of the original non-aged binder. Nonetheless, it has been demonstrated that not all the rejuvenators have the same effect at higher quantities and thus not all of these additives have the same optimal dosage.
- As the test temperature increases, the rejuvenators (regardless of their nature) have a weaker effect on the rheological response of the aged asphalt (which indicates that these additives would be less effective in RAP materials that operate at higher service temperatures).
- There are variations in the capacity of the bio-rejuvenators to restore the lost properties of the aged binder. In spite of the fact that the rejuvenator composed of oleyldiamine ethoxylate (R1) is the most effective in restoring the rheological properties of the aged binder, a similar level of performance can be achieved by using R3 composed of plant-based ester resins.

Acknowledgement. The present study has been conducted within the framework of the RTE research project (Desarrollo de Pavimentos Reciclados Sostenibles de Larga Duración, RTC-2015-3833-4), funded by the Ministry of Economy and Competitiveness of Spain, inside the National Plan RETOS COLABORACIÓN 2015, and co-funded by the FEDER founds.

References

Al-Qadi, I.L., Abauwad, I.M., Dhasmana, H., et al.: Effects of various asphalt binder additives/modifiers on moisture-susceptible asphaltic mixtures title. Research Report FHWA-ICT-14-004, Illinois Center for Transportation (2014)

Dony, A., Colin, J., Bruneau, D., et al.: Reclaimed asphalt concretes with high recycling rates: changes in reclaimed binder properties according to rejuvenating agent. Constr. Build. Mater. **41**, 175–181 (2013). https://doi.org/10.1016/j.conbuildmat.2012.11.031

Durand, G., Lorserie, A., Gueit, C.: Cold recycling process using renewable resources: contribution of rheological and chromatographic methods for rejuvenating mechanism and duration assessment. In: 6th Eurasphalt & Eurobitume Congress, Prague, Czech Republic, 1–3 June 2016

Huang, S., Qin, Q., Grimes, W.R., et al.: Influence of rejuvenators on the physical properties of RAP binders. J. Test. Eval. **43**(3), 594–603 (2014)

Miró, R., Martínez, A.H., Moreno-Navarro, F., et al.: Effect of ageing and temperature on the fatigue behaviour of bitumens. Mater. Des. **86**, 129–137 (2015)

Mogawer, W., Bennert, T., Austerman, A., Ericson, C.: Investigating the aging mitigation capabilities of rejuvenators in high RAP mixtures using black space diagrams, binder rheology and mixture tests. J. Assoc. Asphalt Paving Technol. **85** (2015)

Moghaddam, T.B., Baaj, H.: The use of rejuvenating agents in production of recycled hot mix asphalt: a systematic review. Constr. Build. Mater. **114**, 805–816 (2016)

Ongel, A., Hugener, M.: Impact of rejuvenators on aging properties of bitumen. Constr. Build. Mater. **94**, 467–474 (2015)

Rowe, G.M., Sharrock, M.J.: Cracking of asphalt pavements and the development of specifications with rheological measurements. In: 6th Eurasphalt & Eurobitume Congress, Prague, Czech Republic, June 2016. https://doi.org/10.14311/ee.2016.215

Valentin, J., Valentová, T., Soukupová, L., et al.: Impact of selected WMA additives on normal PMB used for a standardized SMA. In: 6th Eurasphalt & Eurobitume Congress, Prague, Czech Republic, 1–3 June 2016

Zaumanis, M, Mallick, R.B., Frank, R.: Determining optimum rejuvenator dose for asphalt recycling based on Superpave performance grade specifications. Constr. Build. Mater. **69**, 159–166 (2014a). https://doi.org/10.1016/j.conbuildmat.2014.07.035

Zaumanis, M., Mallick, R.B., Frank, R.: 100% recycled hot mix asphalt: a review and analysis. Resour. Conserv. Recycl. **92**, 230–245 (2014b). https://doi.org/10.1016/j.resconrec.2014.07.007

Chemo-Mechanical Characterization of Bituminous Materials: Multiphase Analysis of Binders

Effect of Recycled Materials on Intermediate Temperature Cracking Performance of Asphalt Mixtures

Wei Cao[1(✉)], Louay Mohammad[2], and Peyman Barghabany[3]

[1] Louisiana Transportation Research Center, Louisiana State University, Baton Rouge, USA
wcao@lsu.edu

[2] Department of Civil and Environmental Engineering and Louisiana Transportation Research Center, Louisiana State University, Baton Rouge, USA
louaym@lsu.edu

[3] Louisiana State University, Baton Rouge, USA
peymanb88@gmail.com

Abstract. With increasing use of recycled materials in paving asphalt mixtures, durability against fatigue cracking has been one of the major concerns in pavement design, construction, and performance. This study was aimed to evaluate the intermediate-temperature fracture/fatigue crack resistance of asphalt mixtures using three test methodologies, namely, semi-circular bend (SCB), Texas overlay, and indirect tension (IDT) tests. The obtained parameters are critical strain energy release rate (J_c), fatigue life ($N_{f,OT}$), and dissipated creep strain energy (DCSE), respectively. Three mixture groups were utilized containing reclaimed asphalt pavement (RAP) of up to 40% by recycled binder ratio (RBR). Analysis of variance (ANOVA) was performed on the parameters determined for each mixture group. Mixtures with increased RAP content exhibited reduced fracture/fatigue crack resistance as measured from parameters of tests considered. Further, SCB test was the only one that statistically distinguished all materials from each of the three mixture groups evaluated.

Keywords: Crack resistance · Recycled asphalts · Semi-circular bend Texas overlay · Indirect tension

1 Introduction

Sustainability has been one of the important concerns in pavement design and construction. An overall trend of increasing the amount and variety of recycled materials used in pavement was seen in the past decades. During the 2016 construction season, approximately 76.9 million tons of reclaimed asphalt pavement (RAP) was placed in asphalt pavements according to a recent industry survey conducted by National Asphalt Pavement Association (NAPA) (Hansen and Copeland 2017). The impact of incorporating RAP into asphalt mixtures is generally found to improve stiffness and thus rutting resistance while maintaining a satisfactory or reduced moisture sensitivity (Williams et al. 2011). Yet, the oxidized asphalts may reduce the stress relaxation capability of the mixtures, and thus increase the propensity to fatigue cracking (Daniel et al. 2013).

L. D. Poulikakos et al. (Eds.): RILEM 252-CMB 2018, RILEM Bookseries 20, pp. 229–235, 2019.
https://doi.org/10.1007/978-3-030-00476-7_36

Currently, there are a number of performance evaluation tests that have been developed and used by researchers and state agencies for fatigue characterization of asphalt mixtures as a complement to the Superpave mix design system. Each test uses a different specimen geometry and loading condition to address either one or both phases of crack initiation and propagation. Additionally, data analysis for arriving at the fatigue parameter is based on different theories and hypotheses. This paper presents the experimental effort in fatigue evaluation of RAP mixtures using three selected test methods: semi-circular bend (SCB), Texas overlay, and indirect tension (IDT) tests. The effort presented is part of the pooled fund project titled *Develop Mix Design and Analysis Procedures for Asphalt Mixtures Containing High-RAP and/or RAS Contents*.

2 Objectives

The objectives of this study are to:

- Evaluate the effect of RAP on the intermediate-temperature fatigue crack resistance of asphalt mixtures; and
- Ascertain if different test methodologies would provide consistent fatigue characterization observations.

3 Materials

To achieve the research objectives, a total of seven plant produced hot-mix asphalt mixtures from three participating agencies were utilized. The RAP content was expressed in terms of recycled binder ratio (RBR), which is defined as the percentage of binder from recycled materials in the total asphalt binder. As shown in Table 1, the mixtures provided by the Colorado DOT contained 0% and 18% RAP, and the mixtures supplied by the Florida DOT utilized 17% and 36% RAP. The materials sampled

Table 1. Asphalt mixture composition

Mixture source	Mixture designation	RAP RBR, %	Design asphalt binder content, %	PG of base binder
Colorado DOT	CO-V	0	5.3	64–22
	CO-18RAP	18	5.3	58–28
Florida DOT	FL-17RAP	17	5.2	58–22
	FL-36RAP	36	5.0	52–28
FHWA ALF	ALF-L1-V	0	5.0	64–22
	ALF-L6-20RAP	20	5.0	64–22
	ALF-L5-40RAP	40	5.0	64–22

NOTE: RAP = reclaimed asphalt pavement; RBR = recycled binder ratio; PG = performance grade.

during the construction of the FHWA Accelerated Loading Facility (ALF) test lanes (i.e., Lanes 1, 6, 5) incorporated 0%, 20%, and 40% RAP. Note that for the Colorado and Florida mixtures, increase in the RAP content was addressed with a lower grade base asphalt. Within each group of mixtures, the aggregate type and RAP source were kept constant.

4 Test Methods

This section briefly discusses methodologies employed in the three tests. Plant-produced loose mixtures were compacted in the laboratory using a Superpave gyratory compactor and then trimmed to obtain the required geometries with target air voids of 7.0 ± 0.5%. Specimens were then long-term oven aged at 85 °C for 120 h in accordance with AASHTO R30, Standard practice for mixture conditioning of hot mix asphalt. The IDT test was conducted using a closed-loop servohydraulic system, while the other two were performed using the Asphalt Mixture Performance Tester (AMPT).

4.1 Semi-circular Bend Test

The SCB test was conducted at 25 °C in accordance with ASTM D8044, Standard test method for evaluation of asphalt mixture cracking resistance using the semi-circular bend (SCB) test at intermediate temperatures. Three notch depths were used: 25.4 mm, 31.8 mm, and 38.1 mm, and for each of four semi-circular specimens (150 mm diameter by 57 mm thick) were prepared. During testing, specimen was loaded monotonically with a displacement rate of 0.5 mm/min. until fracture in a three-point bending configuration. This test has been favored by many researchers and practitioners due to the ease of sample preparation and simple testing procedure.

4.2 Texas Overlay Test

The Texas overlay test was conducted at 25 °C in accordance with TxDOT Tex-248-F, Test procedure for Overlay Test. For each mixture, five oval bricks were prepared with 38-mm thickness, 76-mm width, and 150-mm length. The specimen was loaded in the displacement control mode using a triangular waveform at a frequency of 0.1 Hz with a maximum displacement of 0.6 mm. The Texas overlay test addresses both phases of crack initiation and propagation, and has been employed to investigate reflective cracking and fatigue cracking (Germann and Lytton 1979; Zhou et al. 2007).

4.3 Indirect Tension Test

The indirect tension test consisted of two steps: dynamic modulus test and fracture test. This test was conducted at 10 °C in accordance with the well-established procedures documented in the literature (Kim et al. 2004; Roque and Buttlar 1992; Buttlar and Roque 1994). For each mixtures, four disk specimens with 150-mm diameter and 38-mm thickness were prepared and tested. During testing, two sets of extensometers were mounted on both faces of the specimen for the measurement of horizontal and vertical deformations. The dynamic modulus test was conducted at a frequency of 10 Hz; the load level was adjusted to yield a nominal strain in the horizontal direction within 50 ± 5 microstrain. The fracture test was controlled in a monotonic mode with an actuator-displacement rate of 50 mm/min.

5 Results and Discussions

This section outlines the data analysis procedures and presents the fatigue parameters obtained from each test. Discussions are then provided with respect to material composition, i.e., RAP content.

5.1 Semi-circular Bend Test

The collected load-displacement data were processed based on fracture mechanics principle to obtain the parameter referred to as the critical strain energy release rate:

$$J_c = -(1/b)(dU/da) \tag{1}$$

where J_c is the critical strain energy release rate (kJ/m^2), b is specimen thickness (m), a is notch depth (m), and U is strain energy (kJ) determined as the area under the load-displacement curve up to peak load. J_c measures the strain energy required to form a unit area of new surface that is fractured in a medium. A larger J_c value represents higher toughness, and thus is desired for crack resistant mixtures (Cooper et al. 2014; Mohammad et al. 2016).

The obtained J_c results from the SCB test are presented in Fig. 1. The test variability according to the calculated coefficient of variation (CoV) varied between 10% and 14%, suggesting a satisfactory repeatability. The analysis of variance (ANOVA) was performed within each material group, and the results are indicated by the letters A, B, and C, representing the statistically distinct crack resistance from best to worst. It is seen that the SCB test was able to distinguish the mixtures within each group as per the RAP content; the crack resistance was significantly reduced with increase in the RAP dosage.

5.2 Texas Overlay Test

For the Texas overlay test, fatigue life, denoted by $N_{f,OT}$, was determined at the 93% drop in the peak tensile load. $N_{f,OT}$ was averaged from three replicates that were selected out of the five to yield the minimum coefficient of variation (CoV) as suggested by Walubita et al. (2012).

The $N_{f,OT}$ results are given in Fig. 2. The CoV ranged from 7% to 39% with an overall average of 19%. According to the statistical grouping results, the Texas overlay test was able to discriminate the Colorado and Florida mixtures as per the RAP content, but had a difficulty with the three ALF materials. This observation can be attributed to the high variability typically associated with cycle tests as compared to monotonic tests.

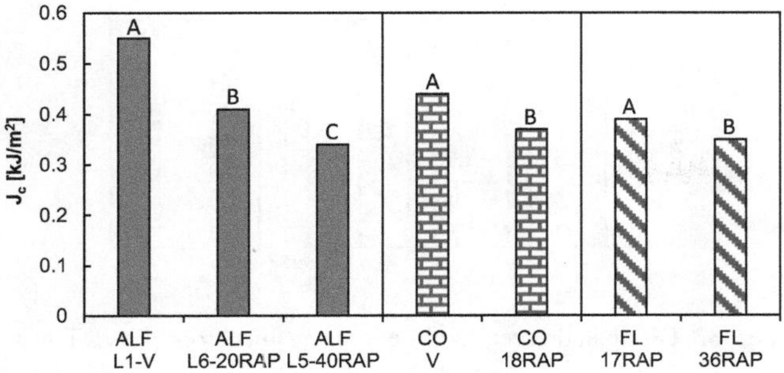

Fig. 1. Critical strain energy release rate results from the SCB test

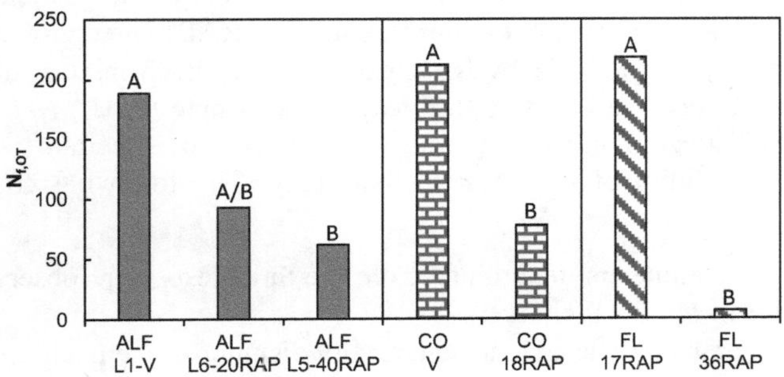

Fig. 2. Fatigue life results from the Texas overlay test

5.3 Indirect Tension Test

Data analysis was based on the concept of the energy threshold for healable microcrack damage: dissipated creep strain energy (DCSE) (Zhang et al. 2001):

$$\text{DCSE} = \text{FE} - \text{EE} \quad \text{with} \quad \text{EE} = 0.5S_T^2/E \tag{2}$$

where DCSE is dissipated creep strain energy (kJ/m^3), FE is fracture energy (kJ/m^3) determined as the area under the load-displacement curve of the fracture test up to peak stress, S_T is tensile strength (kPa), and E denotes dynamic modulus (kPa). Calculation of the dynamic modulus and DCSE was performed following the procedures provided elsewhere (Kim et al. 2004; Roque and Buttlar 1992; Buttlar and Roque 1994).

The obtained DCSE results are provided in Fig. 3. The CoV varied between 8% and 30% with an overall average of 16%. Based on the statistical grouping results, the IDT test was able to distinguish the Colorado mixtures as per the RAP content, but failed to do so for the ALF and Florida mixtures. Nevertheless, DCSE was reduced when more RAP was introduced within each mixture group.

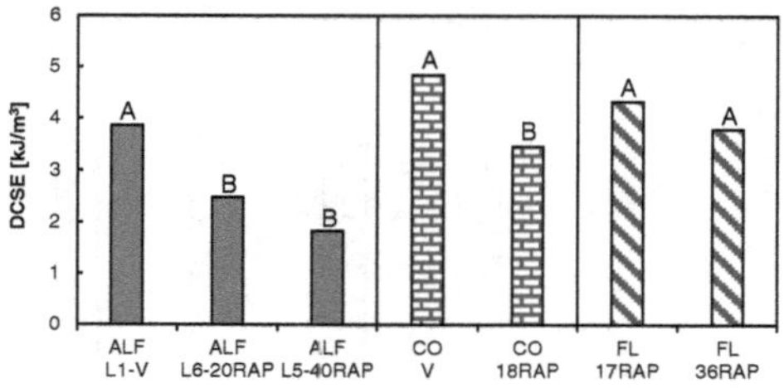

Fig. 3. Dissipated creep strain energy results from the IDT test

6 Summary and Conclusions

This paper presents an experimental effort in characterizing intermediate-temperature fatigue crack resistance of asphalt mixtures containing RAP. Three groups of mixtures with RAP content of up to 40% by RBR were utilized. Each mixture was evaluated using the SCB, Texas overlay, and IDT tests to determine the J_c, $N_{f,OT}$, and DCSE parameters. Statistical comparison was performed to ascertain the mixture-discriminating capability of each test methodology. The following conclusions are drawn:

- In general, all parameters determined from the three tests were observed to reduce with increase in the RAP content;
- The SCB J_c parameter was able to statistically distinguish the materials within each of the three mixture groups, while the Texas overlay and IDT tests had difficulties with one or two mixture groups.

References

Buttlar, W.G., Roque, R.: Development and evaluation of the strategic highway research program measurement and analysis system for indirect tensile testing of asphalt mixtures at low temperatures. Transp. Res. Rec. **1454**, 163–171 (1994)

Cooper Jr., S.B., Mohammad, L.N., Elseifi, M.A.: Laboratory performance of asphalt mixtures containing recycled asphalt shingles. Transp. Res. Rec. **2445**, 94–102 (2014)

Daniel, J.D., Gibson, N., Tarbox, S., Copeland, A., Andriescu, A.: Effect of long term aging on RAP mixtures: laboratory evaluation of plant produced mixtures. J. Assoc. Asphalt Paving Technol. **82**, 327–365 (2013)

Germann, F.P., Lytton, R.L.: Methodology for predicting the reflection cracking life of asphalt concrete overlays. Report No. FHWA-TX-79-09 + 207-5. Texas State Department of Highways and Public Transportation, Austin, TX (1979)

Hansen, K.R., Copeland, A.: Asphalt Pavement Industry Survey on Recycled Materials and Warm-Mix Asphalt Usage: 2016. National Asphalt Pavement Association, Lanham (2017)

Kim, Y.R., Seo, Y., King, M., Momen, M.: Dynamic modulus testing of asphalt concrete in indirect tension mode. Transp. Res. Rec. **1891**, 163–173 (2004)

Mohammad, L.N., Kim, M., Challa, H.: Development of performance-based specifications for Louisiana asphalt mixtures. Report No. FHWA/LA.14/558. Louisiana Transportation Research Center, Baton Rouge, LA (2016)

Roque, R., Buttlar, W.G.: The development of a measurement and analysis system to accurately determine asphalt concrete properties using the indirect tensile mode. J. Assoc. Asphalt Paving Technol. **61**, 304–332 (1992)

Walubita, L.F., Faruk, A.N., Das, G., Tanvir, H.A., Zhang, J., Scullion, T.: The Overlay Tester: a sensitivity study to improve repeatability and minimize variability in the test results. Report No. FHWA/TX-12/0-6607-1. Texas Transportation Institute, College Station, TX (2012)

Williams, R.C., Cascione, A., Haugen, D.S., Buttlar, W.G., Bentsen, R.A., Behnke, J.: Characterization of hot mix asphalt containing post-consumer recycled asphalt shingles and fractionated reclaimed asphalt pavement. Final Report, Iowa State University, Ames, IA (2011)

Zhang, Z., Roque, R., Birgisson, B., Sangpetgnam, B.: Identification and verification of a suitable crack growth law for asphalt mixtures. J. Assoc. Asphalt Paving Technol. **70**, 206–241 (2001)

Zhou, F., Hu, S., Chen, D.H., Scullion, T.: Overlay tester: simple performance test for fatigue cracking. Transp. Res. Rec. **2001**, 1–8 (2007)

Investigation of the Calculation Modeling of Asphalt Binder Surface Energy Based on the Atomic Force Microscope (AFM)

Rong Chang[1(✉)], Erhu Yan[1], Jian Xu[1], and GaoChao Wang[2]

[1] Research Institute of Highway Ministry of Transport, Beijing, China
{r.chang, eh.yan, j.xu}@rioh.cn
[2] ChangSha University of Science and Technology, Changsha, China
wang190808@126.com

Abstract. The adhesion between asphalt binder and aggregate is one of the most important properties of asphalt mixture, and the surface energy is a widely used parameter to evaluate the adhesion property. The conventional method of calculating the surface energy is by measuring the contact angle between the reference sample and bituminous materials, then the Young's equation was conducted for the calculation procedure. However, this is a time consumed method. In this paper, a new procedure was proposed combined with the Atomic Force Microscope (AFM) experimental results, two calculation methods, Johnson-Kendall-Roberts (JKR) and DMT model, and the Fowke theory was conducted for this purpose; the conventional method was used as the reference to evaluate and compare the two methods. Results showed that the JKR(S) model has a better result compared with the DMT model, and the experimental procedure is more reliable, hence, this procedure is suggested as a new method to calculate the surface energy of asphalt binder.

Keywords: Asphalt Pavement · Surface energy · Adhesion work
Atomic force microscopy (AFM) · JKR(S) model

1 Introduction

The surface energy and adhesion work between the asphalt binder and aggregates is one of the common methods to evaluate the adhesion of asphalt mixture (Xiao et al. 2007). The current method (Wang 2013) is by using a referenced liquid, then measuring the contact angle between asphalt binder and aggregates, respectively. Next, the Young's equation is applied to calculate the surface energy. In recent years, an emerging micro structure technology, Atomic Force Microscope (AFM), was widely used to investigate the bituminous materials (Wang 2010; Wang et al. 2012).

In this paper, the objective is to propose a new procedure to calculate the surface energy and adhesion between asphalt binder and aggregate. Three base asphalt binders were selected and tested, a multiscale calculation of surface energy was conducted for each of the asphalt binder, respectively. Next, the contact angle was measured based on the AFM devices. And then, two calculation methods were conducted and compared, a

L. D. Poulikakos et al. (Eds.): RILEM 252-CMB 2018, RILEM Bookseries 20, pp. 236–241, 2019.
https://doi.org/10.1007/978-3-030-00476-7_37

conventional method based on Young's equation was used as the reference. Finally, the results were analyzed, and the suggested method was proposed.

2 Materials and Testing

Three asphalt binders were used in this study, two base asphalt binders 70# and 90#, which were graded using penetration method according to the Chinese national standard 'Technical Specifications for Construction of Highway Asphalt Pavements (JTG F40 2004), and one polymer modified asphalt binder.

The sample preparation procedure has a significant influence on the experimental results, according to the authors' previous study, the oven method was selected in this paper. Firstly, the two base asphalt binders and one modified binder were put into the oven until the materials were melt at 135 °C and 150 °C, respectively. Next, tinfoil was taped around the cover glass to form a rectangular sample area, then a few of uniformly mixed asphalt binder sample were dropped on one side, and the samples can be settled on the stand. Then, the samples were then put into the oven, with a settlement scope 25%, for condition 30 min at a constant temperature 130 °C. Finally, the samples can be taken out for the next step.

AFM and the contact angle measurement devices were then performed on the entire set of asphalt binders prepared. The available devices are Agilent 5500 AFM and FTA 1000 in this study, respectively. For the AFM measurement, four representative points were uniformly selected at a contact mode, characteristic parameters were extracted from the mechanical curve. A minimum of two replicates for each binder type, and four replicates for binder sample were tested in this study.

3 Results and Analysis

3.1 Results and Analysis of AFM

Mechanical curve represents a simple tool for understanding the response of surface elasticity, hardness, Hamaker constant, charge density, and adhesion of the bituminous materials. Based on the AFM experimental data, different models, Hertz, DMT, and JKR(S) (Hua et al. 2003), can be used to extract the characteristic parameters from the mechanical curves. For the Hertz model (Hua et al. 2003), the basic assumption is that the material is a pure elastomer and, no adhesion exists on the contact surfaces. And for the DMT model (Shen 2014), only adhesion in the area out of the contact outer ring is considered, the adhesion between the contact surfaces is can be ignored in this model. The application sphere of this method is small curvature radius, low adhesion work and high elastic modulus system. The formula can be expressed as Eq. 1. The assumption of JKR model (Shen 2014) is that partial deformation is acceptable between the contact surfaces; the short-range surface adhesion is only employed at the contact area instead of the outer range of the contact area. Hence, this model is suitable for a system with large radius of curvature, high adhesion work and low elastic modulus. This model is expressed as Eq. 2.

$$F_{ad} = -2\pi\omega R \tag{1}$$

$$F_{ad} = -1.5\pi\omega R \tag{2}$$

where F_{ad} is the adhesion; ω is the adhesion work; R is the contact radius.

The scale of AFM based method is between the DMT and JKR models (Huang 2007); hence, in the purpose of evaluation the better one, these two models were calculated and compared with the conventional method, respectively. The force-displacement curve represents the change of the interaction force between the probe and the asphalt surface during the process of penetration and taking out. The maximum adhesion occurs when the vertical difference between the minimum value of the curve and the initial position of the probe. During the maximum adhesion testing, several external conditions, such as elasticity coefficient temperature, and humidity, maybe effect the experimental results. These external conditions should maintain in a narrow range; otherwise, the mechanical curve is unable to fit well. The mechanical curve make sense, only when the elasticity coefficient between 1 N/m and 10 N/m. Therefore, the experiment conditions should be controlled quite carefully.

The surface energy can be decomposed into two parts: a dispersive component and a non-dispersive component. The relationship is shown in Eq. 3:

$$W_{ad} = 2\sqrt{\gamma_a^d \gamma_b^d} \tag{3}$$

where W_{ad} is the adhesion work between two materials a and b; γ_a and γ_b are surface energy of the two materials, respectively. In the Fowkes model, the relationship exists $\gamma_a = \gamma_a^d$, $\gamma_b = \gamma_b^d$, when one of the materials is non-polar material, Asphalt is a nonpolar material (Huang 2007). Since γ_b can be obtained by calibration the probe, the adhesion work and the surface energy can be calculated, and the results are displayed in Table 1.

Table 1. The calculated results of adhesion and asphalt surface energy by different models

Materials	#70	#90	SBS I-D
Adhesion (nN)	8.47	7.43	7.97
JKR adhesion work (mJ/m^2)	224.67	197.09	211.41
JKR surface energy (mJ/m^2)	28.03	21.57	24.82
DMT adhesion work (mJ/m^2)	168.51	147.82	155.56
DMT surface energy (mJ/m^2)	15.77	12.13	13.44

3.2 Results and Analysis of Contact Angle Test

According to the physical and chemical theory of the material surface, the surface energy of the liquid-solid interface and the polarity of London dispersive force γ^d and hydrogen bond γ^p are taken into consideration, Equation can be calculated by combining Young's equation (Owens and Wndt 1969):

$$\frac{1+\cos\theta}{2}\frac{\gamma_l}{\sqrt{\gamma_l^d}} = \sqrt{\gamma_s^p}\sqrt{\frac{\gamma_l^p}{\gamma_l^d}} + \sqrt{\gamma_s^d} \quad (3)$$

where θ is the contact angle; γ_s, γ_s^d and γ_s^p are the surface energy, dispersion component and polar component of the solid material, respectively; and $\gamma_s = \gamma_s^d + \gamma_s^p$; γ_l, γ_l^d and γ_l^p are the surface energy, dispersion component and polar component of the liquid material, respectively; and $\gamma_l = \gamma_l^d + \gamma_l^p$. By using the given liquids in Table 2(a), the values of γ_l, γ_l^d and γ_l^p as shown in Table 2(a). Two types of liquids were used to calculate the contact angle, results were the listed in Table 2(b); And the corresponding surface energy were computed and shown in Table 2(c).

Table 2. **(a)** Given liquid surface energy (mJ/m^2)

	γ_l	γ_l^d	γ_l^p
Water	72.8	21.8	51.0
Ethylene glycol	64.0	34.0	30.0
Glycerol	63.4	37.4	26.0

Table 2. **(b)** Contact angle between liquid and the asphalt binder sample

	#70	#90	SBS I-D
Water	94.6	102.9	92.1
Glycerol	82.9	91.5	82.1

Table 2. **(c)** Contact angle method Asphalt surface energy and component

	γ_s	γ_s^d	γ_s^p
#70	30.18	28.30	1.88
#90	24.64	23.72	0.92
SBS I-D	25.75	22.37	3.38

The surface energy comparison between the conventional contact angle method and the ones from AFM technology are shown in Fig. 1. It can be concluded that the surface energy calculated by using the contact angle is higher than the ones based on AFM. The reason maybe be like this: the Fowke's model did not consider the forces caused by polar bonds whereas bituminous materials have a small amount of polar bonds as organic matter; hence, the surface energy obtained by AFM is smaller than that using a contact angle test. However, the results of contact angle test are similar to those of JKR model, which the minimum difference is only 4%. The deviation from the data measured by the DMT model is relatively large. The difference among the results of three asphalt tests can be as high as 50%. Therefore, when the AFM technology is used to calculate the surface energy of asphalt binder, the JKR model is recommended.

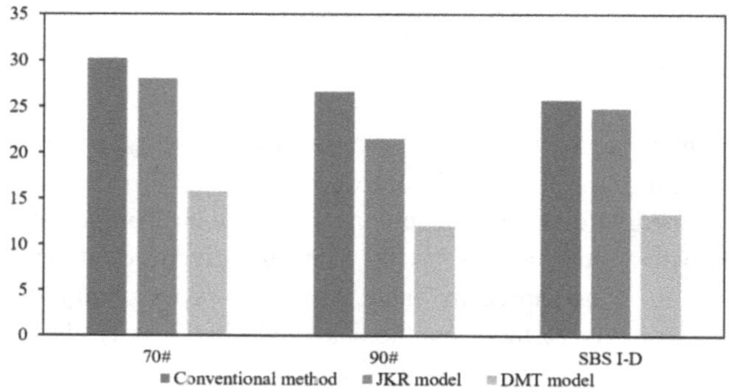

Fig. 1. Comparison of surface energy between different methods

4 Conclusion

(1) The surface energy obtained based on AFM can be well fitted with the result from conventional method; therefore, this method is suggested as a test and calculation procedure to calculate the surface energy of the asphalt binder.
(2) When the AFM technology is applied to calculate the surface energy of asphalt binder which abstained from contact angle method, JKR(S) model shown better results compared with the DMT method.
(3) The advantage of AFM includes: small human error, high test accuracy, good fitting f mechanical curve, less amount of sample during the test procedure, and easy to operation. The nanoscale microscopic technology show potential a new evaluation technology in the aspect of the microscopic performance investigation on bituminous materials.

References

Hua, J., Xu, S.M., Xu, J.M.: Principles of force-distance curve determination by atomic force microscopy. Met. Mine **1**, 25–30 (2003)

Huang, H.Z.: Surface Chemical Analysis. East China University of Science and Technology Press, Shanghai (2007). (in Chinese)

JTG F40: Technical Specifications for Construction of Highway Asphalt Pavements. Ministry of Communications of the People's Republic of China, Beijing (2004). (in Chinese)

Owens, D.K., Wendt, R.C.: Estimation of the surface free energy of polymers. J. Appl. Polym. Sci. **13**(8), 1741–1747 (1969)

Shen, Q.: Polymer Surface Chemistry. Science Press, Beijing (2014). (in Chinese)

Wang, L.L.: Based on the surface free energy of asphalt and mineral materials adhesion effect. Master thesis, Chang'an University (2013). (in Chinese)

Wang, Y.: Study on Adhesion between Asphalt and Aggregates using Surface Energy Theory. Ph.D. thesis, Hubab University (2010). (in Chinese)

Wang, Y.Y., Ling, T.Q., Shi, Z.H.: Research on the adhesion of asphalt and aggregates based on surface energy. New technology and application of special concrete and asphalt concrete, pp. 274–276 (2012)

Xiao, Q.Y., Hao, P.W., Xu, O.M., Wang, H.N., Feng, X.J.: New method for evaluating adhesion between asphalt and aggregate. J. Chang'an Uni. (Natur. Edi.) **1**, 19–22 (2007). (in Chinese)

Qualitative Detection of the Presence of Gilsonite in the Bituminous Blends Based on Thin Layer Chromatography

Michalina Makowska(✉) and Terhi Pellinen

Department of Civil Engineering, Aalto University, Espoo, Finland
michalina.makowska@aalto.fi

Abstract. The natural asphalts, of which Gilsonite is a representative, are modifiers used to increase the stiffness modulus of the asphalt concrete. For quality control purpose, recognizing if such material is present in the final blend is of an interest. The thin layer chromatography using a flame photometric detector (FPD), in addition to the typical flame ionization detector, was demonstrated hereby as a viable analytical tool for this problem. Gilsonite also contains the material soluble in solvent used in the development of the fraction referred to as aromatics. However, for straight run bitumen and Gilsonite the color of that fraction is different as well as their mobility. Due to the lower mobility of Gilsonite molecules on the stationary phase, the elution stops at different position than for fresh bitumen, convoluting the signal in the chromatogram region typically assigned for resins. The presence of Gilsonite is identifiable visually after the second development bath, but also with FPD from the final chromatogram.

Keywords: Thin Layer Chromatography · Flame Photometric Detector Gilsonite

1 Introduction

There are specific applications in paving industry where the high stiffness modulus of asphalt is preferred such as the bound base courses of the heavily loaded roads or the highly rut resistant surface layers. The stiffness modulus of asphalt concrete relates to the stiffness modulus of bitumen (Pellinen et al. 2007). In order to improve resistance to the permanent deformation of asphalt concrete and to increase its modulus, different bitumen modifiers have been investigated over the years, e.g. natural asphalts (NA) (Widyatmoko and Elliott 2008), recycled asphalt shingles (RAS), and recycled asphalt pavements (RAP). However, when certain modifiers such as NA mix with the bitumen, the detection of the modifier from the bitumen blend becomes challenging. Considering three blends with such modifiers at an equal stiffness of the binder, polymer modification (e.g. through RAS) is identifiable and quantifiable using infrared spectroscopy (Zofka et al. 2015), while sadly distinguishing between NA and aged bitumen poses challenge.

L. D. Poulikakos et al. (Eds.): RILEM 252-CMB 2018, RILEM Bookseries 20, pp. 242–247, 2019.
https://doi.org/10.1007/978-3-030-00476-7_38

However, there is a chemical difference between the bitumen and NA, as both the straight run bitumen and the NA differ in polarity as witnessed by their Hansen Solubility Parameters (HSP) (Redelius 2000; Hansen 2007), at least as reported for Gilsonite and various bitumen types. The HSP determination methodology does not allow for quantification of the amount of material, which becomes miscible in a chosen solvent, thus having lower applicability for blend analysis. Potentially, Thin Layer Chromatography (TLC) could allow differentiation between the two materials and even their blends, as it incorporates both information about solubility and polar interaction with the stationary phase (Spangenberg et al. 2011), while the signal can be quantified by the use of detectors. Some of the most applied detectors are based on the light absorbance because the chromatography deals with separation of the mixtures of the colorful compounds (chroma – *gr.* color). Since the separated compounds are often organic, there are also the flame ionization detectors (FID) coupled with the technique, which scintillate the carbon cations using the change in conductivity of the exhaust gas.

The equipment typically used in the bituminous materials industry to perform the TLC is IATROSCAN (Ogasawa et al. 2002). It can be equipped in the flame ionization detector (FID) or with Flame Photometric Detector (FPD). The FPD is in principle an FID, extended by a filter and a photomultiplier tube (Ogasawa et al. 2002). Due to the increasing interest in sulfur (S) distribution in bituminous fractions and their effect on bitumen properties (Makowska et al. 2017; McKenna et al. 2013) the FPD with a filter at 394 nm wavelength is explored (S-FPD).

The bitumen divides into fractions based on the polarity of the molecules with various techniques, and the Corbett technique is the most standard division into Saturates, Aromatics, Resins and Asphaltenes (SARA) (Lehto 1988). Currently, the TLC-FID based SARA fractionation technique is dominant due to its simplicity and low solvent requirement. However, the fractions of the column chromatography (two types of stationary phase) are not exactly the same as those obtained in TLC-FID (one stationary phase) and it was proposed that the products of those separations be named differently (Masson et al. 2001). Additionally, Jiang et al. (2008) pointed out that the fraction obtained in column chromatography, when eluted in TLC-FID program can contribute to more than one peak.

Moreover, comparing the SARA fraction results from TLC-FID performed in different laboratories is challenging due to the use of the differing solvents and times of elution (often not reported in literature) (Masson et al. 2001; Higuerey et al. 2002; Tabatabaee and Kurth 2017; Paliukaite et al. 2014). The laboratories choose the method for the inside quality control of the product, or for the comparison between the products in development stage, based on inside correlations (Lehto 1988; Teugels and Zwijsen 1991). The elution regime is stable within a laboratory to assure the good repeatability, but the economic factors seem to play role in the choice of the parameters in the method (e.g. cost of the rods, length of the process).

Nevertheless, the technique is still providing information about the solubility and the polarity, which is quantifiable. As an example, hereby article was inspired by an observation of a different elution pattern from samples of bitumen modified with Gilsonite (GIL) in the consecutive development steps of typical SARA fractioning through TLC technique. In the development, in so-called second bath, by n-heptan/toluene (80:20% vol.), which for straight run bitumen elutes to the height of about 5 cm from origin, the GIL sample was staining the rods only close to the original spot (approximately to the height of 2–3 cm). It also had a different color (yellow for 70/100 and brown-red for GIL). This observation could be helpful during forensic analysis of the asphalt concrete samples in terms of identification of the presence of GIL as a modifier. Thus, this article investigates how the incorporation of the S-FPD supports qualitative identification of Gilsonite as a modifier in the bituminous blends.

2 Materials and Methodology

The natural asphalt, namely Gilsonite (GIL), and paving grade bitumen 70/100 were used in the analysis. Gilsonite is allowed in the Finnish Asphalt Specifications (Finnish Pavement Technology Advisory Council 2017) as a modifier of asphalt concrete to improve resistance to permanent deformation. It is added at about 0.5% wt. per mass, which makes the approximate ratio between bitumen and Gilsonite in the asphalt concrete about 90:10 by weight and such blend was prepared as a reference for investigation (BLEND).

A solution of approximately 1% vol. of material in chloroform was prepared. 2 μl drop was spotted on each rod. The material studied was then spotted on the rods and after that dried in nitrogen flow for 10 min. One rod in the series stays clean and acts as a reference to evaluate if the full solvent evaporation occurred (Fig. 1).

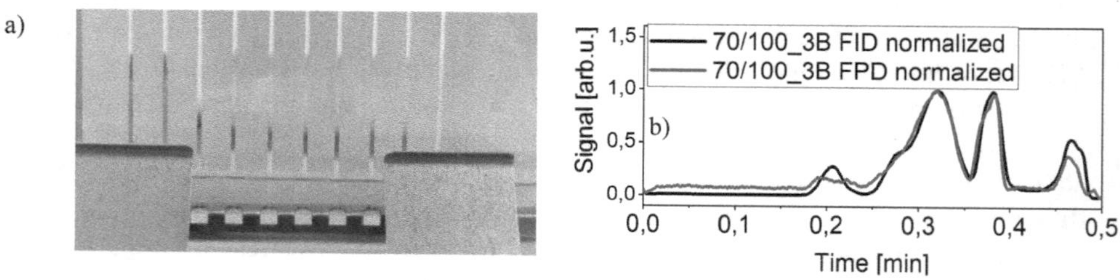

Fig. 1. (a) The picture taken during the development in the third solvent (from the left): GIL (rods no. 1–3), bitumen 70/100 (rods no. 4–6), bitumen BLEND (rods no. 7–9), reference (rod no. 10); (b) an example of the signals obtained from FID and FPD detectors simultaneously from one rod after normalization

A series of 3 baths was prepared: (1) n-heptan – 30 min (9 cm), (2) n-heptan/toluene (80:20% vol.) – 10 min (5 cm), (3) dichloromethane/methanol (95:5% vol.) – 3 min (2.5 cm). The investigated materials were measured in three series (Fig. 2). The first was analyzed according to the classical TLC-FID methodology where bath 1, 2 and 3 were used (3B); the second group was developed only in bath 1 and 2 (2B) and the third group was developed in bath number 1 only (1B) (inspired by the works of Lu et al. (2008). Three repetitions were performed for each sample and each series. Unfortunately, one of the pure Gilsonite repetitions in 3B failed (Fig. 1). Between immersions in each bath and afterwards, rods were removed from development tanks and dried in pure nitrogen flow for a minimum of 5 min. After the development the full pyrolysis scan (30 s/rod) was applied on the rods for all of the groups, using Iatroskan MK-6/6S, equipped in the typical Flame Ionisation Detector, as well as Flame Photometric Detector focused on the detection of the sulfurous compounds.

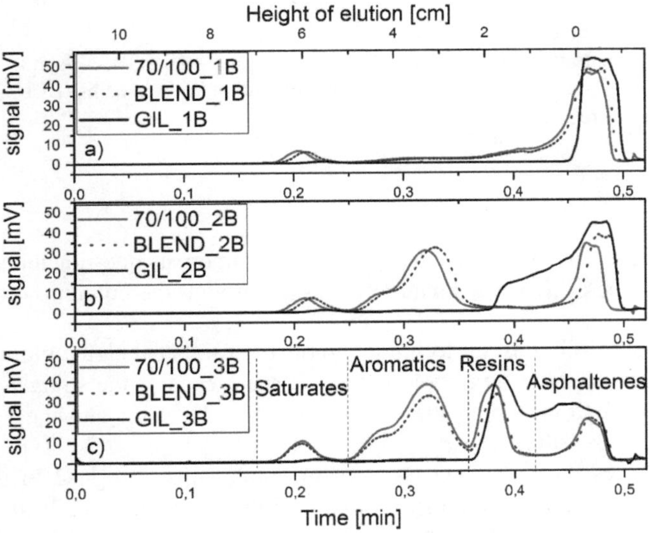

Fig. 2. The chromatogram from TLC-FID of 70/100, GIL and BLEND after (a) 1B, (b) 2B and (c) 3B with visual division of SARA fractions typically assigned in the procedure

3 Results and Discussion

As presented in Fig. 2c the final chromatogram is dividable into four major peaks representing the saturates, aromatics, resins and asphaltenes, or areas where such peaks should be observed. However, due to the overlap of the GIL signal from second bath (Fig. 2b), the position of the valley between the resin and asphaltene characteristic peak is altered for GIL and its blends. The phenomenon demonstrated in Fig. 2 suggests that the molecules soluble in the same solvent of second bath from GIL are expressing lower mobility on the stationary phase (higher polar interaction with solid phase or different size).

As can be seen the majority of the sulfur signal is coming from the peak formed as a result of 2B (Fig. 1) (Ogasawa et al. 2002). Incidentally, it is the peak imposed in the mentioned valley region. The blend of the GIL and the fresh bitumen is indeed giving a slight peak by the S-FPD in that region (Fig. 3), which could prove useful for the qualitative recognition of GIL presence.

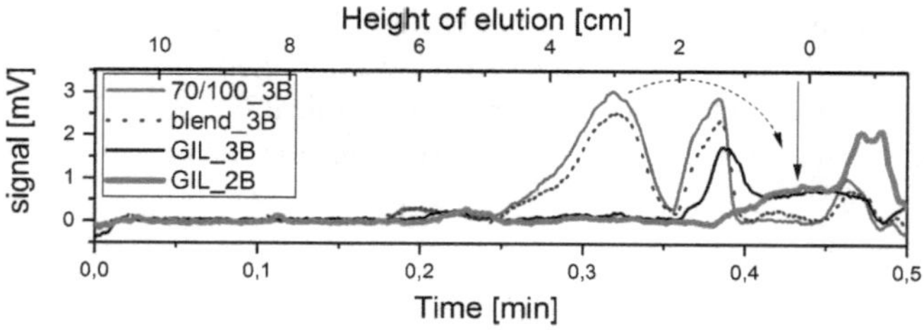

Fig. 3. The chromatogram from TLC-FPD (Sulphur) of 70/100_3B, Gilsonite (GIL_3B and GIL_2B) and blend of the two in the 90:10 ratio after elution in three baths (BLEND_3B), where the arrows indicate region of interest

4 Conclusion

The thin layer chromatography separation of bitumen is questionable in terms of comparison to the SARA fractionation obtained through the column chromatography technique.

However, it is still a useful technique to distinguish between the bitumen samples, as demonstrated on the example of analysis of the blends of the straight run bitumen and Gilsonite, employing both the Flame Ionization Detector and the Flame Photometric Detector (394 nm).

The molecules developed by the solvent combination assigned to the signal of aromatics in the straight run bitumen, solubilized the molecules of Gilsonite. However, their elution upwards the rod was hindered due to their interaction with stationary phase of TLC rods, imposing their signal in the region typically assigned to the resin signal. This fact is observable by an operator due to the difference in color, but also as demonstrated by the utilization of S-FPD detector. The research demonstrated how this analytical technique is applicable for the qualitative recognition of the Gilsonite presence in the bituminous blends.

References

Finnish Pavement Technology Advisory Council: Finnish Asphalt Specifications 2017. Premedia Helsinki Oy, Helsinki (2017)

Hansen, C.: Hansen Solubility Parameters - A user's Handbook. CRC Press, Boca Raton, Fl (2007)

Higuerey, I., Orea, M., Pereira, P.: Estimation Of visbroken and selective catalytic steam cracked product stability using iatroscan TLC-FID. Fuel Chem. Div. Prepr. **47**(2), 656–658 (2002)

Jiang, C., Larter, S., Noke, K., Snowdown, L.: TLC-FID (Iatroscan) analysis of heavy oil and tar sand samples. Org. Geochem. **39**, 1210–1214 (2008)

Lehto, E.: Bitumin fraktiointi TLC/FID-menetelmällä [translation: Bitumen fractionation by TLC-FID method]. University of Oulu, Oulu (1988)

Lu, X., Kalman, B., Redelius, P.: A new test method for determination of wax content in crude oils, residues and bitumens. Fuel **87**, 1543–1551 (2008)

Makowska, M., Hartikainen, A., Pellinen, T.: The oxidation of bitumen witnessed in-situ by infrared spectroscopy. Mater. Struct. **50**, 189 (2017)

Masson, J.-F., Price, T., Collins, P.: Dynamics of bitumen fractions by thin-layer chromatography/flame ionization detection. Energy Fuels **15**, 955–960 (2001)

McKenna, A.M., Marshall, A.G., Rodgers, R.P.: Heavy petroleum composition. 4. asphaltene compositional space. Energy Fuels **27**, 1257–1267 (2013)

Ogasawa, M., Tsuruta, K., Arao, S.: Flame photometric detector for thin-layer chromatography. J. Chromatog. A **973**(1–2), 151–158 (2002)

Paliukaite, M., Vaitkus, A., Zofka, A.: Evaluation of bitumen fractional composition depending on the crude oil type and production technology, Vilnus, Lithuania, 22–23 May 2014

Pellinen, T., Zofka, A., Marasteanu, M., Funk, N.: Asphalt mixture stiffness predictive models (with discussion). J. Assoc. Asphalt Paving Technol. From Proc. Tech. Sessions **76**, 575–626 (2007)

Redelius, P.: Solubility parameters and bitumen. Fuel **79**, 27–35 (2000)

Spangenberg, B., Poole, C., Weins, C.: Quantitative Thin-Layer Chromatography (A practical Survey). Springer, Berlin Heidelberg (2011)

Tabatabaee, H., Kurth, T.: Analytical investigation of the impact of a novel bio-based recycling agent on the colloidal stability of aged bitumen. Road Mater. Pavement Des. **18**(sup2: EATA 2017), 131–140 (2017)

Teugels, W., Zwijsen, M.: Analysis of the generic composition. Appraisal of the Iatroskan method. Paper presented at the International Symposium Chemistry of Bitumens, Rome (1991)

Widyatmoko, I., Elliott, R.: Characteristics of elastomeric and plastomeric binders in contact with natural asphalts. Constr. Build. Mater. **22**, 239–249 (2008)

Zofka, A., Maliszewska, D., Maliszewski, M., Boratynski, J.: Application of FTIR ATR method to examine the polymer content in the modified bitumen and to assess susceptibility of bitumen to ageing. Road Bridges **14**, 163–174 (2015)

Resistance to Moisture-Induced Damage of Asphalt Mixtures and Aggregate-Binder Interfaces

Jorge Lucas Júnior(✉), Lucas Babadopulos, and Jorge Soares

Centro de Tecnologia, Universidade Federal do Ceará, Fortaleza, Brazil
{j.lucas.j,lucasbaba,jsoares}@det.ufc.br

Abstract. The aggregate-asphalt binder adhesiveness is considerably affected by moisture. This property may accelerate distresses in asphalt pavement surface courses when aggregate-binder compatibility is not adequate. Numerous tests have been developed to measure this property, such as the Moisture-Induced Damage test (AASHTO T283) and the Asphalt Bond Strength test (AASTHO TP 91). The former measures the indirect tensile strength of dry and moisture-conditioned asphalt mixtures. The latter measures the tension stress needed to remove a pullout stub with asphalt binder from a solid substrate. Although these are distinct tests, both may be used as a means to analyze how moisture affects a certain combination of mineral aggregate and asphalt binder. The objective of this work is to correlate the results of these two tests and to analyze if the tested combinations of materials follow the same trend under both circumstances. Chemical compositions of the tested aggregates are also investigated. Results from Moisture-Induced Damage tests show that the use of an antistripping agent may reduce the impact of moisture-damage on tensile strength and that the aggregate with a higher content of calcium oxide performed better. The 12 h moisture conditioning in the Asphalt Bond Strength test had a strong correlation with the dry mixes, which was not noticed for the mixtures conditioned for 3 and 6 h. Asphalt mixture's moisture sensitivity results correlated well to the loss of adhesion in the aggregate-binder interface.

Keywords: Adhesion · Moisture-induced damage · Asphalt Bond Strength Chemical composition

1 Introduction

The aggregate-binder adhesiveness within asphalt mixtures is appreciably affected by moisture (Hicks 1991). Low adhesion can lead to defects in pavement asphalt surface courses. According to Sebaaly et al. (2015), two types of additives may be used to prevent such distress: (i) antistripping agents, and (ii) lime (which contains CaO). The Chemical Reaction Theory attempts to explain how the aggregate-binder adhesion is affected by the chemical composition of the aggregates and the asphalt binder. Chaturabong and Bahia (2016) account the chemical nature of mineral fillers are significant factors in the moisture damage resistance of mixture asphalt. Bagampadde

L. D. Poulikakos et al. (Eds.): RILEM 252-CMB 2018, RILEM Bookseries 20, pp. 248–253, 2019.
https://doi.org/10.1007/978-3-030-00476-7_39

et al. (2005) reported that aggregates which alkali metallic elements like sodium (Na_2O) and potassium (K_2O) exhibited high moisture sensitivity.

Several authors try to compare results of different adhesion tests to predict the behavior of materials (either asphalt binder or asphalt mixture properties). Kanitpong and Bahia (2005) compared pull-off tensile strength (POTS) results with the cohesion test performed in a Dynamic Shear Rheometer. There was a strong correlation between these two tests' results. Huang et al. (2010) mentioned that the Dynamic Modulus and the Resilient Modulus of the moisture conditioned mixtures presented similar trends when compared with the results of tensile strength ratio (TSR) measured by the modified Lottman test (AASHTO T823).

The objective of the present work is to study the effect of moisture on the adhesiveness measured in compacted asphalt mixtures as well as on the corresponding aggregate-binder interface, relating these two behaviors. Two types of aggregates with pure or modified asphalt binder (with antistripping additive) are investigated. The chemical composition of each aggregate was obtained by X-ray Fluorescence, and the adhesiveness was measured by Moisture-Induced Damage test on asphalt mixtures and by Asphalt Bond Strength test on the aggregate-binder interface.

2 Materials

Two types of aggregate were used - (i) Phonolite ("Ph") and (ii) Granite ("G") and two asphalt binders - one pure asphalt binder ("P"), and this same binder modified by an antistripping agent ("A"). To determine the chemical composition of the aggregates, X-Ray Fluorescence was used. Table 1 presents a summary of the characterization for materials investigated.

Table 1. Basic characterization results of the tested aggregates and asphalt

		Al_2O_3	SiO_2	K_2O	CaO	MnO	Fe_2O_3	Others
Aggregates	Phonolite [%]	12.64	49.37	14.71	5.58	1.05	12.12	4.53
	Granite [%]	6.89	49.09	5.88	11.13	0.60	11.49	15.92
Pure asphaltic binder	Penetration, 25 °C [dmm]		Softening point [°C]		Temperature susceptibility		Rotational viscosity, 135 °C [cP]	
	50		49		−1.4		395	

Antistripping agent: amine-based liquid anti-stripping additive

3 Moisture-Induced Damage

For the Moisture-Induced Damage test (AASHTO T283), the asphalt mixtures were produced with 7 ± 0.5% air voids. Specimens contained 5% in mass of asphalt binder, according to the Superpave design method, targeting 4% air voids for 100 gyrations on a Superpave gyratory compactor. Three specimens were submitted to indirect tensile strength (ITS) test and three others were conditioned to moisture. The three conditioned specimens were submitted to 70–80% saturation and were cooled to –18 ± 3 °C for 16 h. They were then taken to a water bath at 60 ± 1 °C for 24 ± 1 h. In the subsequent step, the specimens were placed in a water bath at 25 ± 1 °C for 2 to 3 h and, then, conditioned indirect tensile strength (ITS_C) was obtained. The indirect tensile strength ratio (*TSR*), in percentage, is determined by "Eq. 1".

$$TSR = \frac{ITS_C}{ITS} \cdot 100 \tag{1}$$

Figure 1(a) and (b) show that the mixture with granitic aggregate is more resistant to moisture damage than the mixture with phonolytic aggregate. For the mixtures with pure binder, the Ph-P blend had TSR equal to 30%, whereas for the G-P mixture it was 55%. For proper moisture resistance, standards recommend TSR of at least 80%. The observed difference between Ph mixtures and G mixtures was maintained even after adding the antistripping agent, with TSR of 80% for Ph-A and 89% for G-A. The better moisture damage resistance of the mixture with granitic aggregate when compared to the phonolytic one can be explained by their respective contents of CaO and K_2O (Table 1 and Fig. 1(b)).

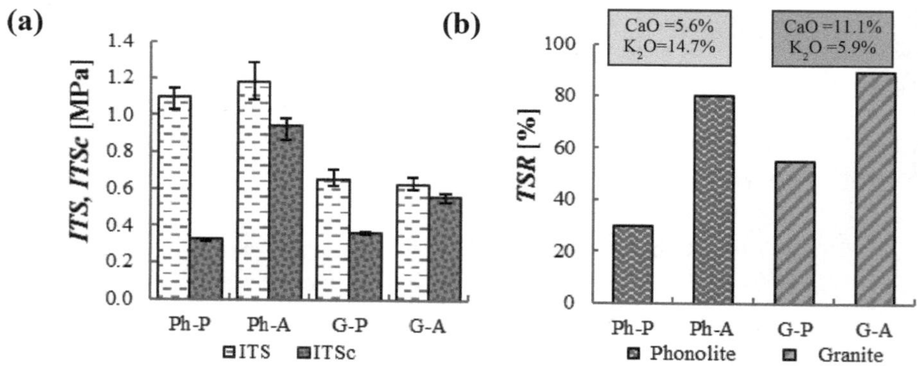

Fig. 1. Moisture-induced damage of the tested asphalt mixtures

4 Asphalt Bond Strength (ABS)

Asphalt bond strength was obtained according to AASHTO TP 91-11. The mineral substrates were sawn and taken to an ultrasonic bath at 60 °C for 60 min to ensure surface cleaning. The substrate and a mass of 0.4 ± 0.05 g of asphalt binder were heated at 150 °C for 30 min. The substrate was then placed in another oven at 60 °C for additional 30 min, while the binder was placed in the silicone mold at room temperature for the same period. The 20 mm diameter metal stubs were heated at 150 °C for 30 min. The binder was added to the lower part of the metal stub, which is attached to the substrate. The pull-off tensile strength (POTS) is determined using the ABS test, which applies increasing pull-off pressure until separating metal stubs from the substrate.

This is done under four types of conditioning, 24 h dry and 3, 6 and 12 h wet. The equipment used has a hydraulic pressure system and reports POTS automatically.

The bond strength ratio (R_{POTS}), in percentage, is determined by "Eq. 2".

$$R_{POTS} = \frac{POTS_{WET}}{POTS_{DRY}} \cdot 100 \tag{2}$$

Analyzing Fig. 2(a), it is noticed that the POTS results from the wet conditions (POTS WET) at conditioning times of 3 and 6 h have no strong correlation with POTS results from the dry condition (POTS DRY). Theoretically, the POTS results should follow the following tendency: Dry > Wet-3 h > Wet-6 h > Wet-12 h, but this not occur in any of the mixtures analyzed. It is speculated that some mixtures suffer more strongly the moisture damage, with adhesion as a preponderant mechanism, while other mixtures suffer more interference from the viscoelastic effects of the asphalt binder, which would cause cohesive disrupt. On the other hand, POTS WET at 12 h of conditioning showed strong correlation (R^2 of 99%) with POTS DRY. Therefore, $POTS_{WET}$ [12 h] was used to determine R_{POTS}.

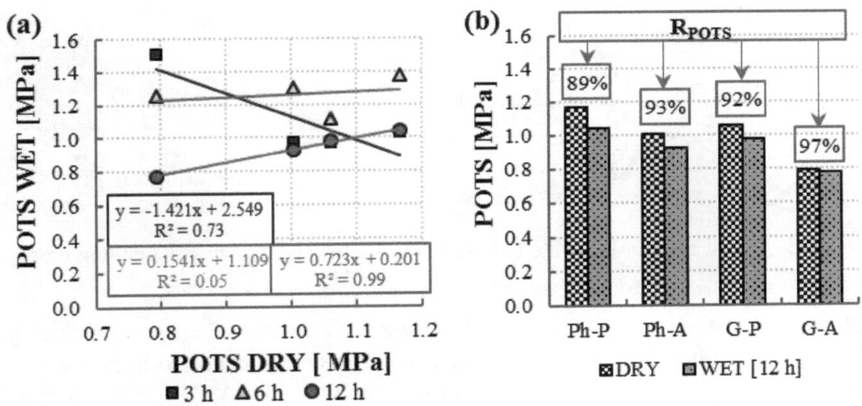

Fig. 2. Relationships between different test results: (a) POTS WET as a function of POTS DRY and (b) POTS WET [12 h] and POTS DRY for the different aggregate-binder combinations.

5 Relationship Between Moisture-Induced Damage and Asphalt Bond Strength

Figure 3(a), (b) and (c) presents the observed correlation between R_{POTS} and TSR. It is noted that the increase in conditioning time (from 3 h to 12 h) increases the correlation. Figure 3(d) shows that, despite a few points, only G-A showed a strong correlation with conditioning time, with R^2 of 99%, leading to the conclusion that different aggregate-binder combinations are affected at different conditioning times, i.e. there must be a minimum conditioning time in which the samples begin to be damaged by moisture more evenly.

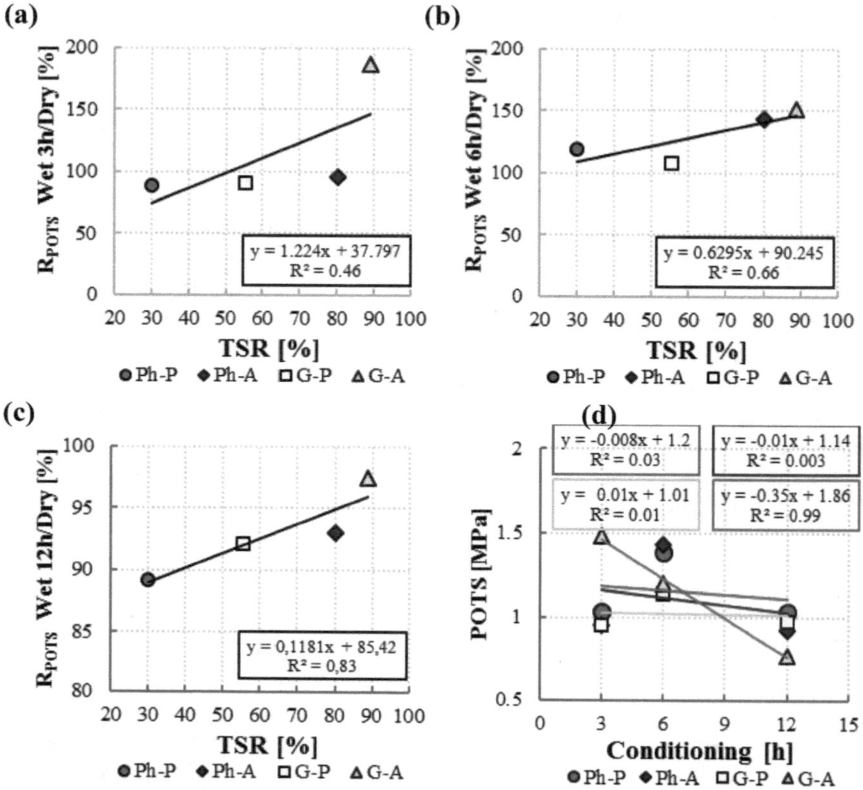

Fig. 3. R_{POTS} relationship with TSR and POTS relationship with conditioning time

It is noted that the highest CaO and the lower K_2O content of the granitic aggregate compared to the phonolytic aggregate (cf. Table 1) was associated to higher values of R_{POTS} (Wet 12 h/Dry) and TSR (G-P > Ph-P and G-A > Ph-A).

6 Conclusions

Moisture-Induced Damage (modified Lottman) test results showed that the amine-based antistripping agent was effective to increase moisture damage resistance for mixtures with both tested aggregates. Before binder modification, the mixtures were unfit to paving applications, becoming fit upon modification. The mixture with granitic aggregate had better moisture resistance than the one with phonolytic aggregate. As shown by the X-Ray fluorescence results, the granitic aggregate has higher calcium oxide and lower potassium oxide content: CaO and K_2O contents in the mixture are important variables for moisture damage resistance, as also suggested in the literature. This result indicates that an increase in moisture damage resistance can be achieved by the incorporation of lime as filler in the asphalt mixture, as reported by the open literature. When it comes to the Asphalt Bond Strength tests, results with 3 and 6 h conditioning did not follow the same trend as results with dry samples. However, a higher conditioning time such as 12 h, indicate that there is a strong correlation between moisture damage resistance on the asphalt mixture (modified Lottman test) and on the aggregate-binder interface (ABS test). A asphalt mixture's moisture sensitivity is associated to the loss of adhesion in the aggregate-binder interface. More research is needed on the fundamental aspects (e.g. viscoelastic effects) of the Asphalt Bond Strength test.

Acknowledgments. The authors thank Fundação Cearense de Apoio ao Desenvolvimento Científico e Tecnológico – FUNCAP and Brazilian Federal Agency for Post-graduate Education (CAPES).

References

AASHTO TP 91-11: Test Designation: Determining Asphalt Binder Bond Strength by Means of the Asphalt Bond Strength (ABS) Test. AASTHO (2013)

AASHTO T283: Standard method of test for resistance of compacted asphalt mixtures to moisture-induced damage. Washington, DC (2014)

Bagampadde, U., Isacsson, U., Kiggundu, B.M.: Influence of aggregate chemical and mineralogical composition on stripping in bituminous mixtures. Int. J. Pavement Eng. **6**(4), 229–239 (2005). https://doi.org/10.1080/10298430500440796

Chaturabong, P., Bahia, H.U.: Effect of moisture on the cohesion of asphalt mastics and bonding with surface of aggregates. Road Materials and Pavement Design (2016)

Hicks, R.G.: Moisture damage in asphalt concrete. NCHRRP. Synthesis of Highway Practice 175, Transportation Research Board, Washington (1991)

Huang, B., Shu, X., Dong, Q., Shen, J.: Laboratory evaluation of moisture susceptibility of hot-mix asphalt containing cementitious fillers. J. Mater. Civil Eng. **22**, 667–673 (2010). https://doi.org/10.1061/ASCEMT.1943-5533.0000064

Kanitpong, K., Bahia, H.U.: Relating adhesion and cohesion of asphalts to the effect of moisture on laboratory performance of asphalt mixtures. Transp. Res. Rec. J Transp. Res. Board **1901**, 33–43 (2005). https://doi.org/10.3141/1901-05. National Research Council, Washington, D.C.

Sebaaly, P.E., Hajja, E.Y., Sathanathana, T., Shivakolunthara, S.: A comprehensive evaluation of moisture damage of asphalt concrete mixtures. Int. J. Pavement Eng. **18**(2), 169–182 (2015). https://doi.org/10.1080/10298436.2015.1065404

Study on Effects of Aging on SBS Modified Asphalt Based on GPC and Rheological Methods

Daisong Luo[1,2], Meng Guo[3(✉)], Yiqiu Tan[1(✉)], and Yafei Li[2]

[1] School of Transportation Science and Engineering, Harbin Institute of Technology, Harbin 150090, China
luodaisongl@163.com, tanyiqiu@hit.edu.cn

[2] Engineering Technology and Materials Research Center, China Academy of Transportation Sciences, Beijing 100029, China
liyafei0927@163.com

[3] Key Laboratory of Urban Security and Disaster Engineering of Ministry of Education, Beijing University of Technology, Beijing 100124, China
gm@bjut.edu.cn

Abstract. In order to study the effects of aging on rheological and molecular properties of styrene-butadiene-styrene (SBS) modified asphalt, rheological and gel permeation chromatography (GPC) experiments were conducted on unaged and aged asphalt binders. The results show that the penetration of SBS modified asphalt droped with aging. The softening point of SBS modified as-phalt rises with the aging progress. The molecular oxidative condensation effect exceeds the down regulation effect of SBS degradation, leading to an increasing viscosity. After aging, ductility at 5 °C sharply attenuates and asphalt shows brittleness and no longer meets the requirements for high-grade highway pavement. SBS decomposition can inhibit the aging effect of asphalt molecules. This can restrain strengthening of high-temperature performance and reduction of low temperature anti-crack property of SBS modified asphalt.

Keywords: Asphalt · SBS · Aging · Rheology · GPC

1 Introduction

How to make efficient use of reclaimed asphalt pavement (RAP) has always been one technical focus and difficulty in the highway industry of China. Researchers have done a lot of research work respectively in the aspects of the physical properties decay law, component change law and molecular structure changes. However, these were mainly done through indoor aging tests, and many factors affecting asphalt aging remains to be further studies, such as influence mechanism of water on asphalt aging, the relationship and difference between simulated and natural aging, aging performance and mechanism of modified asphalt. Therefore, we also need to further study the asphalt aging performance and mechanism using advanced modern analytical and test methods and various aging test methods to provide a theoretical basis for asphalt pavement recycling.

L. D. Poulikakos et al. (Eds.): RILEM 252-CMB 2018, RILEM Bookseries 20, pp. 254–259, 2019.
https://doi.org/10.1007/978-3-030-00476-7_40

Ruan et al. (2003) studied the effect of long-term aging on rheological properties of polymer modified asphalt binders. They found that polymer modification resulted in increased asphalt complex modulus at high temperatures, decreased asphalt complex modulus at low temperatures, broadened relaxation spectra, and improved ductility. Yu et al. (2009) investigated the effect of organo-montmorillonite (OMMT) on aging properties of asphalt. They found that the viscosity aging index (VAI) and softening point increment (AS) of OMMT modified asphalt decrease significantly due to introduction of OMMT. Sengoz and Isikyakar (2008) studied the effect of styrene-butadiene-styrene (SBS) modifier on asphalt using fluorescent microscopy and conventional test methods. They found that SBS modification improved the conventional properties (penetration, softening point, etc.). Wu et al. (2009) explored the influence of aging on the evolution of structure, morphology and rheology of base and SBS modified bitumen. They found that more carbonyl and sulphoxide groups but less chain segments of butadiene were available after aging. Ouyang et al. (2006) used Fourier Transform Infrared (FTIR) spectroscopy to evaluate the aging properties of base asphalt and styrene-butadiene-styrene tri-block copolymer (SBS) modified asphalts (PMA). They found that the modification extent of the modifiers increases in the order: oil, zinc dibutyldithiocarbamate, zinc dialkyldithiophosphate. Siddiqui et al. (2002) used X-ray diffraction to assess the aging pattern of asphalt fractions. They found that the source and chemistry of asphalt are responsible for the aging behavior of its components. Guo et al. (2016) developed a test to simulate the long term aging of asphalt binder. Using this method, they prepared the artificial RAP aggregate. They further investigated the interaction of an asphalt binder with the surface of simulated RAP aggregates. Zhang et al. (2017) studied the three types of aging methods on performance of SBS modified asphalt, including TFOT, PAV and ultraviolet (UV) radiation. They found that SBS modified asphalt with penetration 90 had a higher retained penetration and ductility as well as the lower viscosity aging index compared to SBS modified asphalt with penetration 70. Kim et al. (2016) estimated the service-life reduction of asphalt pavement due to short-term ageing measured by gel permeation chromatography (GPC) from asphalt mixture. They found that the ageing level of binders in the warm-mix asphalt mixture was significantly lower than that of binders in hot-mix asphalt (HMA) after short-term aging.

The objective of this study was to investigate the mechanisms of aging of SBS modified asphalt by conducting comprehensive performance tests and physical-chemistry experiments, and further to improve the aging resistance of SBS modified asphalt.

2 Materials and Methods

2.1 Materials

SBS modified asphalt was used in this research. Its basic properties before aging and after aging are shown in Table 1.

Table 1. The properties of SBS modified asphalt before and after aging

Items	Penetration (25 °C)	Penetration index (PI)	Softening point	Kinematic viscosity (135 °C)	Ductility (5 °C)	Elastic recovery (25 °C)
Unit	0.1 mm	–	°C	Pa·S	cm	%
Requirements	40–60	≥0	≥60	≤3	≥20	≥75
Before aging	48	0.12	69.4	2.4	29.88	85
After aging	29	0.18	81.3	4.9	1.28	6

2.2 Laboratory Tests

In this research, penetration, softening point, ductility and viscosity tests are conducted to evaluate the effect of aging on rheological of SBS modified asphalt on the macro level.

Gel permeation chromatography (GPC) has low requirements for flow phase and features moderate test conditions, good repeatability and fast analysis speed. It is currently the most widely used method for determining the molecular weight distribution of high polymer materials. Its separation basis is the different hydrodynamic volumes of solute molecules in solution. The molecular elution volume of solute depends on the physical parameters such as molecular dimension, filler aperture, porosity and column volume. Two replicates were tested in this research, and the data shown in this paper was the average of the two replicates.

3 Results and Discussion

In order to find the rule of SBS modified asphalt aging, the laboratory adopted I-D finished modified asphalt meeting the requirements of Technical Specification for Construction of Highway Asphalt Pavements (JTGF40-2004) as a sample and put it in an RTFOT at 163 °C for aging for 5 h; then put the asphalt aged by RTFOT in a PAV at 100 °C for aging for 20 h. Finally, the changes in performance indicators, functional groups, components and molecular weight of the aged SBS modified asphalt were analyzed.

As a response to asphalt aging, performance indicators vary depending on different periods. In order to explore the changes of SBS modified asphalt performance in different aging periods, original SBS modified asphalt was aged with RTFOT and PAV successively, and then the penetration, ductility, softening point, kinematic viscosity at 135 °C and elastic recovery test of the aged SBS modified asphalt were conducted. The results are shown in Table 1. It can be seen from Table 1 that the penetration of SBS modified asphalt drops with aging. the softening point of SBS modified asphalt rises with the aging progress. As the space network structure of SBS modified asphalt gradually disintegrates, the shielding nad hindering effect of the huge benzene ring pendant groups in the polystyrene segments of SBS on the oxidizing reaction on principal chain is weakening. After aging, ductility at 5 °C sharply attenuates and asphalt shows brittleness and no longer meets the requirements for high-grade highway pavement. The molecular oxidative condensation effect exceeds the down regulation effect of SBS degradation, leading to an increasing viscosity.

The physical and mechanical properties of asphalt are closely related to the average molecular weight and also associated with molecular weight distribution. The discussion above has proved that asphalt viscosity is closely related to the molecular structure and intermolecular forces of asphalt, while the viscosity of molten asphalt is strongly dependent on the molecular weight of asphalt. The test results of SBS modified asphalt of different aging periods show that the higher the aging degree of asphalt is, the higher the viscosity will be. In order to study the effects of molecular weight on the properties of asphalt, gel permeation chromatography was used to study the effect of aging on the molecular weight distribution of SBS modified asphalt. Results are shown in Fig. 1 and Table 2.

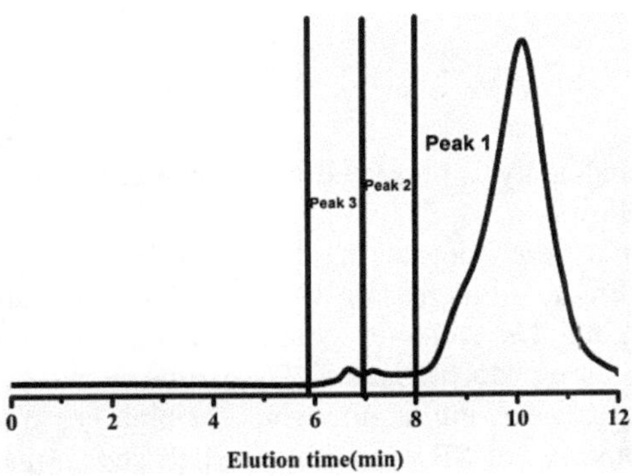

Fig. 1. Molecular weight distribution of SBS modified asphalt

Table 2. Effects of aging on the molecular weight distribution of SBS modified asphalt.

Molecular weight distribution of SBS modified asphalt before aging								
Peak 1			Peak 2			Peak 3		
Mn	Mw	Mz	Mn	Mw	Mz	Mn	Mw	Mz
215	221	227	112	115	118	808	1701	3867
Molecular weight distribution of SBS modified asphalt after aging								
Peak 1			Peak 2			Peak 3		
Mn	Mw	Mz	Mn	Mw	Mz	Mn	Mw	Mz
220	230	243	112	115	119	595	2698	9670

In Fig. 1, peak 1 represents the characteristic peak of asphalt, and peaks 2 and 3 are the characteristic peaks of SBS. Combined with Table 2, it can be known that the number-average molecular weight Mn at peak 1 has increased after SBS modified asphalt aging, indicating an increase in the relative molecular mass after asphalt aging. That's because aging effect causes the crosslinking between some of the light components of asphalt, and associated large molecules will also appear. Meanwhile, it also explains the increase in the stiffness of asphalt after aging.

As can be seen from Table 2, the number-average molecular weight at peak 3 has decreased, indicating that aging effect has caused the cracking reaction of SBS. That's because the polybutadiene segments of SBS contain reach carbon-carbon double bonds. In thermal oxygen environment, they may easily have bond breaking, decompose or crack into small molecules, or have oxygen absorption reaction and produce high-polarity substance. SBS has consumed the activation energy provided by light and heat conditions for asphalt aging and produced polarity substance, which inhibited the production of products of asphalt small molecules' oxygen absorbing and aging reaction, and the small molecules produced by cracking have supplemented for the loss of asphalt small molecules, having certain protective effect on the small molecules.

4 Conclusions

Based on the testing and analysis presented in this paper, the conclusions of the study are summarized as follows:

The molecular oxidative condensation effect exceeds the down regulation effect of SBS degradation, leading to an increasing viscosity. The penetration of SBS modified asphalt droped with aging. The softening point of SBS modified asphalt rises with the aging progress. After aging, ductility at 5 °C sharply attenuates and asphalt shows brittleness and no longer meets the requirements for high-grade highway pavement.

Gel chromatography test of SBS modified asphalt shows that, in the early aging period of asphalt, the characteristic peak of SBS reduced sharply and the small molecular peaks and macromolecular peaks of asphalt molecules went close to the middle at a lower speed. It indicates that SBS decomposition consumed much aging activation energy and inhibited the aging effect of asphalt molecules, such as condensation polymerization and decomposition, etc. This can further restrain continuous change of viscosity, strengthening of high-temperature performance and reduction of low temperature anti-crack property.

Actually more advanced tests were conducted to investigate the mechanism of effect of aging-recycling cycle on rheological of SBS modified asphalt on the micro level, including thermogravimetric analysis (TG), hydrogen nuclear magnetic resonance spectrometry (1HNMR), and scanning electron microscope (SEM) and energy dispersive spectrometer (EDS). But due to the limited paper length, the details cannot be shown in this paper.

References

Guo, M., Motamed, A., Tan, Y., Bhasin, A.: Investigating the interaction between asphalt binder and fresh and simulated RAP aggregate. Mater. Des. **105**, 25–33 (2016)

Kim, S., Lee, S.H., Kwon, O., Han, J.Y., Kim, Y.S., Kim, K.W.: Estimation of service-life reduction of asphalt pavement due to short-term ageing measured by GPC from asphalt mixture. Road Mater. Pavement Des. **17**(1), 153–167 (2016)

Ouyang, C., Wang, S.F., Zhang, Y., Zhang, Y.X.: Improving the aging resistance of styrene-butadiene-styrene tri-block copolymer modified asphalt by addition of antioxidants. Polym. Degrad. Stab. **91**(4), 795–804 (2006)

Ruan, Y.H., Davison, R.R., Glover, C.J.: The effect of long-term oxidation on the rheological properties of polymer modified asphalts. Fuel **82**(14), 1763–1773 (2003)

Sengoz, B., Isikyakar, G.: Analysis of styrene-butadiene-styrene polymer modified bitumen using fluorescent microscopy and conventional test methods. J. Hazard. Mater. **150**(2), 424–432 (2008)

Siddiqui, M.N., Ali, M.F., Shirokoff, J.: Use of X-ray diffraction in assessing the aging pattern of asphalt fractions. Fuel **81**(1), 51–58 (2002)

Wu, S.P., Pang, L., Mo, L.T., Chen, Y.C., Zhu, G.J.: Influence of aging on the evolution of structure, morphology and rheology of base and SBS modified bitumen. Constr. Build. Mater. **23**(2), 1005–1010 (2009)

Yu, J.Y., Feng, P.C., Zhang, H.L., Wu, S.P.: Effect of organo-montmorillonite on aging properties of asphalt. Constr. Build. Mater. **23**(7), 2636–2640 (2009)

Zhang, D.M., Zhang, H.L., Shi, C.J.: Investigation of aging performance of SBS modified asphalt with various aging methods. Constr. Build. Mater. **145**, 445–451 (2017)

Chemo-Mechanical Characterization of Bituminous Materials: Other Approaches

How to Evaluate with Relevance the Compactability of Warm Mixes Using the Gyratory Compactor (GC)?

Abdeldjalil Daoudi[1,2](✉), Anne Dony[2], Layella Ziyani[2], Nicolas Picard[3], and Julien Buisson[3]

[1] Ecole de technologie supérieure, 1100 Rue Notre-Dame Ouest, Montréal, QC H3C 1K3, Canada
abdeldjalil.daoudi.1@etsmtl.net

[2] Université Paris Est, Institut de Recherche en Constructibilité, Ecole Spéciale des Travaux Publics, 28 avenue du Président Wilson, 94234 Cachan Cedex, France
{adony,lziyani}@estp-paris.eu

[3] Ingevity, Parc de la Haute Borne, 20 rue Haddock, 59650 Villeneuve d'Ascq, France
{nicolas.picard,julien.buisson}@ingevity.com

Abstract. The development of Warm Mix Asphalt (WMA) requires the control of the performances at laying temperatures to ensure those at service life temperatures. A simple lowering of mixing temperature can reduce the workability and the compactability of asphalt mixture. So, several techniques such as additivation and foaming are proposed to avoid or reduce this laying issue. The shear gyratory compactor (GC) conventionally evaluates the compactability but different approaches can be used to analyse data. This paper aims to assess the relevance of these methods by an experimental program conducted on two WMA (with EVOTHERM® technology WM30 additive developed by Ingevity and foam process) compared with a conventional hot mix asphalt (HMA), at three compaction temperatures. The classic exploitation of GC's data (i.e. considering the void content at a given number of gyrations) is not relevant to evaluate the effect of the decrease in temperature on the compactability. The calculation of compactability indices (CEI), used to obtain a level of compaction (90%, 92%, and 95%), is interesting but takes more time. The calculation of performance ratios is the simplest and most relevant approach. The use of liquid additive seems the most promising towards the compactability.

Keywords: Compactability · Warm Mix Asphalt (WMA)
Compaction energy · Gyratory compaction

1 Introduction

The warm mix asphalts (WMA, i.e. asphalt mixtures manufactured at a temperature of 30 °C lower than the average production temperature of classic Hot Mix Asphalts (HMA)) are developed to meet the expectations of the environmental, economic and

L. D. Poulikakos et al. (Eds.): RILEM 252-CMB 2018, RILEM Bookseries 20, pp. 263–268, 2019.
https://doi.org/10.1007/978-3-030-00476-7_41

sanitary context. These new mixes must guarantee technical performances at least equivalent or even better than HMA in terms of both laying and finished product properties. But a single decrease in manufacturing temperature can lead to a lower quality of the mixing and consequently a loss of workability and/or compactability (two properties that could be connected but are different). The French lowering temperature of bituminous mixes guide IDRRIM guide suggests several techniques such as additivation and foaming to develop WMA. This paper focuses on the comparison of several approaches of compactability evaluation from GC data and the application to processes developed in France (additive and foam).

2 Literature Review

It is known that the temperature of a bituminous mixture directly affects its compaction behavior and therefore the lowering of the temperature causes a decrease in the compaction ability (compactability) of the asphalt mix, influencing air void content and thus mechanical properties.

The French guide for compaction of hot mix asphalts (Gallenne et al. 2003) fixed according to the bitumen grade, recommendations regarding the temperatures of mixing, spreading and end of compaction. For example, for a 35/50 penetration grade bitumen, the temperature of mixing is about 150 to 170 °C, the minimal temperature of spreading is 130 °C and the minimal temperature of compaction is 110 °C.

The difference between manufacturing temperature and the minimum compaction temperature is about 40 ° to 60 °C. This value covers the cooling due to possible storage, transport, spreading and compaction and is not influenced by bitumen grade; but could we apply the same delta temperature in the case of WMA?

The compactability can conventionally be assessed using the gyratory compactor (GC) test. This test is essential to evaluate a grading formulation (French Béton Bitumineux Semi-Grenu, Grave Bitume…). The classic analysis of the results consists of responding to a level of compaction with respect to a number of gyrations (EN 12697-10); for example, the density of an EB (Enrobé Bitumineux) 0/10 must be between 90% to 95% at 60 gyrations. However, according to several studies (Dony et al. 2010), the GC is not very sensitive to temperature variations between 160 and 110 °C.

Several approaches were developed to evaluate the compactability at different temperatures, using the data of GC.

First, the calculation of the CEI (Compaction Energy Index) (Bahia et al. 1998) shows the area below the compaction curve from eight gyrations until reaching 92% of compaction (Fig. 1). It simulates the work performed during the compaction period on site. According to (Bahia et al. 1998), the first eight gyrations simulate the force applied by a typical paver during the asphalt laying action (so called pre-compaction).

In the National Cooperative Highway Research Program (Bonaquist 2011), the author compared the number of gyrations needed to reach a fixed level of compaction, then calculated the gyration ratio (Fig. 1), thus comparing mixes manufactured at different temperatures. According to NCHRP, for mixes which have the similar compactability, the

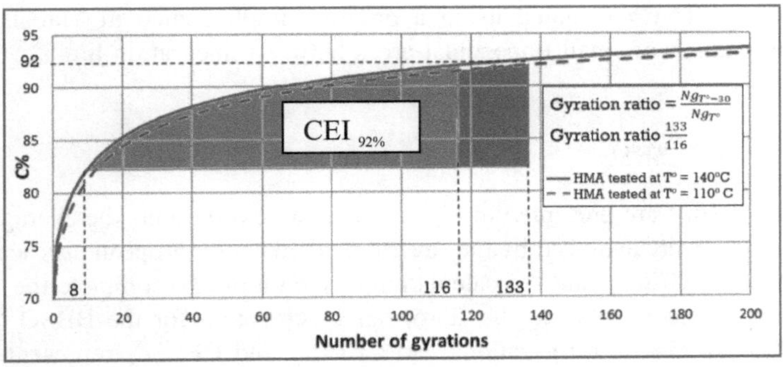

Fig. 1. Calculation of the CEI and the gyration ratio

gyration ratio must be more or less equal then 1.25. Compared to the CEI method, this method is simpler, with a high sensitivity to variations in the test temperature.

Recently, a new parameter named "performance ratio" was defined for an additive or foam WMA process (Smith and James 2016). The authors compared the numbers of gyrations for a fixed density at the same temperature (Eq. 1), in the case of two WMA mixtures with and without additive or new process. The performance ratio allowed the evaluation of a WMA technique (additivation, foaming…) for compactability at one temperature.

$$\text{Performance ratio} = \left(\frac{Ng_{HMA}}{Ng_{WMA}}\right)_{T^\circ ref} \tag{1}$$

3 Objective and Methodology

The main objective of this study was to compare several indicators calculated from the experimental GC data for different asphalt mixtures: a conventional HMA, a WMA with a liquid chemical additive (Evotherm® WM30 from Ingevity) and a WMA produced with foam process.

Compactability tests were performed with the GC press on three types of mixes: a reference HMA (mixing temperature at 165 °C), a WMA with WM30 additive (mixing temperature at 135 °C) and a foam WMA (performed by adding 2.5% of water compared to the mass of bitumen, mixing temperature at 135 °C). The same formulation (BBSG 0/10), the same bitumen type (35/50) and the same bitumen content (5.6 ppc) were used for the study.

Based on the compaction table for 35/50 bitumen in HMA (Table 1), three target temperatures were chosen: 140 °C (optimum laying temperature), 110 °C (minimum compaction temperature), 80 °C (minimum compaction temperature, extrapoled for WMA by fixing the same Δ temperature of 30 °C). After the manufacture of mixes at the controlled temperature and filling of pre-heated GC molds, the samples were left until the correct test temperature was reached to start the tests.

CEI indices were calculated using a program implemented in Matlab® (MathWorks) (summation of small trapezoidal areas between integration limits).

4 Results

The GC tests results are presented in Fig. 2, each curve represents the average of three tests. The results were analyzed by the several methods: the European classical one, the calculation of the $CEI_{92\%}$ and the calculation of the ratios. To complete the study, the CEI index was also adapted to the European specification for the BBSG 0/10 (C% between 90% to 95% at 60 gyrations), so CEI $_{90\%}$ and CEI $_{95\%}$ represent the areas below the compaction curve from eight gyrations until reaching 90% and 95% of compaction, and CEI 60 $_{\text{gyrations}}$ represents the area below the curve of compaction from eight gyrations until reaching 60 gyrations. Figure 2 presents the evolution of density with the gyration number for all mixes. The several indicators are calculated in Table 1.

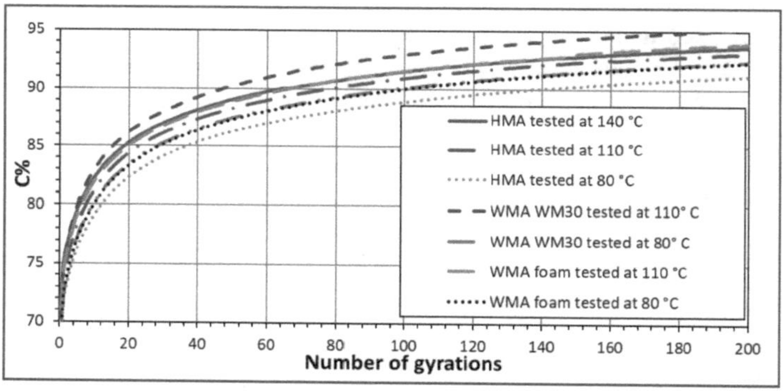

Fig. 2. Gyration compaction results of WMA and HMA at several temperatures

Table 1. Levels of compaction at 60 gyrations, CEI, gyration and performance ratio of mixes

	(T° C) test	C (%) 60 gyrations	CEI $_{60}$ gyrations	CEI 92 %	CEI 90 %	CEI 95 %	Gyration ratio		Performance ratio	
							110 °C	80 °C	110 °C	80 °C
HMA	140	89.7 ± 0.4	290	834	323	2,497				
	110	89.0 ± 0.3	303	1,096	459	3,104	1.20	1.77	1.00	1.00
	80	87.0 ± 0.6	304	2,132	1,101	5,972				
WM30 WMA	110	91.0 ± 0.3	311	473	201	1,769	0.70	1.28	1.60	1.40
	80	88.2 ± 0.3	313	1,531	678	3,949				
Foam WMA	110	89.6 ± 0.2	295	819	348	2,466	1.00	1.30	1.20	1.40
	80	88.1 ± 0.3	311	1,684	700	4,045				

The classical approach for a BBSG 0/10 determines the density of mixes at 60 gyrations. We note that the WM30 WMA tested at 110 °C presents the highest level of density (C = 91.0% ± 0.3). The WMA foam tested at 110 °C and the HMA tested at 140 °C have the same level of compaction.

Moreover, the difference between level of compaction of HMA tested at 140 °C and HMA tested at 110 °C is not very significant (about 0.7%). This finding is consistent with the literature review. All the mixes meet the large European specifications (borderline for WM30 WMA).

The analysis of $CEI_{60gyrations}$ (Table 1) confirms that the classical approach is not very sensitive to the variation of test temperature: no significant difference is noted between mixes.

Regarding the calculation of CEI, for a same type of mix, CEI increases with decreasing the test temperature, whatever the targeted compaction level (92%, 90% or 95%): for example, for the HMA, an increase about 31% of $CEI_{92\%}$ between 140 °C and 110 °C and 156% between 140 °C and 80 °C. The lowest value of CEI was found with WM30 WMA tested at 110 °C with a decrease about 45% of $CEI_{92\%}$ compared to the HMA tested at 140 °C and 60% compared to the HMA tested at 110 °C. The CEI of the foam WMA tested at 110 °C and the HMA tested at 140 °C are comparable. The same tendency is observed by analyzing $CEI_{90\%\ and\ 95\%}$, but it is noted for a same type of mix, at one temperature a large difference between $CEI_{90\%}$ and $CEI_{95\%}$. It is also important to fix the level of compaction for each mix to evaluate the energy level.

The calculation of gyration ratio and performance ratio are presented in Table 1. This calculation is more rapid than CEI without calculation of areas. We note that the gyration ratio of the WM30 WMA at 110 °C was less than 1, so the compactability of the WM30 WMA was better compared to the HMA at 140 °C. Moreover, the foam WMA at 110 °C showed a similar compactability as the HMA at 140 °C: the gyration ratio equals 1. At 80 °C, WMA show a compactability very similar to the HMA at 110 °C. This is explained by a gyration ratio close to 1.25 (1.28 for WM30 WMA and 1.3 for foam WMA). The performance ratio calculation showed that the additive WM30 was more efficient than foaming at 110 °C and both WMA have a similar performance at 80 °C.

5 Conclusions

The objective was to evaluate indicators calculated with GC data to compare several techniques of WMA. If the classical European approach is insensitive to changes in temperature test, CEI calculation allows evaluating the effect of temperature but is time-consuming. The choice of targeted compaction level is important. The ratio approach has the advantage of being fast and relevant. Finally, our study shows an improvement in the compactability with the use of the additive WM30.

References

Bahia, H.U., et al.: Optimization of constructibility and resistance to traffic: a new design approach for HMA using the superpave compactor. J. Assoc. Asph. Paving Technol. **67**, 189–232 (1998)

Bonaquist, R.: Mix Design Practices for Warm Mix Asphalt, National Cooperative Highway Research Program (NCHRP) Report 691, TRB, Washington DC (2011)

Dony, A., et al.: Laboratory assessment of Warm Mixes Asphalt by means of two mix design methods. Paper presented in the 16th World Meeting International Road Federation, Lisbon (2010)

Gallenne, M.L., et al.: Compactage des enrobés hydrocarbonés à chaud, Guide technique. LCPC, Paris (2003)

Smith, M., James, A.: New compactability parameter for comparing warm mix additives. In: Eurasphalt & Eurobitume Congress, Prague (2016)

Hybrid Approach to Characterize Reflective Cracking in Airport Pavements

Tirupan Mandal[1(✉)], Mesbah Ahmed[1], Hao Yin[1], and Richard Ji[2]

[1] Gemini Technologies, Jamison, New Jersey, USA
{tirupan.mandal,mesbah.ahmed,hao.yin}@gemitek.com
[2] Federal Aviation Administration, Washington, D.C., New Jersey, USA
richard.ji@faa.gov

Abstract. In airfields, reflective cracking is one of the major distresses in asphalt overlays which are placed on rigid pavements, cracked asphalt pavements, and even in asphalt layers with cement-treated base pavements. Recently, a comprehensive laboratory testing program to characterize reflective cracking was conducted at the Federal Aviation Administration (FAA) NextGen Pavement Materials Laboratory. All tests were conducted on field extracted hot mix asphalt (HMA) cores. Test data from the customized overlay tester (OT) showed an excellent correlation between the Dissipated Energy (*DE*) and fracture parameters for specimens tested with different maximum crack opening displacement (or displacement rate). Parallel to the laboratory study, a finite element (FE) modeling effort was also carried out. The developed 2D FE model was not only capable of simulating the OT test but also accurately compute the fracture parameters, which consequently can be used to predict the fatigue performance of HMA specimens using empirical equations developed from laboratory testing.

Keywords: Reflective cracking · FEM · Dissipated energy
Customized overlay tester

1 Motivation and Objective

Texas Overlay Tester [OT] (TxDOT 2017) represents the best laboratory method to truly simulate horizontal joint movements in the joint/crack vicinity of concrete pavements. Recently, this test has been customized to mimic the full-scale testing conditions to study reflective cracking at the National Airport Pavement Test Facility (NAPTF) (Mandal et al. 2017; Ji et al. 2018), and the laboratory testing results correlated well to the full-scale test data (Mandal et al. 2018). However, the OT testing is extremely time-consuming. In this study, an analytical procedure was developed to predict fracture and fatigue performance of hot mix asphalt (HMA) specimen using a mixed pool of OT test data from laboratory and a 2D finite element (FE) simulation. Figure 1 shows the flowchart of research objective.

L. D. Poulikakos et al. (Eds.): RILEM 252-CMB 2018, RILEM Bookseries 20, pp. 269–274, 2019.
https://doi.org/10.1007/978-3-030-00476-7_42

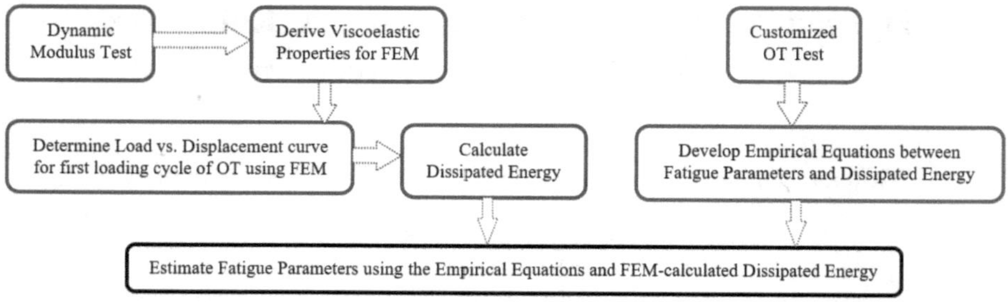

Fig. 1. Flowchart of research objective

2 Materials and Methods

HMA cores of standard FAA P-401 Performance Grade 64-22 (PG 64-22) were extracted from the Phase V test pavement and fabricated for the customized OT test. Three maximum crack opening displacements (sets of specimens) were chosen – 0.1524 mm (0.002 mm/s, low), 0.3048 mm (0.004 mm/s, medium), and 0.4572 mm (0.006 mm/s, high) – to study the effect of displacement rates. The medium set (0.3048 mm) simulated the full-scale testing at the NAPTF. All overlay tests were conducted at 0 °C with haversine loading of 150 s followed by a 150-s rest period. Details about the test method can be found in Mandal et al. (2017).

A Finite Element Model (FEM) was also developed in ABAQUS to simulate the OT test. A 2D geometry was adopted due to the symmetry of test assembly. Generalized Maxwell Model (GMM) was used to characterize the linear viscoelastic behavior of HMA. Dynamic modulus (DM) test was first conducted to determine these GMM parameters. The g_i input values were: 0.114, 0.013, 0.124, 0.111, 0.177, 0.169, 0.154, 0.087, 0.037, 0.008, and 0.003 at reduced time (in seconds) of 2E-5, 2E-4, 2E-2, 2E-1, 2, 2E+1, 2E+2, 2E+3, 2E+4, and 2E+5 at a reference temperature of 0 °C. The Young's modulus was then used as the relaxation modulus at infinite time. A constant Poisson's ratio of 0.35 was assumed.

3 Results and Analysis

3.1 Customized Overlay Tester

Customized OT tests were conducted with at least three replicates for each maximum opening displacement. One specimen was also tested at a higher displacement rate (0.008 mm/s [0.6096-mm opening displacement]), which failed completely after three loading cycles. Following the method described in Ma (2014), the fatigue parameters were determined. The average values are shown in Table 1. As the testing was stopped once the crack was clearly visible on the surface, $Nf_{(final)}$ parameter (i.e., the number of cycles until the test was stopped) was different for each specimen; the number of cycles till 75% load reduction ($Nf_{(75\ \%)}$) was used to compare the sets. In case of the 0.1524-mm opening displacement, the specimens were stopped testing after ~64–69% load reduction (~2 weeks); the number of cycles at 64% load reduction was reported instead. The values for $Nf_{(NLC)}$ (i.e., number of cycles at the failure point), $Nf_{(crack)}$ (number of cycles to the point where the load-displacement curve drops again after the failure point), and $Nf_{(czone)}$ (i.e., number of cycles between $Nf_{(crack)}$ and $Nf_{(NLC)}$) could not be determined for 0.1524-mm opening displacement as the crack was not able to propagate to the surface. As would be expected in Table 1, higher the crack opening displacement, lower is the fatigue life of the specimen (Walubita et al. 2012; Sheng and Ping 2016, and Walubita et al. 2013).

Table 1. Fatigue parameters for different displacement rates

Opening displacement (mm)	$Nf_{(75\%)}$	$Nf_{(NLC)}$	$Nf_{(crack)}$	$Nf_{(czone)}$
0.1524	2320*	N/A	N/A	N/A
0.3048	118	99	152	54
0.4572	16	12	21	9

* Value determined at $Nf_{(64\%)}$

Fracture parameters (critical fracture energy [G_f] and Dissipated Energy [DE] from the first cycle) were also computed for all the specimens and were found to be increasing with increase in opening displacement. The average values for each set are presented in Fig. 2(a). Further, when comparing the three sets, strong relationships ($R^2 > 0.92$) were found between the fatigue parameters and fracture parameter (DE) as seen in Fig. 2(b).

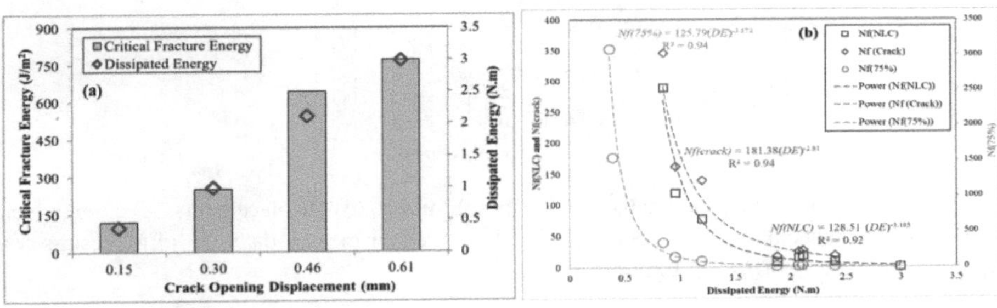

Fig. 2. Results of customized overlay test: (a) fracture properties, (b) DE vs. fatigue parameters

3.2 Finite Element Analysis

The 2D FEM consisted of a 150 × 38.1-mm HMA specimen. Datum points were plotted in the model to separate the 2-mm gap at the center of the specimen. The boundary conditions are shown in Fig. 3(a), where the bottom nodes of the left side of HMA were restricted from translation while a prescribed ramp motion was applied to the right side of the specimen. The same displacement used the laboratory test was applied to the movable face, as shown in Fig. 3(a). For the sake of computational time, only the first loading cycle was simulated. A quad-dominated free mesh was used with

5,992 elements and 6,156 nodes, and the region was modeled using a 4-node bilinear plane stress quadrilateral, CPS4 elements. A finer mesh was generated near the center of the specimen where a steep variation of stress was expected (Ahmed et al. 2015) (see Fig. 3(b)). An example of both strain and displacement distribution at the maximum displacement are provided in Fig. 3(c) and (d), respectively. The strain contour is symmetric with a very high value near the gap. During the laboratory test, it was observed that the crack was initiated at this location due to high strain concentration. Distribution of the simulated strain is qualitatively similar to the laboratory observation and literature (Gu et al. 2015). Figure 3(e) shows an example plot of load vs. displacement data at 0.3048-mm displacement for both FEM and experiment data. Using this load and displacement data from FE simulation, dissipated energy (DE_{FEM}) was calculated for different opening displacements. Figure 3(f) shows the comparison between simulated and measured *DE* parameter. An excellent agreement was observed except for the displacement of 0.6096 mm.

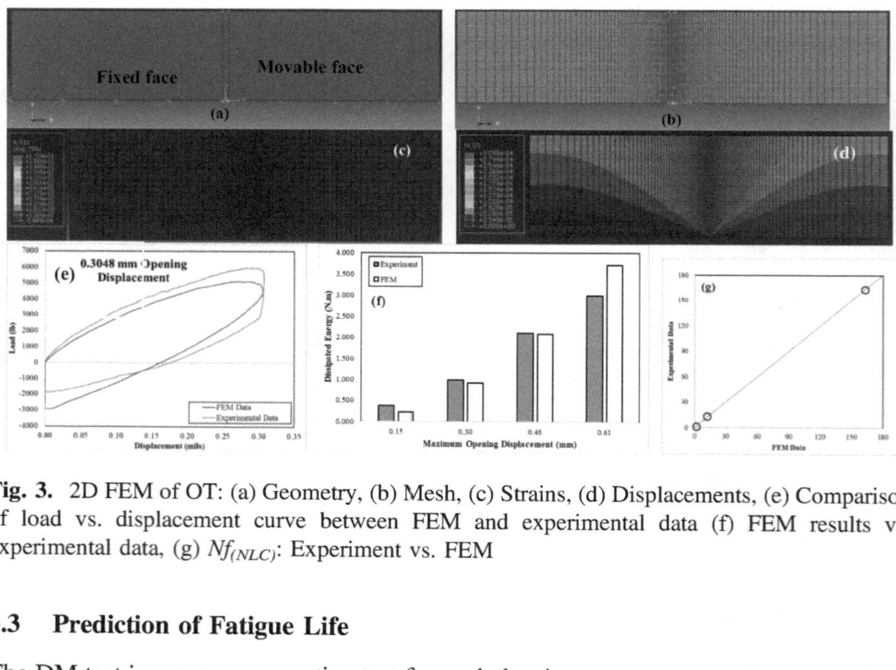

Fig. 3. 2D FEM of OT: (a) Geometry, (b) Mesh, (c) Strains, (d) Displacements, (e) Comparison of load vs. displacement curve between FEM and experimental data (f) FEM results vs. experimental data, (g) $Nf_{(NLC)}$: Experiment vs. FEM

3.3 Prediction of Fatigue Life

The DM test is a very easy routine test for asphalt mixtures as compared to customized OT (due to testing time and specimen fabrication). Viscoelastic properties obtained from the DM test data can be directly input into the FEM. The FEM can then be used to determine the load vs. displacement curve for the first loading cycle; and hence, DE_{FEM} can be calculated under the hysteresis loops (load vs. displacement) generated from the FEM simulation. Using this DE_{FEM}, fatigue parameters can be estimated using the empirical equations developed from laboratory testing of the customized OT (Fig. 2 (b)). The equations for these relationships are shown below:

$$Nf_{(75\%)} = a_1 DE_{FEM}^{-b_1} \tag{1}$$

$$Nf_{(crack)} = a_2 DE_{FEM}^{-b_2} \tag{2}$$

$$Nf_{(NLC)} = a_3 DE_{FEM}^{-b_3} \tag{3}$$

where $Nf_{(75\%)}$ is the number of cycles to 75% load reduction, $Nf_{(crack)}$ and $Nf_{(NLC)}$ are fatigue parameters, DE_{FEM} is the dissipated energy of the first loading cycle calculated from the FEM (N.m), and a_1, a_2, a_3, b_1, b_2, b_3, n_1, and n_2 are laboratory derived regression constants. Figure 3(g) shows an example for the comparison of $N_{f(NLC)}$ determined based on laboratory testing and FEM simulation. It is observed that these values follow a linear trend. In addition, they are very close to the line of equality. It indicated that the FEM simulated *DE* can be used to predict the fatigue parameters.

4 Conclusions and Recommendations

This paper investigates the reflective cracking mechanism of airport pavements through a laboratory testing suite and FEM. The customized OT successfully evaluated the different crack opening displacements using fatigue and fracture parameters. From the experimental data, the fracture parameter (*DE*) was found to have a strong correlation ($R^2 > 0.92$) to fatigue parameters. A 2D FEM was developed to simulate the first loading cycle of the customized OT and a good agreement between the calculated and measured *DE* values were observed. As OT testing can be time-consuming (for e.g., over 2 weeks for 0.1524-opening displacement); using this FEM and the empirical equations developed from the laboratory testing, the fatigue performance of the specimens can be easily predicted. Validating the model for different OT parameters and mix design variables is recommended for future research.

References

Ahmed, M.U., Rahman, A.S.M.A., Islam, M.R.: Combined effect of asphalt concrete cross-anisotropy and temperature variation on pavement stress–strain under dynamic loading. Constr. Build. Mater. **93**, 685–694 (2015)

Gu, F., Luo, X., Zhang, Y., Lytton, R.: Using overlay test to evaluate fracture properties of field-aged asphalt concrete. Constr. Build. Mater. **101**, 1059–1068 (2015)

Ji, R., Mandal, T., Yin, H.: Laboratory characterization of temperature induced reflection cracks. J. Traffic Transp. Eng. (2018)

Ma, W.: Proposed Improvements to Overlay Test for Determining Cracking Resistance of Asphalt Mixtures. MS Thesis, Auburn University, USA (2014)

Mandal, T., Yin, H., Ji, R., Rutter, R.: Laboratory simulation of extreme cooling effects on the propagation of reflection cracks using customized Texas overlay tester. In: Proceedings of International Conference on Highway Pavements and Airfield Technology, Philadelphia, Pennsylvania, USA, pp. 81–93 (2017)

Mandal, T., Yin, H., Ji, R.: Correlating laboratory and full-scale reflective cracking tests for airfield pavements. Constr. Build. Mater. **169**, 47–58 (2018)

Sheng, B., Ping, W.V.: Evaluation of Florida Asphalt Mixes for Crack Resistance Properties using the Laboratory Overlay Test Procedure, FDOT Report BDV30 TWO 977-06 (2016)

TxDOT: Tex-248-F: Test Procedure for Overlay Test, Texas Department of Transportation Standard (2017)

Walubita, L.F., Faruk, A.N., Das, G., Tanvir, H.A., Zhang, J., Scullion, T.: The Overlay Tester: A Sensitivity Study to Improve Repeatability and Minimize Variability in the Test Results, Report No. FHWA/TX-12/0-6607-1 (2012)

Walubita, L.F., Faruk, A.N., Koohi, Y., Luo, R., Scullion, T., Lytton, R.L.: The Overlay Tester - Comparison with Other Crack Test Methods and Recommendations for Surrogate Crack Tests, FHWA/TX-13/0-6607-2 (2013)

Kinetic Analysis of the Thermal Behavior of the Sap of the Petroleum Plant for Producing Bio-Binders

Lilian Medeiros Gondim[1(✉)], Sandra de Aguiar Soares[2], and Suelly Helena de Araújo Barroso[2]

[1] Federal University of Cariri, Juazeiro do Norte, Brazil
lilian.gondim@ufca.edu.br
[2] Federal University of Ceara, Fortaleza, Brazil
sas@ufc.br, suelly@det.ufc.br

Abstract. The Petroleum Plant Sap is a bio-material that could be used for producing bio-binder in order to surrogate asphalt materials. Low amounts of this sap (up to 10%) were applied to an asphalt binder, resulting in very small changes of physical and rheological properties, what would indicate the degradation of some compounds of the sap. The present paper has the main purpose of analyzing the thermal degradation of the sap of the Petroleum Plant, what could point out the safe temperature range for applying the sap on paving applications. For that, thermogravimetry was performed on three heating ratios (5 °C/min, 10 °C/min and 40 °C/min) and the kinetic analysis of the data was processed. The results indicated that the temperature applied form the binder modification and aging simulating process were higher than the maximum safe temperature for the sap, that would be around 140 °C. It was also observed that the sap contains compounds that were oxidized during the heating process. The Thermogravimetric technique showed to be important for characterizing thermal behavior of bio-materials that are candidates for the formulation of bio-binders.

Keywords: Thermogravimetry · *Euphorbia Tirucalli* · Kinetic analysis

1 Introduction

The search for a bio material that could surrogate the asphalt binders has been a response to the increase of world's environmental awareness and to the unpredictability of the petroleum markets. In this sense, many studies have been conducted to develop new products such as bio-binders derived from bio-mass of different kinds (Çelik and Atasağun 2012; Williams et al. 2014; Fini et al. 2011; Mills-Beale et al. 2014; Peralta et al. 2013), vegetal materials (Leite et al. 2012; Feitosa et al. 2015; Zofka and Yut 2012), algae materials (Audo et al. 2012), wasted cooking oils (Wen et al. 2012; Leite et al. 2012) and also by the mixtures of resins and fatty acids (Vasconcelos 2010; Ballie 2004).

As pointed out by Kluttz (2012) one of the main challenges of producing bio-binders are due to the different behavior and chemical composition of these materials when compared to petroleum-based ones. This may cause them to fail to meet the

L. D. Poulikakos et al. (Eds.): RILEM 252-CMB 2018, RILEM Bookseries 20, pp. 275–280, 2019.
https://doi.org/10.1007/978-3-030-00476-7_43

specifications that were formulated specifically for asphaltic materials, even though they may not be necessarily inappropriate for pavement applications.

In this context, the sap of the Petroleum Plant (scientific named as *Euphorbia Tirucalli*) has been studied as a possible material to replace portions of the asphalt binder, producing bio-binders able to reduce the consumption of petroleum-based materials. Gondim et al. (2017) modified an asphalt binder with different contents of this sap (up to 10%) and analyzed the changes on its physical and rheological properties. The results indicated that the addition of the sap occasioned diverse trends of modification on the binder behavior in the different temperature ranges: it hardened the asphalt binder at 25 °C, but it had a solvency effect on the high in-service temperature (led to higher phase angles and lower complex moduli) and more pronounced effects at mixing and compaction temperatures, reducing them in about 7 °C. Nevertheless, most of these changes showed very small magnitude, what would suggest that part of this sap could be degraded during the modification process, due to high temperature: the modification was processed at 160 °C.

Chemical composition and thermal behavior of this bio-material should be analyzed to find out the limitations of application, such as the right temperature of mixture and compaction, changes of material states within the in-service temperature range and also how they would behave when submitted to the aging processes. One of the main techniques for analyzing thermal behavior of samples is Thermogravimetry (TGA) followed by its kinetics analysis.

The Thermogravimetry technique consists on measuring the variation of the mass of a sample due to heating. It makes possible to evaluate the thermal stability of a material, to determine the activation energy necessary to the occurrence of certain degradation and to verify the volatile and moisture content of a given sample.

As a development of the research of Gondim et al. (2017), this work has the main purpose of analyzing the thermal degradation of the sap of the Petroleum Plant, what could point out the safe temperature range for applying the sap on paving applications.

2 Material and Testing Methods

The sample evaluated on this research was the Sap of the Petroleum Plant, harvested and prepared as described by Gondim et al. (2017), where it was used to modify an asphalt binder for paving applications.

To verify the thermal stability and the kinetic analysis of the degradation reactions of the sap of the Petroleum Plant, this sample was submitted to Thermogravimetry Analysis (TGA). Approximately 7 mg of the dehydrated sap were heated at a rate of 5 °C/min, 10 °C/min, and 40 °C/min, over the temperature range of 20 °C to 700 °C, in an inert atmosphere (N_2) and in an oxidative atmosphere (O_2).

The kinetic analysis consisted on calculating the Thermal Activation Energies by the Flynn and Wall method, in which the activation energy (Ea) for a given conversion (mass loss) can be determined from the angular coefficient of the natural logarithm graphic of the ratio of heating (β) as a function of the inverse of temperature (1/T) in Kelvin, according to Eq. 1.

$$Ea = \frac{-R}{b}\left(\frac{d\ln\beta}{d(1/T)}\right) \tag{1}$$

Where Ea is the activation energy (KJ/mol), b is a constant (equal to 1), R is the gas constant (8.314 J/mol * K) and the term in brackets is the slope of the graph line which relates the ratio of heating to the inverse of temperature, as presented on Fig. 1.

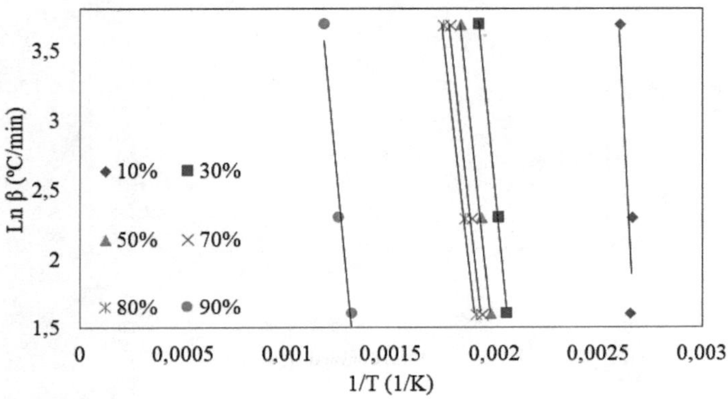

Fig. 1. Ratio of heating (β) as a function of the inverse of temperature (1/T), for a given mass conversion

3 Results and Discussion

The results of the thermogravimetry analysis of the sap of the Petroleum Plant are summarized at Table 1. The first three events in oxidative atmosphere are due to the loss of water and volatiles and are equivalent to the first event observed on N_2 atmosphere. The main event of degradation was the 4th on O_2, and the 2nd on N_2, and it is responsible for the major mass conversion (from 68% to 77%). There is a secondary event on both environments around 365 °C, where mass conversions ranged from 9% on O_2 and 14% on N_2. The last event around 510 °C only occurred on oxidative environment, which would indicate the presence of oxidized compounds due to the heating process.

Table 1. Maximum temperature and percent of degradation of each decomposing event

	O_2						N_2		
Event	1°	2°	3°	4°	5°	6°	1°	2°	3°
T_i (°C)	30	50	80	150	348	460	–	150	332
T_d (°C)	40	60	100	292	365	510	–	292	365
T_f (°C)	50	80	150	348	460	610	–	332	505
Degradation (%)	12,5			68,0	9,0	6,3	3,3	77	14
Residual at 700 °C (%)	4,2						5,2		

The activation energies related to several amounts of mass conversions on oxidative environment are shown in Fig. 2. This Figure is divided into zones representing the degradation events: zone 1 refers to the first three events associated with water and volatile losses; zone 2 represents the main event, which occurs between 150 °C and 348 °C; zone 3 refers to the fifth degradation event, which occurs between 348 °C and 460 °C; zone 4 represents the range in which degradation of oxidized nuclei occurs (460 °C to 610 °C) and zone 5 is characterized as a zone where there is no mass decomposition, and the percentage of sample residue is determined.

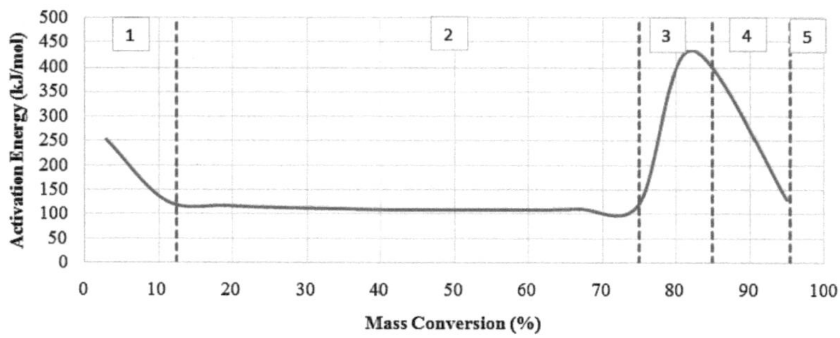

Fig. 2. Activation Energy as a function of mass conversion

The activation energy, defined as the required energy to process a reaction, has an inverse relationship with its speed, so that the higher the Ea is, the slower the associated reaction will be. Thus, it is found that the activation energy in zone 1 reduces with increasing conversion, indicating that a higher energy was required to start the degradation process than to continue, and that decomposition was accelerated with increasing temperature.

In zone 2, which was the main degradation event, there was a maintenance in the value of the energy in almost the entire conversion range, indicating that the reaction was carried out at constant speed, until degradation of approximately 75% of the material. From this percentage, a higher energy was required for the reaction to continue, indicating that the compounds to be degraded at this stage are more thermally stable than the compounds converted at the beginning of this event.

Furthermore, in Zone 2 it was observed the lowest values of Ea for the whole conversion of the sap, indicating that this degradation occurred at a faster rate than all other mass conversion events. It should be noted that this event takes place over a range of temperatures at which the bio-binders were produced at Gondim *et al.* (2017), suggesting that the use of improper temperatures in these processes could easily degrade the sample.

In zones 3 and 4, a constant reduction in the activation energy is observed, which indicates that the speed of these reactions increased with the temperature increases. In zone 5, referring to the residual of Thermogravimetry, there was no decomposition, so there is no associated activation energy.

4 Conclusion

The thermogravimetric analysis of the sap of the petroleum plant indicated a high complexity of this material, once the thermograms on inert and oxidative environments showed differences on both number of events and percentage of mass conversions.

The kinetic analysis of the thermogravimetry showed that the energy required for conducting mass conversion varied to each event, and the energy required for the main event of degradation (that starts at 150 °C) was the lowest one, what indicates the highest speed of reaction.

This observation supports that the use of 160 °C for modification, mixing and aging processes is unsafe for the material stability, since it is higher than 150 °C, and that small variations on temperature on this range would results in major mass conversions. Thus, the temperature to be employed in binder modifications, mixing and simulation of short-term aging of bio-binder based on petroleum plant sap should be lower than 150 °C. Considering the possibilities of variations of temperature on theses process, the maximum safe temperature for this sap is around 140 °C.

At last, the thermogravimetric technique showed to be important for characterizing thermal behavior of bio-materials that are candidates for the formulation of bio-binders.

Acknowledgements. The authors acknowledge FUNCAP for financial support and LUBNOR/Petrobras for the donation of the asphaltic binder samples.

References

Audo, M., Chailleux, E., Bujoli, B., Queffélec, C., Legrand, J., Lépine, O.: Alternative binder from microalgae. In: Alternative Binders for Sustainable Asphalt Pavements – Papers from a Workshop, Transportation Research Circular, Number E-C165, Washington D.C., August 2012

Ballie, M.: COLAS. Liant de Nature Végétale pour la Réalisation de Materiaux pour Le Bâtiment et/ou Les Travaux Publics. Demand de Brevet Europeen, EP 1 466 878 A1, 08 abr 2004

Çelik, O.N., Atasağun, N.: Rheological properties of bituminous binder modified with Nigella pulp liquefied by means of pyrolysis method. In: 2nd International Symposium on Asphalt Pavements & Environment. International Society for Asphalt Pavement – ISAP, Fortaleza (2012)

Feitosa, J.P.M., Alencar, A.E.V., Souza, J.R.R., Soares, J.B., Soares, S.A., Ricardo, N.M.P.S.: Evaluation of carnauba waxes in warm mix asphalt technology. Int. J. Civ. Environ. Eng. IJCEE-IJENS **15**(3), 1–9 (2015)

Fini, E.H., Kalberer, E.W., Shabazi, G.: Application of bio-binder from swine manure in Asphalt Binder. In: Transportation Research Board 90th Annual Meeting, Washington D.C., United States (2011)

Gondim, L.M., Soares, S.A., Barroso, S.H.A., Alecrin, C.M.C.: Chemical and rheological analysis of an asphalt binder modified by the sap of Euphorbia Tirucalli Plant. In: Loizos, et al. (eds.) Bearing Capacity of Roads, Railways and Airfelds. Taylor & Francis Group, London (2017). ISBN 978-1-138-29595-7

Kluttz, R.: Consideration for use of alternative binders in asphalt pavements material characteristics. In: Alternative Binders for Sustainable Asphalt Pavements – Papers from a Workshop, Transportation Research Circular, Number E–C165, Washington D.C., pp. 2–6 (2012)

Leite, L., Chacur, M., Nascimento, L.A., Cravo, M.C., Martins, A.T.: The use of vegetal products as asphalt cement modifiers. In: 5th Euroasphalt & Eurobitume Congress, Istanbul (2012)

Mills-Beale, J., You, Z., Fini, E., Zada, B., Lee, C.H., Yap, Y.K.: Aging influence on rheology properties of petroleum-based asphalt modified with biobinder. J. Mater. Civ. Eng. (2014). https://doi.org/10.1061/(ASCE)MT.1943-5533.0000712

Peralta, J., Silva, H.M.R.D., Williams, R.C., Rover, M., Machado, A.V.A.: Development of an innovative bio-binder using asphalt-rubber technology. Int. J. Pavement Res. Technol. (2013). https://doi.org/10.6135/ijprt.org.tw/2013.6(4)

Vasconcelos, M.A.G.: Formulation D'um Bitume Vert. Rapport de Stage. Laboratoire Central de Ponts et Chaussées de Nantes, Nantes, France (2010)

Wen, H., Bhusal, S., Wen, B.: Laboratory evaluation of waste cooking oil-based bioasphalt as sustainable binder for hot-mix asphalt. In: Alternative Binders for Sustainable Asphalt Pavements – Papers from a Workshop, Transportation Research Circular, Number E-C165, Washington D.C., August 2012

Williams, R.C., Brown, R.C., Tang, S.: Iowa State University Research Foundation, INC. Asphalt Materials Containing Bio-oil and Methods for Production Thereof. United States Patent. Patent Number: US 8.696.806 B2, 15 abr 2014 (2014)

Zofka, A., Yut, I.: Investigation of rheology and aging properties of asphalt binder modified with waste coffee grounds. In: Alternative Binders for Sustainable Asphalt Pavements – Papers from a Workshop, Transportation Research Circular, Number E-C165, Washington D.C., August 2012

Machine Learning Technique for Interpretation of Infrared Spectra Measured on Polymer Modified Binders

Adam Zofka[1(✉)] and Krzysztof Błażejowski[2]

[1] Road and Bridge Research Institute (IBDiM), Warsaw, Poland
azofka@ibdim.edu.pl
[2] ORLEN Asfalt, Płock, Poland
Krzysztof.Blazejowski@orlen.pl

Abstract. Demand and supply of polymer modified binders (PMB) have significantly increased in the past 20 years. These binders are intended to ensure long-lasting performance of asphalt pavements. Depending on the type and magnitude of modification, PMBs can be engineer to maintain superior mechanical properties over a wide range of service temperatures. PMBs can be also used to enhance pavement surface properties such as texture, noise absorption and skid resistance, and they can be applied in the modern surface treatments for pavement maintenance. This paper utilizes one of the machine learning techniques, namely neural network, in order to classify PMBs based on their infrared (IR) signature. Such a classification is particularly justifiable when dealing with highly modified asphalt (HiMA) which is a premium PMB material with a polymer content of at least 7.5 wt%. This paper demonstrates IR measurements and interpretation of 22 different unaged asphalt binders from two producers. Based on the advanced analysis it is shown that machine learning technique can very accurately differentiate between various asphalt binder groups.

Keywords: Polymer modified binder · Machine learning
Highly modified asphalt · Fourier transform infrared spectroscopy

1 Motivation and Objectives

Application of polymer modified binders (PMB) is constantly increasing due to their superior properties and plausible benefit to cost analysis (Bernier et al. 2012; Kluttz et al. 2012; Yut and Zofka 2014; Paliukaite et al. 2015). In the recent years, a new addition to PMB material group has been introduced called highly modified asphalt (HiMA) (Kluttz et al. 2009; Kluttz et al. 2013; Błażejowski et al. 2015; Błażejowski et al. 2016). A distinct feature of HiMA material is its polymer content that exceeds 7 wt%. Such a high binder content is intended for mitigation of pavement cracking and plastic deformations as well as to increase the fatigue resistance of asphalt courses (Timm et al. 2012; Timm et al. 2013). At 7 wt% or higher there is a phase reverse in binder/polymer matrix resulting in continuous polymer network (polymer phase) with a binder inclusions. Figure 1 compares micrographs of typical PMB with

L. D. Poulikakos et al. (Eds.): RILEM 252-CMB 2018, RILEM Bookseries 20, pp. 281–286, 2019.
https://doi.org/10.1007/978-3-030-00476-7_44

HiMA. It can be easily observed that indeed HiMA is a mixtures of asphalt dispersed within a polymer phase.

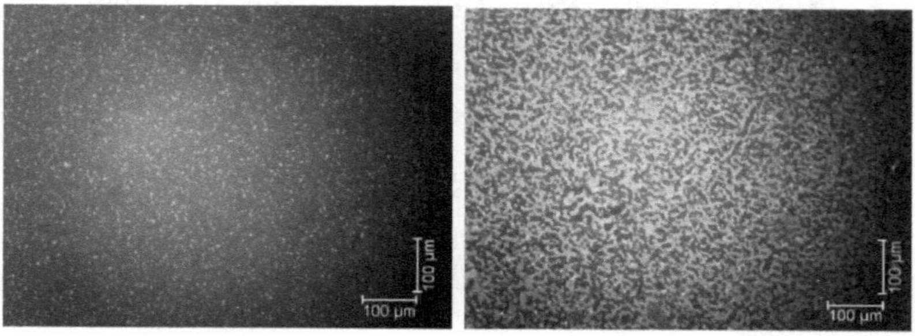

Fig. 1. Micrographs from fluorescence microscope: (a) polymer modified binder (PMB) (PMB 25/55-60), (b) highly modified asphalt (HiMA) (PMB 45/80-80 HiMA)

There is a number of interesting aspects related to HiMA materials, including their production, laboratory performance and pavement performance, but this paper focuses on the issue related to the quality assurance (QA). Typically, in order to differentiate between various asphalt binders, mechanical testing is performed that indicates certain property which is unique for a particular group of materials. For example, a ductility testing is often used as indictor of asphalt modification. In this paper, a procedure based on infrared (IR) signature of asphalt materials is proposed for fingerprinting asphalt binders. Proposed procedure goes beyond a simple IR measurements and uses advanced chemometric analysis based on the artificial neural networks (ANN). ANNs are supervised machine learning techniques that are often used for data classification, regression and/or clustering depending on the objective. In this paper objectives are as follows:

1. Perform IR measurements on various asphalt binders including unmodified binders, PMBs and HiMAs, and determine characteristic IR indices.
2. Perform ANN analysis in order to classify various asphalt binder groups.

2 Materials

This study uses 22 asphalt binders from two producer (Table 1). All binders are combined into three distinct groups based on their penetration values and other standard properties declared by the producers. All material samples were collected directly from the producers and further prepared for the IR measurements in one laboratory. Proper planning was incorporated to ensure that all samples experienced the same treatment in terms of reheating and oven handling. Two reheating steps in the conventional oven were used to prepare all specimens and no other laboratory aging was implemented in the specimen preparation process. Complete Randomized Design (CRD) experiment was employed for the IR measurements as explained in the next section.

Table 1. Binder penetration classes and groups

Penetration classes	Producer	Group
20/30 35/50 50/70 70/100	**A** **B**	**NM** (unmodified)
PMB 10/40-65 PMB 25/55-60 PMB 45/80-55 PMB 45/80-65	**A** **B**	**PM** (modified)
PMB 25/55-80 PMB 45/80-80 PMB 65/105-80	**A** **B**	**HM** (highly modified)

3 Methods

IR measurements were performed in the attenuated total reflection (ATR) mode according to the procedure developed earlier and available in the literature (Yut and Zofka 2011; Zofka et al. 2015). IR spectra were collected with a diamond ATR crystal within a broad absorbance spectrum of wavenumbers from 7500 cm^{-1} up to 375 cm^{-1} at the resolution of 4 cm^{-1}. Based on the previous studies, at least 5 replicate specimens were tested for each binder which resulted in over 110 IR spectra. Raw IR data interpretation comprised the following steps: (1) baseline correction, (2) data smoothing, (3) ATR correction and atmospheric compensation. In the next step, IR spectra was numerically integrated within three regions of interest (ROI) associated with polymer (SBS) presence, i.e. around 700, 725 and 965 cm^{-1}. Example IR spectra within these ROI for all three binder groups (NM, PM, HM) are presented in Fig. 2. In the final step, calculated ROI areas were normalized with the sum of areas under entire IR spectra, and resultant values of normalized indices (A700, A725 and A965) were passed to ANN analysis.

ANN chemometric analysis was performed on the individual values of A700, A725 and A965 obtained from all replicates and all 22 asphalt binders. To increase model accuracy, one additional input parameter was employed in the ANN: it was a first component (PC1) from the principal component analysis (PCA) performed on the normalized indices A700, A725 and A965. ANN architecture was selected based on the previous studies (Zofka and Yut 2012; Yut and Zofka 2014). It consisted of one hidden layer and 10 neurons. For the ANN development, dataset was randomly divided for three phases: training (70%), validation (15%) and testing (15%). Typically, classification ANN results are presented in terms of confusion matrix separately for each phase but this paper shows only testing matrix and overall matrix as they practically summarize the ANN performance.

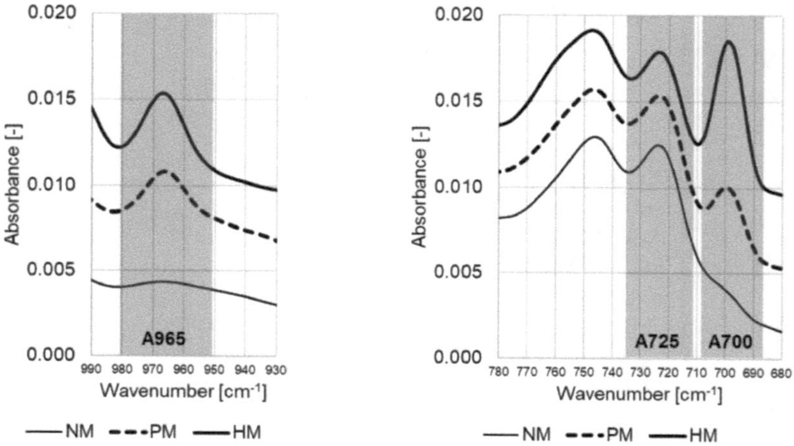

Fig. 2. Example IR spectra within ROI (A965, A725 and A700) for three binder groups (NM, PM, HM)

4 Analysis

Results are presented in two parts: the first part shows the normalized IR indices (A700, A725 and A965) for different asphalt groups whereas the second part shows the performance of ANN model.

4.1 Interpretation – Normalized IR Indices

Figure 3 shows the normalized IR indices separately for three asphalt binder groups, i.e. NM, PM and HM. Since these IR regions (A700 and A965) are associated with polymer presence, Fig. 3 confirms the group assignment and verifies the interpretation procedure together with applied spectra corrections. High variance for PM and HM groups indicates different polymer content and/or different production technology between two producers. Overlapping of the normalized indices hinders the distinction between PM and HM and thus there is a need for more robust classification procedure such as machine learning and more specifically ANN.

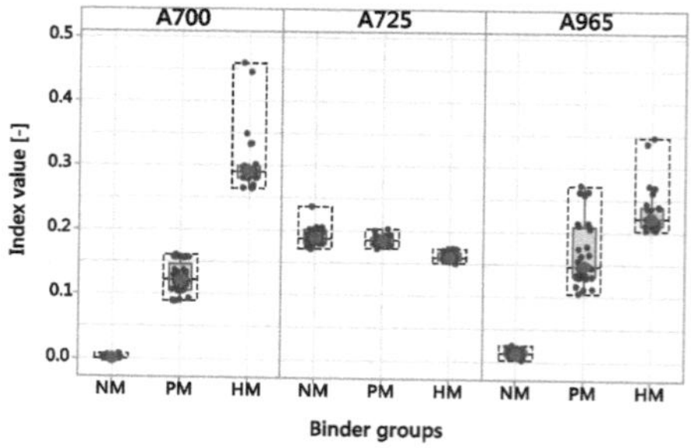

Fig. 3. Index values: A700, A725 and A965 for different binder groups

4.2 Machine Learning (ANN)

Classification results from ANN model are presented in Table 2. Since true positive predictions are located along the diagonal, it can be concluded that ANN model demonstrates 100% accuracy on both testing and overall dataset which proves applicability of proposed approach as classification tool.

Table 2. ANN confusion matrices: (a) testing phase, (b) overall; 1 – NM, 2 – PM, 3 – HM

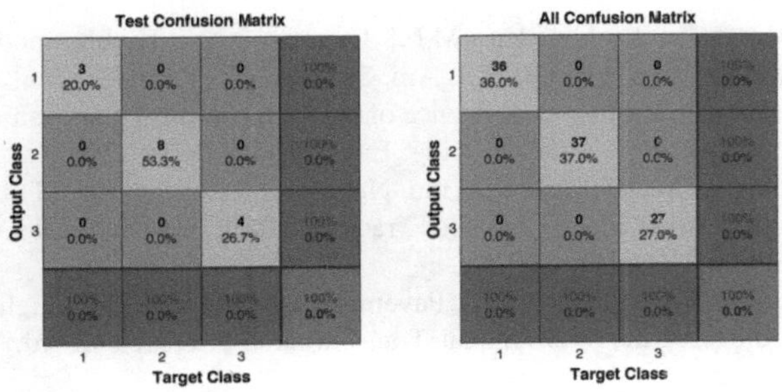

5 Summary

This paper utilizes one of the machine learning techniques, namely artificial neural network (ANN), in order to classify asphalt binders based on their IR signature. FT-IR ATR measurements were performed on 22 different binders and three distinct normalized indices were determined from each spectra. IR indices were combined with the first PCA component and such inputs were used in the development of ANN model. Results showed an excellent performance of ANN based on input values with 100% accuracy on the testing dataset that was not used for model training.

This paper proposed a practical and quick procedure for QA that is based on the direct assessment of intrinsic features of asphalt binders. Proposed procedure requires fairly small workload and it can be performed in the field conditions. It can be of particular interest for investors who deal with high volumes of premium materials such as HiMA.

References

Bernier, A., Zofka, A., Yut, I.: Laboratory evaluation of rutting susceptibility of polymer-modified asphalt mixtures containing recycled pavements. Constr. Build. Mater. **31**, 58–66 (2012)

Błażejowski, K., Wójcik-Wiśniewska, M.: Highly modified bitumen in perpetual pavements. In: Czech National Asphalt Conference Asfaltove Vozovky, Ceske Budejovice, 24–25 November 2015

Błażejowski, K., Wójcik-Wiśniewska, M., Peciakowski, H., Olszacki, J.: The performance of a highly modified asphalt for long-lasting asphalt pavements. Transp. Res. Procedia **14**, 679–684 (2016). Transport Research Arena, Warsaw 2016

Kluttz, R., Richard Willis, J., Molenaar, A., Scarpas, T., Scholten, E.: Fatigue performance of highly modified asphalt mixtures in laboratory and field environment. In: 7th RILEM International Conference on Cracking in Pavements (2012)

Kluttz, R.Q., Molenaar, A.A.A., van de Ven, M.F.C., Poot, M.R., Liu, X., Scarpas, A., Scholten, E.J.: Modified base courses for reduced pavement thickness and improved longevity. In: Proceedings of the International Conference on Perpetual Pavement, Columbus, OH, October 2009

Kluttz, R.Q., Jellema, E., Woldekidan, M.F., Huurman, M.: Highly modified binder for prevention of winter damage in OGFCs. Am. Soc. Civ. Eng. (2013)

Paliukaite, M., Vaitkus, A., Zofka, A.: Influence of bitumen chemical composition and ageing on pavement performance. Balt. J. Road Bridg. Eng. **10**(1), 97–104 (2015)

Timm, D.H., Robbins, M.M., Willis, J.R., Tran, N., Taylor, A.J.: Field and Laboratory Study of High-Polymer Mixtures at the NCAT Test Track. Draft Report, National Center for Asphalt Technology, Auburn University (2013)

Timm, D., Powell, R., Willis, J., Kluttz, R.: Pavement rehabilitation using high polymer asphalt mix. In: Proceedings of the 91st Annual Transportation Research Board, Washington, DC (2012)

Yut, I., Zofka, A.: Attenuated total reflection fourier transform infrared spectroscopy of oxidized polymer-modified bitumens. Appl. Spectrosc. **65**(7), 765–770 (2011)

Yut, I., Zofka, A.: Correlation between rheology and chemical composition of aged polymer-modified asphalts. Constr. Build. Mater. **62**, 109–117 (2014)

Zofka, A., Yut, I.: Prediction of Asphalt Creep Compliance Using Artificial Neural Networks. Arch. Civ. Eng. **58**(2), 153–173 (2012). https://doi.org/10.2478/v.10169-012-0009-9

Zofka, A., Maliszewska, D., Maliszewski, M., Boratynski, J.: Application of FTIR ATR method to examine the polymer content in the modified bitumen and to assess susceptibility of bitumen to ageing. Roads and Bridges - Drogi i Mosty **14**(3), 163–174 (2015)

Meso- to Macroscale Homogenisation of Hot Mix Asphalt Considering Viscoelasticity and the Critical Role of Mortar

Johannes Neumann[1(✉)], Jaan-Willem Simon[2], and Stefanie Reese[2]

[1] Aachen Association of Highway Engineering GmbH, Pascalstraße 6, 52076 Aachen, Germany
neumann@isac-gmbh.com

[2] Institute of Applied Mechanics, RWTH Aachen University, Mies-van-der-Rohe-Str. 1, 52074 Aachen, Germany

Abstract. Computational homogenisation is a useful tool to predict the macroscale response of composite materials without the cumbersome experimental programme. However, careful verification is required for computational methods as well. A recently proposed method to create synthetic 3D mesoscale models of hot mix asphalt uses Voronoi polyhedra to represent the mineral aggregate. The agreement of these synthetic shapes with data from XRCT scanning is investigated. Typical shape measures are compared, and good agreement is found. The generalised Maxwell model is used to describe the viscoelasticity of the mortar. However, the previously used experimental data is found to be inadequate. This is attributed to the hitherto employed mortar design, which is revised, and a review of the relevant literature is conducted. Reported experimental data is critically assessed in terms of usefulness for homogenisation schemes. First-order strain driven homogenisation is then carried out in frequency domain in order to obtain the macroscale response, which is compared to macroscale experimental data. A satisfactory agreement is found.

Keywords: Hot mix asphalt · Computational homogenisation
Mortar

1 Introduction

Synthetic modelling of composite materials in general and of asphalt concrete (AC) in particular can provide a useful addition and/or alternative to experimental methods. However, the complexity of the material requires great care during the design phase in order to achieve sensible results. From a morphological point of view, AC is a random heterogeneous material with irregularly shaped inclusions. From a materials science perspective, complexities include viscoelasticity, plasticity, anisotropy, temperature dependence, and many more.

Several researchers have identified the need for 3D modelling of AC since 2D approaches typically underestimate the stresses (Chen et al. 2015; Ozer et al. 2016). Here, a recently proposed synthetic 3D mesoscale model of hot mix asphalt is employed (Neumann et al. 2017). It uses a nested Voronoi tessellation to mimic the

L. D. Poulikakos et al. (Eds.): RILEM 252-CMB 2018, RILEM Bookseries 20, pp. 287–292, 2019.
https://doi.org/10.1007/978-3-030-00476-7_45

rocks which are larger than a variable size threshold and also captures the particle size distributions. In a yet unpublished article, XRCT data of an SMA13 type AC has been analysed. Now, the agreement of several typical shape measures in road engineering is checked between real and synthetic particles.

The bituminous mortar[1] is obtained by removing the larger rock fractions from the mixture. The motivation to test mortar instead of mixture properties is fuelled by the following considerations: it is easier, faster and cheaper than the testing of mixture samples.

2 Material Modelling

A regularised linear least-square method is used to identify material parameters of the generalized Maxwell model. Details are given in (Neumann et al. 2017). Further investigations have revealed that the mortar used there is rather soft due to a high bitumen content of 21.1 m.-%. Other researchers have taken the amount of bitumen ab- and adsorbed by the larger rock fractions into account and designed mortars with significantly lower bitumen content of 8–13 m.-%. For reasons of comparability, all master curves taken from literature sources have been shifted horizontally to a common reference temperature of $\Theta_r = 20$ °C. The shifted master-curve data is shown in Fig. 1. The first two mortars, FAM 76-16 and FAM 64-22 are from (Gudipudi and Underwood 2015). The next five (CFL-9 to VFFS-2) are reported in (Underwood and Kim 2013). Due to uncertainties about the void ratio in mortar, (Karki et al. 2015) investigated mortar designs with 1 and 5.5% air void. The mortar hitherto used by the author is termed "Exp FAM SMA" and indicated by black crosses. It can be seen that it is softer than the reported mortar data by a decade, even in the high frequency regime. For comparison, data of a pure 20/30 binder (Schüler et al. 2013) and the same binder reinforced with filler (Schüler et al. 2016) are presented.

3 Geometry Modelling

The notion of a representative volume element is of pivotal importance for computational homogenisation of random heterogeneous materials. Volume elements become representative with increasing size once the quantities of interest have converged to a value that does not change significantly when the size is increased further. Strictly speaking, this is only achieved in the limit case of infinite volume. Thus, the pragmatic way to evade this dilemma is to accept quantities as converged once their scatter drops below a certain threshold. For reasons of computational complexity, one is interested in RVEs which are as small as possible. The concept of ergodicity is important here, since it allows to average over several small volumes instead of one, or a few, large ones.

However, it needs to be guaranteed that the mean is not biased, otherwise properties will not converge to realistic values. Naturally, for a volume element to be

[1] Often termed fine aggregate matrix (FAM) in the relevant literature.

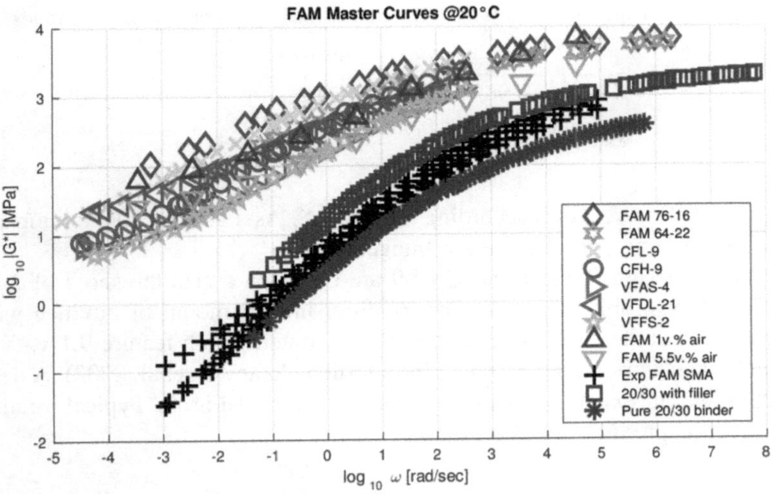

Fig. 1. Master curves of complex shear modulus of different mortar designs available in literature

representative, it must contain enough particles so that the complete particle size distribution is contained within one sample. This requirement is automatically satisfied by the algorithm of (Neumann et al. 2017). Furthermore, particle representation must be accurate. The present algorithm uses Voronoi polyhedra to represent the real aggregate particles. Recently, shape analysis of XRCT data of mixture SMA13 has been undertaken by the author. There, particle shape is assessed via three orthogonal linear dimensions: length L, width W, and thickness T. These dimensions are derived from a principal component analysis of the inertia tensor of each particle.

A comparison between XRCT scanned real and synthetic Voronoi grains shows high agreement, please see Table 1. Two shape measures are employed. The sphericity $\Psi = \sqrt[3]{(WT/L^\wedge 2)}$ is a nondimensional measure which yields unity for spheres and approaches zero with increasing aspect ratios (Krumbein 1941). Furthermore, the shape index (Anon 2015) counts the mass[2] of those particles for which $L/T > 3$. For closed porous top layers, the shape index must not exceed 20 m.-%. On the downside, Voronoi polyhedra are strictly convex, where real aggregate particles are not. The topology has not been investigated, yet.

Depending on the maximum size of the particles in the mortar, the size threshold of explicit geometrical modelling is chosen. For European mixtures, an obvious choice is 2 mm. However, for mixtures based on other standards and guidelines, particularly from the US, other thresholds are used. Choosing a small threshold is advantageous from an experimentalist's perspective because smaller samples can be used. However, a small threshold can even be prohibitive for computational homogenisation since the

[2] Or volume in case of uniform density.

Table 1. Shape measures of XRCT scanned and synthetic Voronoi particles

	Ψ	Shape index [m.-%]
XRCT (n = 98)	0.6782	13.02
Voronoi (n = 129)	0.6658	12.44

number of particles, N, grows according to $N \propto S^{-3}$. As a general rule, mixtures with a high number of large particles are advantageous.

The mortar designs CFL-9 and CFH-9 are based on a size threshold of 2.36 mm and are therefore usable. The two mortars differ in the amount of bitumen with "L" indicating low and "H" indicating high bitumen content. Both feature 9.1 v.-% voids. The particle size distribution was taken from (Allen Cooley Jr. et al. 2002). A depiction of the PSD and an example volume element is given in Fig. 2. Typical volume elements contain approximately 200 grains.

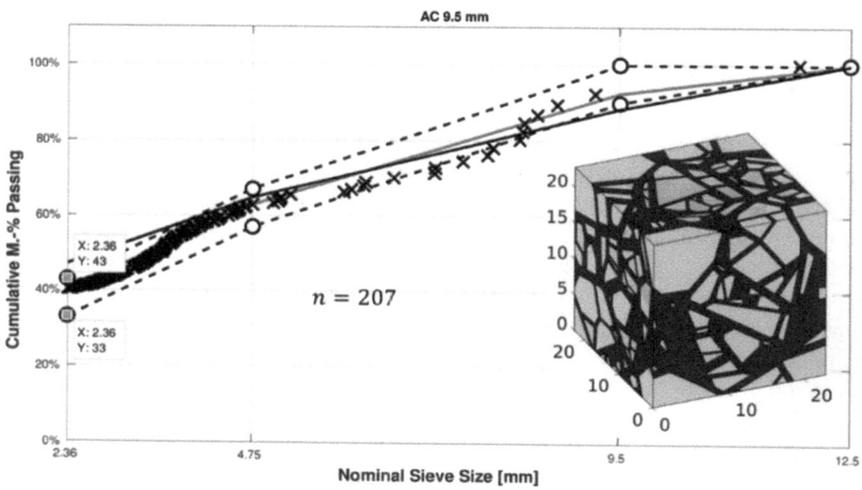

Fig. 2. Particle size distribution and synthetic volume element of AC 9.5 after (Allen Cooley Jr. et al. 2002)

4 Results and Discussion

First order strain driven computational homogenisation is used in conjunction with periodic boundary conditions to minimise boundary effects. In addition to previous works (Neumann et al. 2017), the homogenisation is now carried out directly in frequency space using steady-state dynamics. Preliminary results using only one volume element are shown in Fig. 3. On the left, the CFL-9 mortars stiffness is rather close to mix stiffness. Therefore, the macroscale prediction overshoots the experimental mix data. On the right, however, the homogenised result matches the mix stiffness quite well. This result requires additional investigations in terms of mesh density, RVE size and number of samples. Nevertheless, mortar design turns out to be critical for computational homogenisation. Unfortunately, there is no consensus about the right mortar design at the moment.

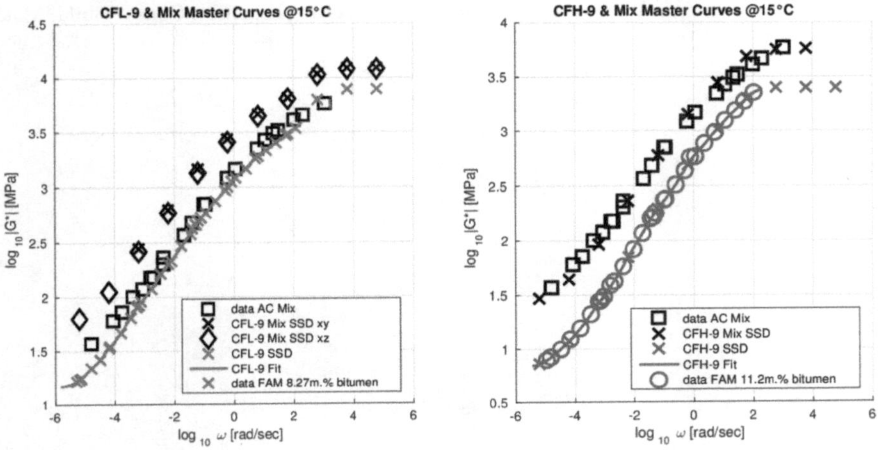

Fig. 3. Mortar data, parameter fit, validation, results of computational homogenisation, and mixture data for CFL-9 (left), and CFH-9 (right)

References

Allen Cooley Jr., L., Prowell, B.D., Brown, E.R.: Issues pertaining to the permeability characteristics of coarse-graded superpave mixes, s.l.: s.n. (2002)

Anon: DIN EN 933-4 Tests for geometrical properties of aggregates - Part 4: Determination of particle shape - Shape index; German version EN 933-4:2008. s.l.:s.n. (2015)

Chen, J., Wang, H., Li, L.: Virtual testing of asphalt mixture with two-dimensional and three-dimensional random aggregate structures. Int. J. Pavement Eng. **18**(9), 824–836 (2015). https://doi.org/10.1080/10298436.2015.1066005

Gudipudi, P., Underwood, B.S.: Testing and modeling of fine aggregate matrix and its relationship to asphalt concrete mix. Transp. Res. Rec. J. Transp. Res. Board **2507**(1), 120–127 (2015)

Karki, P., Kim, Y.-R., Little, D.N.: Dynamic modulus prediction of asphalt concrete mixtures through computational micromechanics. Transp. Res. Rec. J. Transp. Res. Board **2507**, 1–9 (2015)

Krumbein, W.C.: Measurement and geological significance of shape and roundness of sedimentary particles. J. Sediment. Res. **11**, 64–72 (1941)

Neumann, J., Simon, J.-W., Mollenhauer, K., Reese, S.: A framework for 3D synthetic mesoscale models of hot mix asphalt for the finite element method. Constr. Build. Mater. **148**, 857–873 (2017)

Ozer, H., Ghauch, Z.G., Dhasmana, H., Al-Qadi, I.L.: Computational micromechanical analysis of the representative volume element of bituminous composite materials. Mech. Time Depend. Mater. **20**(3), 441–453 (2016)

Schüler, T., Jänicke, R., Steeb, H.: Nonlinear modeling and computational homogenisation of asphalt concrete on the basis of XRCT scans. Constr. Build. Mater. **109**, 96–108 (2016)
Schüler, T., et al.: Multi-scale modelling of elastic/viscoelastic compounds. ZAMM J. Appl. Math. Mech./Z. Für Angew. Math. Und Mech. **93**, 126–137 (2013)
Underwood, B.S., Kim, Y.R.: Effect of volumetric factors on the mechanical behavior of asphalt fine aggregate matrix and the relationship to asphalt mixture properties. Constr. Build. Mater. **49**, 672–681 (2013)

Novel Application of the Falling Weight Deflectometer Test: Detection of Surface and Subsurface Distresses

Anirban Chatterjee(✉) and Yichang(James) Tsai

Georgia Institute of Technology, Atlanta, GA, USA
a.chatterjee@gatech.edu

Abstract. Falling Weight Deflectometers (FWD) and Traffic Speed Deflectometers (TSD) have become increasingly popular for monitoring the structural performance of roadway pavements. However, many transportation agencies measure the functional performance of the pavement (given by surface distresses such as cracking, rutting and raveling) for maintenance purposes. This study explored the feasibility of estimating surface and subsurface distresses using FWD tests. It was found that the presence of pavement cracking can be detected through FWD tests but not pavement rutting. A random forest classifier was used to predict the presence of cracking between layer interfaces below the pavement surface. The classifier achieved an accuracy of 92.3%. With the methodology presented in this paper, FWD tests can be used to measure the structural as well as functional performance of pavements. The methodology also serves as a proof-of-concept for using TSDs to measure the functional performance of pavements at high speeds. This provides an alternative to unsafe and laborious on-foot survey practices.

1 Introduction

The emergence of new sensor technologies has opened opportunities for innovative integrations to assess road pavement conditions. Falling Weight Deflectometers (FWD) are non-destructive test devices which measure pavement deflection on the application of a fixed load. FWDs are widely used for the backcalculation of pavement layer moduli and to design suitable overlays to achieve a desired design life.

Historically, FWDs have been mainly used as a non-destructive test for determining the structural properties of the pavement. Diefenderfer (2008), Ali and Khosla (1987), Xu et al. (2002), Zhang et al. (2003), Hoffman (2003), Kim and Park (2002), Sharma and Stubstad (1980) are some examples. However, some transportation agencies like the Georgia Department of Transportation (2007) and Florida Department of Transportation (2015) use the functional properties of the pavement for evaluating pavement performance. These functional properties, which affect user experience, are present on the surface of the pavement, such as cracking and rutting. Although directly using the FWD to find surface distresses is pointless, this research serves as a proof-of-concept that deflection measuring devices, such as Traffic Speed Deflectometers (TSDs), can be used to detect surface distresses. TSDs can then be used to detect pavement distresses

L. D. Poulikakos et al. (Eds.): RILEM 252-CMB 2018, RILEM Bookseries 20, pp. 293–299, 2019.
https://doi.org/10.1007/978-3-030-00476-7_46

at much higher speeds. Moreover, the correlation can be used in the other direction, allowing us to understand the effect of pavement distresses on FWD readings, and subsequently, the layer moduli. Another possibility is to use the FWD test to predict the presence of subsurface distresses, which can lead to functional performance degradation in the future.

Hence, the first objective of this project was to study the effect of pavement distresses on FWD backcalculated layer moduli and vice versa. Clearly other factors affect the FWD readings as well, such as traffic loading and age. This study examines the effect of pavement distresses at that particular age and traffic loading. The second objective of this project was to test the feasibility of subsurface distress prediction using only the FWD test and observable surface properties.

2 Methodology

FWD readings recorded on ten 152 m (500 ft.) sections (five in each direction) of US Route 80 near Savannah, GA in February 2016 were used for this research. The presence of pavement distresses was recorded for each of the sections. A total of 56 FWD readings were taken, mostly on the right wheel path.

Coring samples were taken to identify the asphalt layer thicknesses at 26 locations, although it is possible to calculate the layer thicknesses by non-destructive means: from the construction and maintenance history of the section. These coring samples also provided information about the presence of cracks, crack width and crack depth. It was also observed that in some coring samples, the layers of asphalt had separated at one or more of the interfaces.

2.1 Processing of Coring Samples

First, the layer thicknesses were estimated using coring samples (Fig. 1) and plans. The pavement often consisted of multiple asphalt concrete (AC) layers which were added up to give a total AC layer thickness. The AC layer was followed by a granular aggregate base (GAB) layer in almost all the sections. The thickness of the GAB layer was determined using engineering drawings for the pavement profile. One section contained a coring sample with a cement-treated base. This section was ignored since it was not possible to determine which FWD readings lay on the cement-treated base. Below the GAB layer was the subgrade.

In order to allocate a thickness for each FWD test, the average thickness of all the cores were calculated for each section. This average thickness was used for all FWD tests in that section.

Apart from layer thicknesses, the coring samples also provided crack information (Fig. 1b). The crack width and depth was measured in each of the samples. It was also noted down whether there were any breaks in the interface between two asphalt layers (Fig. 1c). This was a subsurface distress which was not observable without the destructive coring.

(a) (b) (c)

Fig. 1. Coring samples

2.2 Processing of FWD Data

The thicknesses obtained from the coring samples and plans served as the input for the backcalculation of layer moduli using the deflection basin fit method developed by Hossain and Zaniewski (1991). The asphalt layer temperature at the middle of the asphalt layer was also required for backcalculation. This was calculated by using the BELLS2 model, developed by Stubstad et al. (1994), which takes the asphalt surface temperature and mean air temperature on the previous day as the input. The asphalt surface temperature was measured by the FWD and the mean air temperature of the previous day was found using local weather data.

3 Results and Analysis

The correlation between the layer moduli and the presence of pavement distresses, namely cracking and rutting, was determined. The layer moduli were used because they can have better correlation with the pavement distresses than the deflection, since they are intrinsic properties of the pavement.

3.1 Rutting

The plot of layer moduli for FWD points with and without rutting is shown in Fig. 2a. Rutting was assumed to be present if the rut depth exceeded 3.2 mm (0.25 in.). The AC layer, GAB layer and subgrade are represented by E1, E2 and E3 respectively. The dotted lines connect the mean moduli for each layer. No significant change occurred in the layer moduli due to the presence of rutting.

3.2 Cracking

The plot of layer moduli for FWD points with various associated crack widths is shown in Fig. 2b. The dotted lines represent the trendline for each layer. A clear increase in elastic modulus in the first two layers is seen when cracks start to appear. It appears that the layer moduli slightly fall as the crack width increases from 2 mm to 5 mm, but the presence of the three FWD points with very high moduli may have skewed the results. What remains conclusive is that the presence of cracks causes a clear rise in elastic

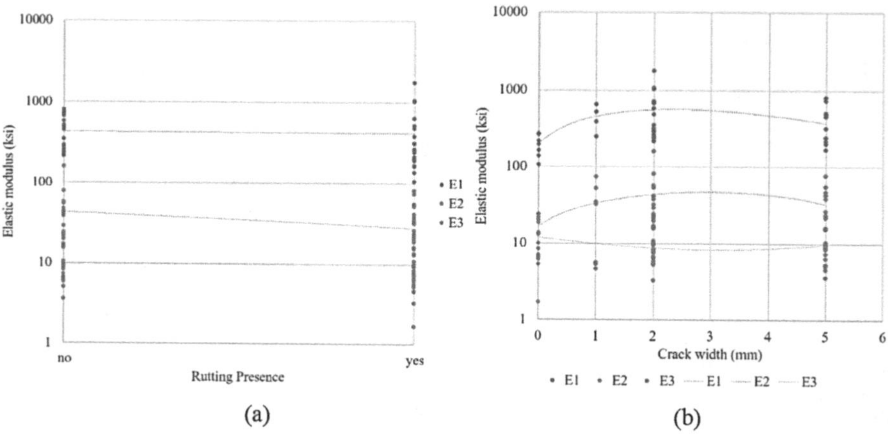

Fig. 2. Effect of (a) rutting and (b) cracking on layer moduli

moduli of the AC and GAB layers. This can be attributed to the loss of load transfer efficiency that occurs when cracks appear. This means the load applied by the FWD is not propagated further, causing lower deflections at the farther geophones (deflection measuring devices). Lower deflections correspond to higher stiffness, hence causing the elastic modulus to increase.

3.3 Interface Cracking

Separation of asphalt layers was observed in the coring samples. This separation could result in the development of potholes, an important surface distress that hampers the functional performance of pavements. However, this subsurface distress is difficult to detect without the use of destructive tests such as coring or specialized sensors such as ground-penetrating radars.

It is possible that such a cracking along layer interfaces would affect load-deflection relationships. The possibility of detecting layer separation using the FWD test was explored. This was achieved by training a random forest classifier on attributes that would be available through non-destructive testing: total AC layer thickness (from plans), crack presence, crack width and layer moduli (from FWD test). Based on these inputs, the classifier had to predict whether or not there were broken interfaces between the AC layers. The dataset consisted of 39 FWD points (after outlier removal) with the interface broken in 15 of the points. So the base accuracy (minimum accuracy that can be achieved simply by classifying all points as not broken) is 61.5%. The accuracy was measured using 10-fold cross validation. An accuracy of 92.3% was achieved by the classifier. The crack width was the principal component: achieving an 82% accuracy alone. The confusion matrix for the classifier is given in Table 1.

Table 1. Confusion matrix for interface broken classifier

	Predicted: broken	Predicted: not broken
Ground truth: Broken	13 (TP)	2 (FN)
Ground truth: Not broken	1 (FP)	23 (TN)

4 Extension to Traffic Speed Deflectometers

FWDs are a fast non-destructive test to estimate the structural properties of pavements at the project-level. In this paper, the ability of FWD tests to estimate the functional properties of pavements was demonstrated. FWDs are not suitable for network-level analysis although it has been attempted by Diefenderfer (2008) and Zhang et al. (2003). However, with Traffic Speed Deflectometers (also known as Rolling Wheel Deflectometers) becoming more and more common, the principles and research for FWDs can be applied at a much larger scale. Using the approaches presented in this paper, the functional performance of highways can be measured at highway speeds with TSDs. This provides a strong alternative to unsafe and laborious manual survey practices currently employed by transportation agencies.

5 Conclusion

This paper contributes the following findings:

1. It was found that pavement rutting had no significant impact on the FWD test. Hence, it is difficult to detect rutting based on load-deflection tests alone.
2. The presence of pavement cracking had a clear impact on FWD tests. The back-calculated layer moduli for the asphalt concrete and granular aggregate base layers would increase with the presence of cracks. Hence, FWD tests can be utilized to detect the presence of cracks.
3. Using data obtained by non-destructive means, including the FWD test, a random forest classifier was trained to predict the presence of a break between asphalt layers —a subsurface distress—with a 92.3% accuracy. Therefore, it is possible to use FWD tests to predict the presence of this subsurface distress and the formation of potholes.
4. Traffic Speed Deflectometers operate on the same load-deflection principle as FWDs but at much higher speeds. Hence, the above stated ideas can be in principle employed for high-speed network-level functional performance measurement of road pavements.

There are two future research recommendations. First, a more generalized study can be carried out to establish the relationships between the FWD test and surface distresses for various pavement ages and thicknesses. Second, some FWDs are fitted with geophones on either side of the load to measure load transfer efficiency across concrete joints. FWDs are generally not used when cracking is present in asphalt pavements, since cracking can decrease the load transfer efficiency in the pavement and cause overestimation of layer moduli due to decreased deflections. This shortcoming can be harnessed to study the cracks themselves. It is possible that the load transfer efficiency across a crack is related to the depth of a crack. Then by placing an FWD to measure the load transfer efficiency across a crack, the crack depth can be determined non-destructively.

This paper explored the relation between the functional performance of the pavement and the FWD test. The results laid the groundwork for a safe and efficient approach to functional performance measurement. This paper also opened new research avenues for innovative applications of the FWD test. With the increasing popularity of TSDs, further research into this area can help create safer and more efficient transportation systems.

Acknowledgements. The authors would like to thank the research project sponsored by the Georgia Department and Transportation, "RESEARCH PROJECT 14-05: Study of Georgia's Pavement Deterioration/Life and Potential Risks of Delayed Pavement Resurfacing and Rehabilitation", especially Jewell Stone and Neoma Cole from GDOT for their Falling Weight Deflectometer Tests on SR 26/US 80 near Savannah, GA, and the support provided by Ernay Robison, Eugene and Binh Bui from GDOT. We would like to thank the research team, Yiching Wu, Dr. Ross Wang, Dr. Zhaohua Wang, Geoffrey Price, Georgene Geary, Vincent Cartillier and April Gadsby from Georgia Tech for collecting 3D data on SR 26/US 80.

References

Ali, N.A., Khosla, N.P.: Determination of layer moduli using a falling weight deflectometer. Transp. Res. Rec. **1117**, 1–10 (1987)

Diefenderfer, B.K.: Network-Level Pavement Evaluation of Virginia's Inter-State System using the Falling Weight Deflectometer, Virginia Transportation Research Council, Report No. VTRC 08-R18 (2008)

FDOT: 2015 Flexible Pavement Condition Survey Handbook. Florida Department of Transportation (2015)

GDOT: Pavement Condition Evaluation System. Georgia Department of Transportation (2007)

Hoffman, M.: Direct method for evaluating structural needs of flexible pavements with falling weight deflectometer. Transp. Res. Rec. J. Transp. Res. Board **1860**, 41–47 (2003)

Hossain, A., Zaniewski, J.P.: Characterization of falling weight deflectometer deflection basin. Transp. Res. Rec. **1293** (1991)

Kim, Y.R., Park, H.-G.: Use of Falling Weight Deflectometer Multi-Load Data for Pavement Strength Estimation. North Carolina Department of Transportation, Report No. FHWA/NC/2002-006 (2002)

Sharma, J., Stubstad, R.: Evaluation of pavement in florida by using the falling weight deflectometer. Transp. Res. Rec. **755**, 42–48 (1980)

Stubstad, R., Baltzer, S., Lukanen, E.O., Ertman-Larsen, H.: Prediction of AC mat temperatures for routine load/deflection measurements. In: 4th International Conference, Bearing Capacity of Roads and Airfields, vol. 1 (1994)

Xu, B., Ranji Ranjithan, S., Richard Kim, Y.: New relationships between falling weight deflectometer deflections and asphalt pavement layer condition indicators. Transp. Res. Rec. J. Transp. Res. Board **1806**, 48–56 (2002)

Zhang, Z., Claros, G., Manuel, L., Damnjanovic, I.: Evaluation of the pavement structural condition at network level using falling weight deflectometer (FWD) data. In: 82nd Transportation Research Board meeting, Washington, DC, USA (2003)

Peat as an Example of a Natural Fiber in Bitumen

Hilde Soenen[1(✉)], Patricia Kara De Maeijer[2], Johan Blom[2], and Wim Van den Bergh[2]

[1] Nynas NV, Bitumen Research, Groenenborgerlaan 171, 2020 Antwerp, Belgium
hilde.soenen@nynas.com

[2] Energy and Materials in Infrastructure and Buildings (EMIB), Faculty of Applied Engineering, University of Antwerp, Groenenborgerlaan 171, 2020 Antwerp, Belgium
{patricija.karademaeijer, johan.blom, wim.vandenbergh}@uantwerpen.be

Abstract. In this study, the suitability of a natural peat fiber, as a modifier for bitumen was investigated. Peat fibers were dried, ground, and fractionated into a fiber and a granular fraction. Rheological data on peat modified binders indicated that the granular fraction is acting as a filler mainly stiffening the bituminous binders whilst the fiber fraction mainly increases the elastic behavior, which became especially visible at higher service temperatures. It was observed that by changing the ratio of fibers versus granular particles, the rheological behavior can be regulated. In addition, drainage tests were carried out on stone mastic asphalt (SMA) mixes. The data indicated that peat fibers can effectively reduce drainage, and can be used as an alternative for the currently used cellulose fibers.

Keywords: Peat fibers · Bitumen · Rheology · Drainage

1 Introduction

Peat soil is predominantly decayed organic matter which is formed, under water-logged conditions such as bogs, it is formed from dead plant material and accumulates where rainfall is high and loss of water through evapo-transpiration is low (Bain et al. 2011). In some countries peat is regarded as a slow-renewable material, however, as the rate of extraction and usage of peat far exceeds the rate of reforming, its renewable character has been under debate. Peat's main usage is in energy production, for example, in 2015 in Finland 3% of the annual energy production was provided by peat (Statistics Finland 2016). In the industrial sector, peat is also used for example as an oil absorbent or as an efficient filtration medium for mine waste streams, municipal storm drainage and septic systems (USGS 2017).

Peat has already been applied in asphalt, as a stabilizing additive for peat-based asphalt-concrete mixes, providing high performance to the road surface at a low cost (Kudrjashov et al. 2013). Fibers are used in asphalt to prevent drainage of bitumen, this becomes more important for gap graded mixes like stone mastic asphalt (SMA) which

L. D. Poulikakos et al. (Eds.): RILEM 252-CMB 2018, RILEM Bookseries 20, pp. 300–305, 2019.
https://doi.org/10.1007/978-3-030-00476-7_47

have a high binder content (Hansen et al. 2000). Typically, the optimum fibers content is 0.3%–0.5% by weight of the asphalt mix (Oda et al. 2012). The addition of fibers provides advantages such as increased binder content, increased aggregate binder film thickness, increased mix stability and a reduction in binder drainage during transportation and laying.

This study will discuss the initial results of an investigation on the effects of addition of peat and peat fibers into bituminous binders, where peat is seen as an example of a natural additive such as cellulose and many other fiber types.

2 Materials and Methods

Peat fibers were sourced from a Finnish company VAPO Fibers consisting of medium (PM), long (PL) and extra-long (PEL), with approximate lengths of < 8 mm, < 16 mm and > 16 mm, respectively. Each size was obtained in two moisture contents: a high moisture content of around 48–50 wt%, denoted as peat 50, and a low moisture content of around 20–24 wt%, denoted as peat 20. The long peat fibers were ground in a kitchen blender for 3 and 5 min, which resulted in a fine powder mixed with a fiber fraction. Peat fibers were dried for at least 2 h in an oven at 110 °C before using in asphalt/mixing. The bitumen was a standard 70/100 unmodified binder, with a penetration of 72 dmm and a softening point of 47 °C. The mixing procedure of peat and bitumen was as follows: bitumen was heated to 160 °C, various types and percentages of unheated dried peat fibers/powder were added into hot bitumen and manually stirred for at least 1 min before preparing test specimens.

FT-IR Spectroscopy. Fourier Transform Infrared Spectroscopy (FT-IR), combined with attenuated total reflection (ATR) was used. The instrument was a Nicolet IS 10, with a diamond cell (smart-orbit).

Optical Microscopy. A Carl Zeiss Axioskop 40Fl microscope, equipped with a digital camera deltapix DP200 was used. Two light sources are available, a high pressure mercury arc lamp HBO50 which transmits intense light with a wavelength between 450 to 490 nm, to be used in fluorescence mode, or a standard white light bulb, to be used in transmission mode. Magnifications levels of 50x, 100x and 200x are available.

Drainage. The binder drainage test was performed in accordance with NBN EN 12697-18 (2017), using drainage baskets constructed from 3.15 mm perforated stainless steel sheets, in accordance with ISO 3310-2, on side and base to form 100 mm cubes with feet at each corner of the base. The asphalt mixes consisted of 1.1 kg batches of SMA with the following composition: 70.7% crushed porphyry aggregates with the maximum size of 10 mm, 20.4% crushed porphyry sand, 8.90% Duras filler and 6.9% bitumen of standard penetration grade 70/100. Asphalt specimens were placed in baskets on a pre-wrapped tray in the oven at 180 °C for between 3 h and 3 h 15 min and the binder drainage was calculated according to Eq. (1):

$$D = 100 \times \frac{(W_2 - W_1)}{(1100 + B)} \quad (1)$$

where D is the drained material (%), W_1 is the initial mass of the tray and foil (g), W_2 is the mass of the tray and foil with drained material (g), B is the initial mass of binder in asphalt mix (g).

Rheology. Rheological properties were determined by an Anton Paar MCR 500 rheometer using 25 mm parallel plates, with a gap setting of 2 mm. The specimens were first poured in silicon molds and afterwards transferred to the rheometer. Frequency sweeps were performed from 0.01 Hz to 10 Hz, from 40 °C–90 °C, with a strain level of 0.01, in steps of 10 °C. The base binder was also investigated with 8 mm plates, from 0 °C–40 °C, using a gap setting of 2 mm, and a strain of 0.0005.

3 Results and Discussion

Initially, some solubility tests were carried out, using various organic solvents such as toluene and tetrahydrofuran, in order to evaluate the solubility of peat in bitumen. It was concluded that there is no solubility in any of these solvents, and therefore no solubility bitumen is expected.

FT-IR spectroscopy was used to investigate the water content differences between peat 50 and peat 20; this was clearly visible by the broad hydrogen bonding region (3500–3000 cm^{-1}). In Fig. 1, the spectra of the medium fibers containing 50% and 20% of water are shown. The data confirm differences in moisture content between both samples. For each moisture content, 3 repeat tests were performed and they showed little variation. Besides water, the spectra indicated that peat fibers consist of saturated organic groups, alcoholic groups, a very small amount of C = O groups, and probably also inorganic material, indicated by the large signal at 1000 cm^{-1}. Moreover, there were no differences observed in the spectra of the powder and the fiber fraction, indicating that both fractions are chemically identical.

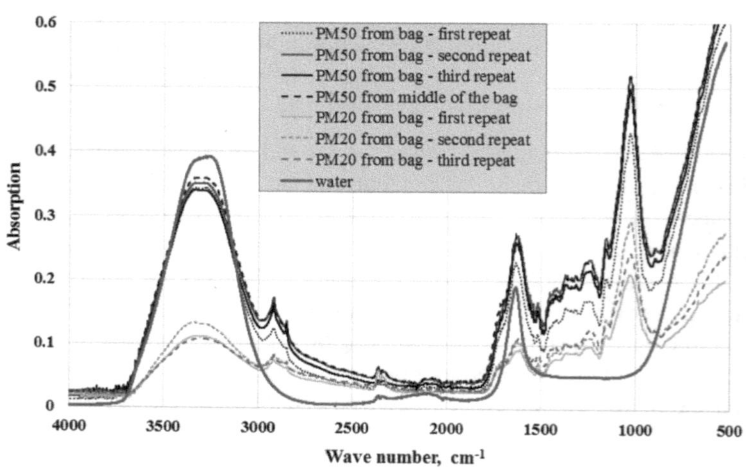

Fig. 1. FT-IR spectra of medium peat fibers containing two water contents

In Fig. 2a, a micrograph of PEL20(3) peat powder & fibers shows the particle size ranges from 0.1 mm up to 1 mm; it can be also observed that fibers can be long up to about 3 cm. In Fig. 2b the fluorescence image, shows that the fibers on a microscopic level are very inhomogeneous, as demonstrated by large differences in fluoresce.

(a) (b)

Fig. 2. (a) PEL20(3) ground peat material, microscopy image (50×); (b) fluorescence image of PM20 (50×)

For the rheological tests, the ground peat fibers PEL20(3) were selected, the PEL20 (3) fiber part could be easily separated from the powder fraction and the fractions were manually mixed with bitumen in different dosages, ranging from 0.1 to 1%, by weight of bitumen. Black plots and frequency sweeps of various combinations of bitumen, peat powder and fibers are presented in Fig. 3.

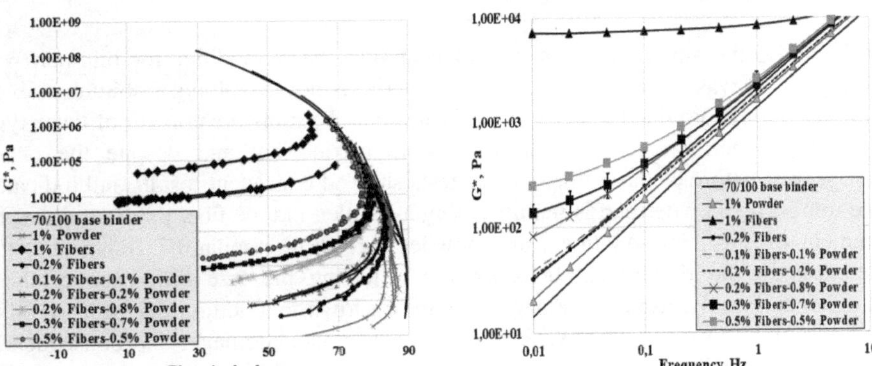

(a) (b)

Fig. 3. (a) black plots of peat PEL20(3) and bitumen and (b) frequency sweeps at 60 °C of peat PEL20(3) and bitumen

The curves clearly demonstrate an increased elasticity by increasing the fibers versus powder content in the base binder. It can be observed for the 1% fiber curve, that the modulus stays at a high level, at decreasing temperature and frequency. For this sample the modulus shows a plateau region at around 10 kPa. Samples with less fibers also show a plateau region in the modulus, but appearing at a lower stiffness level. The addition of only powder, for example the 1% curve, is contributing mainly to a stiffening effect. Only at the highest investigated temperature of 90 °C, some elastic contribution can be observed, which can also be observed in mastics and is related to a particle interaction. In Fig. 3b, for the sample 0.3 fibers & 0.5 powder errors bars, equal to the standard deviation of 5 repeats are added. Repeatability is better for the larger frequencies, because then the effect of the powder is dominating the behavior, even though repeatability is less good at low frequencies, trends are still clear.

It is known that natural fibers are used to prevent drainage, therefore, this test was also performed in this study. At first, the optimal amount of fibers (%) per asphalt mix composition was defined, in order to keep asphalt batch mass homogeneous. The variation of fibers 0.3–1.5% was applied and according to obtained results, it was decided to keep 0.5% of peat fibers per mix. In the first set of experiments peat fibers PM, PM20, PL20 and PEL20 were used. According to the obtained results the amount of drained material for: reference asphalt mix (REF) was 0.32%, PM – 0.03%, PM20 – 0.02%, PL20 – 0.07% and PEL20 – 0.01%. In a second step PM20 and PEL20 peat fibers were grinded for 3 and 5 min and grinded peat material: PM20(3), PM20(5), PEL20(3) and PEL20(5) was obtained. According to the drainage test results, the amount of drained material was: REF – 0.33%, PM20(5) – 0.1% and no drainage was observed for mixes PM20(3), PEL20(3) and PEL20(5). The data show that peat fibers grinded/non-grinded are effective in reducing or even preventing drainage.

4 Conclusions

In this study, the suitability of a natural peat fiber, as a modifier for bitumen was investigated. It was noticed that peat fiber material is not dissolving in bitumen, even not partly. FT-IR provided a quick test to determine the moisture content of these type of fibers. Furthermore, rheological tests were carried out and despite the rather heterogenous nature of these fibers, these tests showed consistent results and indicated some trends; the powder part is mainly acting as a filler and the fiber part introduces an increment of elasticity. As fibers and powder part show similar FT-IR spectra, the differences in the rheological behavior are reflecting the size difference of these components. Finally, asphalt drainage tests have shown that adding dry peat, whether this is ground or not, is effective in reducing the binder drainage, and consequently from this point of view would allow for thicker binder films around the aggregate. These type of fibers, even though they are not dissolving in bitumen show interesting effects. For a definite conclusion, more tests will need to be carried out, such as low temperature fracture tests, an investigation of aging effects and water sensitivity measures, as well as a large-scale evaluation. At this stage, the test results look promising.

References

Bain, C., Bonn, A., Stoneman, R., Chapman, S., Coupar, A., Evans, M., Gearey, B., Howat, M., Joosten, H., Keenleyside, C., Labadz, J., Lindsay, R., Littlewood, N., Lunt, P., Miller, C., Moxey, A., Orr, H., Reed, M., Smith, P., Swales, V., Thompson, D., Thompson, P., Van de Noort, R., Wilson, J., Worrall, F.: IUCN UK Commission of Inquiry on Peatlands. http://www.iucn-uk-peatlandprogramme.org. Accessed 1 Feb 2018 (2011)

Energy in Finland Communication and Information Services Statistics Finland, ISBN 978−952−244−552−0. http://www.stat.fi. Accessed 1 Feb 2018 (2016)

Hansen, R., McGennis, B., Prowell, B., Stonex, A.: Current and future uses of non-bituminous components of bituminous paving mixtures. Transportation in the new millennium. TRB 2A2002 Washington, USA. http://www.trb.org/publications/millennium/00079.pdf (2000)

Kudrjashov, A.P., Kudrjashov, I.V., Kudrjashov, P.A., Germashev, V.G., Jadykina, V.V.: Stabilising additive for peat-based asphalt-concrete mixture (versions) and method of producing structure-forming agent thereof. Patent RU 2479524 C2 (2013)

NBN EN 12697-18 Bituminous mixtures – test methods for hot mix asphalt – Part 18: Binder drainage (2017)

Oda, S., Fernandes, J.L., Ildefonso, J.S.: Analysis of use of natural fibers and asphalt rubber binder in discontinuous asphalt mixtures. Constr. Build. Mater. **26**(1), 13–20 (2012). https://doi.org/10.1016/j.conbuildmat.2011.06.030

U.S. Geological Survey Mineral Commodity Summaries. https://minerals.usgs.gov/minerals/pubs/mcs/2017/mcs2017.pdf. Accessed 1 Feb 2018 (2017)

Promotion of Bitumen-Impregnated Cellulose Fibres from Lightweight Roofing Tiles in Stone Mastic Asphalt

Clara Tamburini[1,2](✉), Layella Ziyani[2], Anne Dony[2], Christophe Rohart[3], and Emanuele Toraldo[1]

[1] Department of Civil and Environmental Engineering, Politecnico di Milano, Piazza Leonardo da Vinci 32, 20133 Milan, Italy
tamburiniclara@gmail.com, emanuele.toraldo@polimi.it
[2] Université Paris Est, Institut de Recherche en Constructibilité, Ecole Spéciale des Travaux Publics, 28 avenue du Président Wilson, 94234 Cachan Cedex, France
{lziyani,adony}@estp-paris.eu
[3] Onduline, Zone Industrielle, 76480 Yainville, France
crohart@ofitech.fr

Abstract. Applied in wearing courses, Stone Mastic Asphalt (SMA) is known to enhance pavement durability, in particular to reduce permanent deformation. The peculiarities of this bituminous mixture consist of its high content of coarse aggregates, its gap-graded gradation and the presence of fibres inside. Traditionally, cellulose fibres extracted from different sources (paper, magazines) are employed. In this research work, bitumen-impregnated cellulose fibres made according to the same manufacturing process as for lightweight roofing tiles were used. These materials are composed of a soft pure bitumen, recycled magazines and old corrugated paper (OCC). The objective of this study was to evaluate the influence of their introduction on the compactability, volumetric and mechanical properties of SMA mixtures. The results showed that adding impregnated fibres decreases binder drainage. Moreover, incorporating a specific fibre content of 0.3% wt. improves compactability, compression and rutting resistances of the mixtures. As a conclusion, bitumen-impregnated cellulose fibres similar to those used in roofing industry are viable for a potential use in SMA.

Keywords: Stone Mastic Asphalt · Bitumen-impregnated fibres Tiles · Formulation

1 Introduction

To limit interventions on roads, asphalt pavement durability is one of the major challenges of road industry, sometimes requiring to turn to new processes or the evolution of existing materials. Among these solutions is listed the Stone Mastic Asphalt (SMA), a mixture composed of aggregates (mainly coarse aggregates, often a gap-graded gradation), bitumen-based mastic and fibres. The use of fibres in road technique, as practiced in the SMA, is known for favouring resistance to the permanent

L. D. Poulikakos et al. (Eds.): RILEM 252-CMB 2018, RILEM Bookseries 20, pp. 306–311, 2019.
https://doi.org/10.1007/978-3-030-00476-7_48

deformations related to traffic loadings. The aim of the use of SMA is to obtain a mix with a longer service life and better resistance to fatigue (Blazejowski 2001). Nowadays, it is current to add additives such as cellulose fibres in SMA, These are incorporated in the bituminous mixes to fix the binder and to stiffen the mastic. They absorb the excess bitumen and hold it in place; they then act as a stabiliser (Putman and Amirkhanian 2004).

Cellulose fibres are also used in corrugated bitumen sheets. These products are found in roofing or waterproofing applications (specifications according to NF EN 534 2010). They usually consist of a cardboard made with cellulose extracted from paper or magazines and impregnated with bitumen. This "absorption" process allows the fibres to be fully covered by bitumen. To our state of knowledge, these kind of fibres have never been introduced in bituminous mixtures. Meanwhile, they can represent an opportunity in road field, in particular when they are recycled from waste production or recovered from tiles at their end of service life.

The objective of this study was to incorporate these bitumen-impregnated fibres in SMA mixtures and to assess their effect on the mechanical properties of SMA. In particular, the influence of fibre content was considered and SMA mixtures were evaluated in terms of compression and rutting resistances. The quality of the mixtures was compared to that of SMA without fibres, in order to evaluate the contribution of fibres.

2 Methodology

The aggregates used during this research (0/2, 2/4 and 6/10) were of diorite nature and were supplied by a French quarry in the West of France, while the fillers employed were of limestone nature. The binder used for SMA formulations was a 35/50 penetration grade pure bitumen, supplied by Total S.A. The cellulose fibres used were recovered from bitumen-impregnated tiles ($\approx$50% wt. of soft bitumen in the total mass of a tile) produced by Onduline company. They contain fibres from recycled newspaper or magazine and OCC (old corrugated container), stronger than traditional cardboard. The impregnated tiles were extracted from a production output (and thus did not face the effect of climate stress). They were crushed in order to obtain <5 mm diameter materials.

A same theoretical grading curve was adopted for the formulations of SMA. The effects of such fibres were investigated preparing in the laboratory two SMA mixes at different fibre contents: SMA0.3 and SMA1 (containing 0.3 and 1% wt. of fibres respectively). For comparative purposes, a SMA mixture without fibres (REF) was also prepared. Table 1 shows the composition of the three mixtures. Moreover, in order to evidence the effect of impregnated fibers exclusively, the binder thickness (according to the "module de richesse" method) was kept constant at 3.96 and the true density at 2.577 g/cm^3 for the SMAs. In the case of SMA0.3 and SMA1, the aggregates were mixed for 20 s. Then, the fillers and the right amount of impregnated fibres were added as received in the mixer and the whole was stirred for 10 s. Finally, bitumen was introduced and the whole was mixed at 60 rpm for 5 min.

First, the three mixtures were characterised in terms of Gyratory Compaction and binder drainage, following the beaker method (NF EN 12697-18 2017). The draindown D calculated corresponds to the ratio of the mass of matter remained in a beaker to the

Table 1. Weight composition of the three SMA mixtures

	0/2	2/4	6/10	Filler	BI fibres	Bitumen
REF	20.70%	7.53%	60.23%	5.65%	0%	5.89%
SMA0.3	20.64%	7.51%	60.05%	5.63%	0.3%	5.87%
SMA1	20.50%	7.45%	59.63%	5.59%	1%	5.83%

BI: bitumen-impregnated. Bitumen content is indicated without taking into account the binder contained in BI fibers.

mass of the mixture introduced, after heating the beaker at 185 °C for an hour. Then, water sensitivity of SMAs was assessed using the French Duriez test (NF EN 12697-12 – Annex B 2008). The compression resistances in water (i), in air (C) and the i/C parameter, which traduces stripping resistance, were determined after the storage of specimens at two different climatic conditions for seven days (20 °C–55% of relative humidity and 40 °C). Finally, 500 mm × 180 mm × 100 mm slabs were prepared. After the application of a roller compaction (strong mode – 90% of density), the slabs were stored for 10 days at room temperature before rutting tests (performed at 60 °C).

3 Results and Discussion

Results of the Gyratory Compaction tests are shown in Fig. 1(a). The void percentage of SMA0.3 is lower than the one of REF throughout the experiment, whereas the curve of SMA1 is above the other two curves. Considering a number of gyrations of 60, the difference in void percentage between REF and SMA0.3 is 1.1%. This value is close to the repeatability of a trial (about 1%). It can be concluded that adding 0.3% of impregnated fibres slightly increases compactability of SMA. Meanwhile, at the same number of gyrations, it is pointed out that SMA1 has a void percentage of 12.5%, i.e. 4% more than what obtained with REF. According to the results, it seems that a given fibre content exists to improve compactability. Fibres are able to fill the voids and contribute, with the mastic, to favour the stone-to-stone contact, until a critical point. Over this point, the fibres produce an excessive increase in mastic's viscosity, reducing the mixture's ability to be compacted.

The results in binder drainage tests depicted in Fig. 1(b) stress the bad draindown effect of REF (>0.9%). When adding fibres, draindown D is below the recommended limit of <0.2% (Blazejowski 2001). This shows the positive effect of the addition of bitumen-impregnated fibres in SMA, despite their pre-impregnation with bitumen (good ability to absorb). Figure 1(b) also illustrates a decrease in percentage of drained bitumen as the amount of fibres incorporated into the mixture increases. Actually, SMA1 shows the lowest drainage value, confirming the excessive increase in the mastic's viscosity observed during the previous test.

The results in water sensitivity tests are presented in Table 2. The geometric voids found for the three mixtures were between 9 and 11%. Consequently, the results in compression resistance are comparable. First, it is noted that at 20 °C and 55% of relative humidity in air, the compression resistances of SMA0.3 and SMA1 are higher

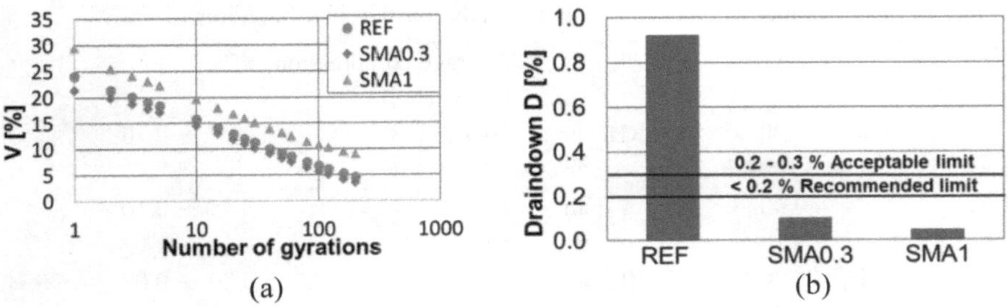

Fig. 1. (a) Void percentages in the three SMA mixtures as a function of the number of gyrations, (b) Draindown results and comparison with requirements for SMA mixtures

than the resistance of REF. In the same climatic conditions, but in water, the resistances are similar (between 10.1 and 10.6 MPa). Comparing water sensitivities of mixtures, it is observed that i/C ratio is the highest for REF. As a conclusion, adding impregnated fibres at 20 °C does not improve water sensitivity, but it increases compression resistance. Moreover, regarding SMA0.3 and SMA1, the i/C ratio at this temperature (0.89 and 0.92 respectively) is above 0.80, the minimum limit value requested for asphalt concrete mixtures (Delorme et al. 2007). Second, at 40 °C, compression resistance of SMA0.3 is the highest (11.9 MPa), whereas those of REF and SMA1 are similar, confirming that there is a critical content of fibres that increases too much the mastic's viscosity during compaction and, thus, increase the compression resistances. The compression resistances of specimens containing fibres and immersed in water are slightly higher than the one from REF, although the standard deviation found for this mixture is a little elevated (0.8 MPa). The i/C ratio is the lowest for SMA0.3. Meanwhile, the value found (0.79) is close to the minimum of 0.80.

Comparing both climatic conditions, compression resistances in air are similar, except for SMA1, for which it is lower at 40 °C. The resistances decrease in water from 20 to 40 °C, although it is noticed that the difference in resistance is weaker for SMA containing fibres. The water sensitivity parameter i/C is lower at 40 °C, whatever the SMA mixture considered.

Finally, results of rutting tests are gathered in Table 3: a limit of nearly 5.0% is reached at 10,000 cycles for REF and SMA1 and at 100,000 cycles for SMA0.3. REF and SMA1 show similar rutting percentages for all the number of cycles monitored. At 10,000 cycles, rutting percentage of SMA0.3 is 1.5% less than the one from REF, and 2.0% less than the one from SMA1. This confirms the results of water sensitivity test, i.e. there is a specific fibre content, which assures rigidity of the mixture and certain mechanical performance. Above a given fibre content, there is a decrease in rutting resistance. Probably above a given amount of fibres, the influence of bitumen contained in them is greater than the influence of the fibres themselves. Considering high temperature of the test (60 °C) and higher bitumen content in the mixture, there is an increase in permanent deformation. This trend is in accordance with a previous study on hot mix asphalt without fibres, for which it was proved than an addition of 0.2% of bitumen increased rutting percentage for around 4% at 30,000 cycles (Delorme et al. 2007).

Table 2. Water sensitivity results for the three mixtures

			Average compression resistance [Pa]	i/C
REF	20 °C, 55% RH	air	10.4 ± 0.8	0.97 ± 0.10
		water	10.1 ± 0.4	
	40 °C	air	10.2 ± 0.4	0.85 ± 0.10
		water	8.6 ± 0.8	
SMA0.3	20 °C, 55% RH	air	11.9 ± 0.5	0.89 ± 0.07
		water	10.6 ± 0.4	
	40 °C	air	11.9 ± 0.2	0.79 ± 0.06
		water	9.4 ± 0.4	
SMA1	20 °C, 55% RH	air	11.2 ± 0.2	0.92 ± 0.07
		water	10.3 ± 0.6	
	40 °C	air	10.5 ± 0.2	0.89 ± 0.65
		water	9.3 ± 0.3	

RH: relative humidity

Table 3. Rutting percentages of the three SMA mixtures

Number of cycles	Average rutting percentage of REF [%]	Average rutting percentage of SMA0.3 [%]	Average rutting percentage of SMA1 [%]
100	2.14 ± 0.41	1.46 ± 0.03	2.85 ± 0.62
300	3.19 ± 1.00	1.54 ± 0.31	3.19 ± 1.08
1,000	3.27 ± 0.45	1.94 ± 0.21	3.59 ± 1.20
3,000	3.80 ± 0.08	2.69 ± 0.22	4.48 ± 1.11
10,000	4.80 ± 0.14	3.22 ± 0.34	5.17 ± 0.96
30,000	/	4.32 ± 0.30	/
100,000	/	5.07 ± 0.24	/

4 Conclusions

The aim of the laboratory investigation described in this paper was the introduction of bitumen-impregnated cellulose fibres from lightweight roofs in SMA mixtures and the assessment of their effect on the compactability, volumetric and mechanical properties of SMA. According to the results from Gyratory Compaction, compression resistance and rutting tests, SMA0.3 shows the best performances. This implies that there is a fiber content for which the mechanical properties of SMA are improved. This study demonstrates also the viability of the bitumen-impregnated fibres in SMA. This is encouraging for the promotion in the future of bitumen-impregnated cellulose materials used for roofing tiles in other applications than roofing.

References

Blazejowski, K.: Stone Matrix Asphalt – Theory and Practice. CRC Press, Taylor and Francis Group, Boca Raton (2001)

Delorme, J.L., De la Roche, C., Wendling, L.: LPC bituminous mixtures design guide (2007)

NF EN 534 Corrugated bitumen sheets. Product specification and test methods (2010)

NF EN 12697-12 Bituminous mixtures. Test method for hot mix asphalt. Determination of the water sensitivity of bituminous specimens (2008)

NF EN 12697-18 Bituminous mixtures. Test methods. Binder drainage (2017)

Putman, B.J., Amirkhanian, S.N.: Utilization of waste fibers in stone matrix asphalt mixtures. Res. Con. Rec. **42**, 265–274 (2004). https://doi.org/10.1016/j.resconrec.2004.04.005

Author Index

L. D. Poulikakos et al. (Eds.): RILEM 252-CMB 2018, RILEM Bookseries 20, pp. 313–315, 2019.
https://doi.org/10.1007/978-3-030-00476-7

Kolibri 360 GmbH
Ferdinand-Jühlke-Straße 7
99095 Erfurt, Deutschland
produktsicherheit@kolibri360.de